Waveform Diversity

Waveform Diversity

Theory & Applications

Unnikrishna Pillai
Polytechnic Institute of NYU, New York

Ke Yong Li
C & P Technologies, Inc., New Jersey

Ivan Selesnick
Polytechnic Institute of NYU, New York

Braham Himed
Air Force Research Lab, Dayton, Ohio

New York Chicago San Francisco
Lisbon London Madrid Mexico City
Milan New Delhi San Juan
Seoul Singapore Sydney Toronto

The **McGraw-Hill** *Companies*

Cataloging-in-Publication Data is on file with the Library of Congress

McGraw-Hill books are available at special quantity discounts to use as premiums and sales promotions, or for use in corporate training programs. To contact a representative, please e-mail us at bulksales@mcgraw-hill.com.

Waveform Diversity: Theory & Applications

1234567890 QFR QFR 10987654321

ISBN 978-0-07-162289-9
MHID 0-07-162289-6

Sponsoring Editor
Wendy Rinaldi

Editorial Supervisor
Jody McKenzie

Project Manager
Madhu Bhardwaj
Glyph International

Acquisitions Coordinator
Joya Anthony

Copy Editor
Manish Tiwari

Proofreader
Nidhi Chopra

Indexer
Valerie Haynes Perry

Production Supervisor
Jean Bodeaux

Composition
Glyph International

Illustration
Glyph International

Art Director, Cover
Jeff Weeks

Cover Designer
Jeff Weeks

About the Authors

Unnikrishna Pillai is a professor of electrical and computer engineering at Polytechnic Institute of NYU in Brooklyn, New York. His research interests include radar signal processing, blind identification, spectrum estimation, data recovery, and waveform diversity. Dr. Pillai is the author of *Array Signal Processing* and coauthor of *Spectrum Estimation and system Identification*, Prof. Papoulis' *Probability, Random Variables and Stochastic processes* (Fourth edition), and *Space Based Radar: Theory & Applications*.

Ke Yong Li is a senior engineer at C & P Technologies, Inc. in Closter, New Jersey. His areas of research include space-time adaptive processing (STAP), waveform diversity and radar signal processing. He is the coauthor of the book on *Space Based Radar: Theory & Applications*.

Ivan Selesnick is an associate professor of electrical and computer engineering at Polytechnic Institute of NYU in Brooklyn, New York. His current research interests are in the areas of digital signal and image processing, wavelet and sparsity based methods for signal restoration, and biomedical signal processing.

Braham Himed is the technical advisor in the RF Technology Branch at the U.S. Air Force Research Laboratory, in Dayton, Ohio. His research interests include radar signal processing, detection, estimation, multichannel adaptive processing, time series analysis, and array processing. He is the coauthor of the book on *Space Based Radar: Theory & Applications*.

"Follow the flagship, and ask (no) questions"

–Ferdinand Magellan (1518)

Contents At A Glance

Contents

List of Abbreviations

AMTI	Airborne Moving Target Indication
AR	Auto Regressive
ARMA	Auto Regressive Moving Average
CNR	Clutter to Noise Ratio
CR	Cramer-Ratio
DFT	Discrete Fourier Transform
GMTI	Ground Moving Target Indication
HTP	Hung-Turner Projection
i.i.d.	Independent and Identically Distributed
MF	Matched Filter
ML	Maximum Likelihood
PRF	Pulse Repetition Frequency
PRI	Pulse Repetition Interval
RCS	Radar Cross Section
SAR	Synthetic Aperture Radar
SBR	Space Based Radar
SINR	Signal to Interference plus Noise Ratio
SNR	Signal to Noise Ratio
STAP	Space-Time Adaptive Processing
SVD	Singular Value Decomposition
UAV	Unmanned Aerial Vehicle
WD	Waveform Diversity

CHAPTER 1
Introduction

1.1 Introduction

Waveform diversity refers to exploiting the shape of the transmitter and receiver waveforms for improving signal detection, interference, and noise suppression across various disciplines, such as radar, sonar, and communications that utilize waveforms in their daily use. The electromagnetic spectrum is becoming crowded and efficient use of bandwidth and power by primarily shaping the waveforms to meet the performance criterion is a goal of waveform design. Modern advances in hardware design offer extra degree of freedom that can be exploited in this context [1–5]. In recent years, there has been a trend toward ever greater degrees of digitization and more widespread use of solid state power amplification that will have a significant impact on the overall system performance [6]. As a result, there are compelling changes in system requirements such as more efficient spectrum usage, higher sensitivities, transmitter/receiver agility, greater information content, improved robustness to errors, etc. This combination of changes is fueling a worldwide interest in the subject of waveform design and diversity [6]. The role of waveform design is effective system resource management because it demands that the waveform used is continually updated and adjusted to meet the system needs [1].

Waveform diversity has its origins in the 1930s, when techniques involving a varying chirp slope were developed [4]. In the 1950s activities occurred as part of a variety of other efforts, including high-power microwave. In the 1960s there was a lot of interest in waveform design for clutter rejection, electromagnetic compatibility, and spread-spectrum techniques for communication and radar [7]. In this context continued interest in ultra wideband application, along with polarization diversity, led to the development of an orthogonally polarized quadraphase electromagnetic radiator for wideband applications [8].

Wireless communications applications increased rapidly in the 1990s, and waveform diversity began to emerge as a distinct technology. Wireless systems utilize diversity in time, space, frequency, and polarization. Waveform diversity techniques are exploited there to provide multiple independent channels for message signals in order to increase redundancy and reduce bit error rates. Besides communications, optimal waveform for radar applications is another area of interest [4].

More recently, knowledge-based radar signal and data processing has undergone considerable activity. Knowledge-based processing takes advantage of known information about the target and interference environment, and alters algorithms, data, and parameters based on this information [9–11].

In the context of radar, with increasing numbers of radio frequency sensors and communication systems, battlefield scenarios have become more complex. The challenge is to use this technology to enhance the overall performance of advanced radar systems operating within available spectrum in these battlefield environments. Technologies essential to this goal include knowledge-based systems for waveform generation' timing and control, and cognitive radar signal processing for detection, parameter estimation, and tracking. Cognition is the process of knowing, along with awareness and judgment. Awareness of the dynamically changing spectral environment and judgment for the selection of optimal waveform parameters and sensor placement can enable improved performance for target surveillance/reconnaissance in addition to enhanced interference mitigation. These factors make waveform diversity-based systems a logical and affordable alternative to classically designed radars [2].

Adaptivity has been extended from the receiver side to the transmitter side as well. With waveform diversity, the transmitted signal is changed to better match the target interference environment [12]. This is classified under "target matched illumination" where prior knowledge about the target and environment is used in the waveform design process for selecting particular responses. Self-learning approaches can be adopted to select the waveform for a given detection or classification task by training against targets that are required to be detected along with the joint design of transmitter and receiver design in the presence of targets, noise, and inference [12–14]. Based on waveform design and knowledge-based activities, a fully adaptive radar is now possible, where the transmitted waveform and the processing of those waveforms is optimized in response to the environment.

Waveform diversity can be exploited spatially using a multiple set of sensors for both transmission and/or reception, and in the time-frequency domain using distinct waveforms of different durations over different spectral bands [15–20]. In addition, other features such as constant envelope property, polarization, energy distribution of various transmit signals along with good pulse compression at the receiver can be used for further optimization. Spatial diversity can be realized from a single platform or multiple platforms for both transmission and reception [21–28]. For example, in monostatic mode the same set of sensors is used for transmission and reception, whereas in the bistatic mode different sensors are used for these two functions. Thus in the bistatic case, a dedicated sensor is used for transmission and a multiple set of sensors can be used as the receivers. In both cases, the platform involved may be stationary or in motion [29–36]. In a typical airborne or space-based radar situation, the platform is moving and both transmitter and receiver may be timesharing the same set of antennas in different configurations giving rise to a monostatic situation. If we envision a set of unmanned aerial vehicles (UAVs) into this situation purely in the receiver mode, then we have a multisensor bistatic situation at hand [37–40].

The objective in such situations also may comprise multiple tasks, such as ground moving target indication (GMTI), airborne moving target indication (AMTI), and synthetic aperture radar (SAR). In the case of GMTI, the task is to improve the overall system design so that slowly moving targets can be detected and identified in the presence of clutter, jamming signals, and noise. Once the transmitter and receiver array configurations and their relative positions are known, the problem is to select various transmitter and receiver waveforms for target return enhancement and simultaneous suppression of interference and noise. The target scene may consist of single or multiple targets which may be stationary or moving under various conditions. The system goal is to detect and identify all of them, and estimate their relevant parameters such as their range, Doppler, elevation, and azimuth angles, etc. Usually target return is buried in competing interference which consists of stochastic return signals from the environment and noise. These random returns are transmit signal-dependent and hence they directly compete with the target returns, usually dominating them. One task of waveform diversity is to distinguish the interference spectral characteristics and design waveforms accordingly to minimize the competing signal-dependent interferences [41–45].

1.2 Organization of the Book and Notations

This book is primarily intended for seniors/graduate students as well as professional engineers who have basic understanding of linear systems and radar fundamentals. Chapter 2 gives an introduction to the joint transmitter receiver waveform diversity. For a given transmit waveform, target (channel) response, and noise scene, it is well known that the classic matched filter gives the optimum receiver characterization. The problem of transmitter optimization and the potential advantages to be gained by transmitting a specific waveform in a particular context—specific target (channel) and interference/noise scene—are addressed in this chapter.

New methods for optimum transmitter and receiver design are introduced in Chap. 3. One such method uses the target characteristics as well as clutter and noise spectral features to complete the transmitter waveform design, and hence it can be used for tasks such as AMTI/GMTI. The second method assumes the target characteristics are unknown and uses the total interference spectrum to complete the transmitter design. Hence, this approach is desirable for SAR applications where the objective is to perform high-resolution imaging rather than specific target detection. Further, the concept of transmitter threshold energy is introduced in this chapter. It is shown that to achieve the maximum realizable *SINR* (i.e., good performance), the transmitted energy must be greater than certain threshold value. The energy-bandwidth trade off concept is also introduced here. It is demonstrated here that by increasing the transmit energy or bandwidth, better performance can be attained. More importantly, performance can be maintained at the same level by increasing one parameter and decreasing the other. This leads to the bandwidth-energy trade off for both designs. The energy-bandwidth trade off concept is further extended to show that simultaneous savings of both energy and bandwidth can be realized by using new waveform design. In other words, the shape of the waveform is used to offset both bandwidth and energy, thus releasing both of these precious commodities for other uses.

One useful feature for transmit signals is the time domain constant envelope property that maximizes the transmit power efficiency. In that case, the time domain waveform will ideally have constant envelope property or in practice possess an approximate constant envelope, thereby distributing the peak transmit power more evenly in time. Chapter 4 reviews various methods for generating constant envelope waveforms in time domain given the frequency magnitude transform. The methods of alternating projections are reviewed here and a projection-based method to generate constant/near-constant envelope signal is also introduced.

Chapter 5 discusses the optimum waveform design problem in presence of noise and clutter in detail. It is shown that in the point target and the white noise case, the optimum transmit signal frequency transform must be inversely proportional to the clutter spectrum. This, of course, assumes that there is no restriction on the transmit signal bandwidth. Thus by increasing the transmit bandwidth *indefinitely*, it is possible to eliminate flat-clutter completely and the *SINR* is only noise-limited [46–51]. However, with finite bandwidth imposed through a square-integrable transmit output filter constraint, increasing the chirp bandwidth eliminates clutter; however, it is no longer possible to achieve maximum *SINR* in that manner even in the white noise case, and selective transmitter-receiver design must be carried out to maximize the output *SINR* [52–57]. Toward this, the optimum transmit waveform design method that takes advantage of the differences between the target, noise, and clutter is introduced in this chapter. It is shown that the optimum transmitter exploits the differences among the target, clutter, and noise spectral shapes and synthesizes a finite duration waveform that further amplifies their difference, thus achieving a better output *SINR* than any other waveform [58–61]. The optimum waveform design in discrete-time domain is carried out in Chap. 6. Chapter 6 also discusses the optimum multichannel waveform design problems from an output *SINR* maximization viewpoint and their solutions.

Chapter 7 discusses the sparsity-based receiver design methods. Recent advances in L_1-based optimization are reviewed and their usefulness in sparse target situations is illustrated. When the underlying data scene is sparse in some domain, such as frequency, space, angle, and Doppler, this information can be exploited using L_1-based optimization methods as opposed to the conventional L_2-based optimization. One advantage of this approach is its effectiveness in small sample support situations and this is particularly attractive in nonstationary data scene.

Finally, a word about notations used in this text. Throughout this book, scalars are represented by regular lowercase or uppercase Roman and Greek letters (e.g., a, A, α, or λ); vectors are denoted by lowercase bold letters *with* or *without* underline (e.g., $\mathbf{a}$, $\underline{\mathbf{a}}$, or $\boldsymbol{\beta}$), or lowercase letters *with* underline (e.g., $\underline{a}$, $\underline{\alpha}$, or $\underline{\beta}$). In a similar manner, matrices are represented by uppercase bold Roman letters and uppercase Greek letters (e.g., $\mathbf{R}$, $\mathbf{U}$, or Λ). In addition, $\bar{\mathbf{A}}$, $\mathbf{A}^T$, $\mathbf{A}^*$, $tr(\mathbf{A})$, $\det(\mathbf{A}) = |\mathbf{A}|$, $\mathbf{A}^{-1}$ represent the complex conjugate, transpose, complex conjugate transpose, trace, determinant, and inverse of the matrix $\mathbf{A}$, respectively. The identity matrix is represented by $\mathbf{I}$ whose size should be clear from the context.

References

[1] C.J. Baker, H.D. Griffiths, "Radar Systems and Waveforms," *Proc. of First International Conference on Waveform Diversity & Design Conference*, Edinburgh, UK, Nov. 8–10, 2004.

[2] M. Wicks, "Waveform Diversity & Knowledge Based Systems," *2010 IEEE International Radar Conference Tutorial*, Washington DC, USA, May 10–14, 2010.

[3] L. Brennan, I. Reed, "Theory of Adaptive Radar," *IEEE Trans. Aerospace and Electronics Systems*, vol. AES-9, no. 2, pp. 237–252, March 1973.

[4] M. Wicks, "A Brief History of Waveform Diversity," *2009 IEEE Radar Conference*, Pasadena, CA, May 4–8, 2009.

[5] *Proce. of First International Conference on Waveform Diversity & Design Conference*, Edinburgh, UK, Nov. 8–10, 2004.

[6] V. Amuso, S. Blunt, "Message from the Chairs," *IEEE International Waveform Diversity & Design Conference*, Orlando, FL, Feb. 8–13, 2009.

[7] J.R. Klauder, A.C. Price, S. Darlington, W.J. Albersheim, "The Theory and Design of Chirp Radars," *Bell System Technical Journal*, vol. XXXIX, no. 4, pp. 745–808, July 1960.

[8] M.C. Wicks, P.V. Etten, "Orthogonally Polarized Quadra-Phase Electromagnetic Radiator," U.S. Patent 5,068,671, Nov. 26, 1991.

[9] P. Antonik, H. Schuman, P. Li, W. Melvin, M. Wicks, "Knowledge-Based Space-Time Adaptive Processing," *Proc. 1997 National Radar Conference*, Syracuse, NY, 1997.

[10] Knowledge Aided Sensor Signal Processing and Expert Reasoning (KASSPER), *Third Annual Workshop*, Clearwater, FL, Feb. 22–24, 2005.

[11] J. Guerci, "Waveform Diversity/KASSPER Connection," *Second Tri-Service Waveform Diversity Workshop*, Verona, NY, Feb. 2–3, 2004.

[12] S.U. Pillai, H.S. Oh, D.C. Youla, J.R. Guerci, "Optimum Transmit Receiver Design in the Presence of Signal-Dependent Interference and Channel Noise," *IEEE Trans. Inform. Theory*, vol. 46, no. 2, pp. 577–584, March 2000.

[13] J.R. Guerci, P. Grieve, "Optimum Matched Illumination Waveform Design Process," U.S. Patent 5,121,125, June 9, 1992.

[14] R.S. Adve, R.A. Schneible, G. Genello, P. Antonik, "Waveform-Space-Time Adaptive Processing for Distributed Aperture Radars," *Proc. 2005 IEEE International Radar Conference*, pp. 93–97, Arlington, VA, May 9–12, 2005.

[15] P. Stoica, J. Li and X. Zhu, "Waveform Synthesis for Diversity-Based Transmit Beampattern Design," *IEEE Transactions on Signal Processing*, vol. 56, no. 6, June 2008.

[16] K.W. Forsythe, D.W. Bliss, "Waveform Correlation and Optimization Issues for MIMO Radar," *Proc. 39th Asilomar Conf. Signals, Systems, Computer*, Pacific Grove, CA, pp. 1306–1310, Nov. 2005.

[17] B. Friedlander, "Waveform Design for MIMO Radars," *IEEE Transactions on Aerospace and Electronic Systems*, vol. 43, no. 3, pp. 1227–1238, Oct. 2007.

[18] Y. Chi, A. Pezeshki, R. Calderbank, S. Howard, "Range Sidelobe Suppression in a Desired Doppler Interval," *IEEE International Waveform Diversity & Design Conference*, Orlando, FL, Feb. 8–13, 2009.

[19] Y. Li, W. Moran, S. Sira, A. Papandreou-Suppappola, "Adaptive Waveform Design in Rapidly Varying Radar Scenes," *IEEE International Waveform Diversity & Design Conference*, Orlando, FL, Feb. 8–13, 2009.

[20] I. Demirkiran, D. Weiner, A. Drozd, I. Kasperovich, "Equivalent Based Analysis for Narrow Weakly Nonlinear Bandpass Systems for Waveform Diversity Applications," *IEEE International Waveform Diversity & Design Conference*, Orlando, FL, Feb. 8–13, 2009.

[21] M. Stuff, B. Thelen, N. Subotic, J. Parker, J. Browning, "Optimization and Waveforms for Compressive Sensing Applications in the Presence of Interference," *IEEE International Waveform Diversity & Design Conference*, Orlando, FL, Feb. 8–13, 2009.
[22] H. Mendelson, P. Zulch, "Mitigation of Radar Range Ambiguities via Waveform Diversity," *IEEE International Waveform Diversity & Design Conference*, Orlando, FL, Feb. 8–13, 2009.
[23] J. Goodman, B. Miller, M. Herman, M. Wai, P. Monticciolo, "Extending the Dynamic Range of RF Receiver with Nonlinear Equalization," *IEEE International Waveform Diversity & Design Conference*, Orlando, FL, Feb. 8–13, 2009.
[24] J. Zhang, A. Papandreou-Suppappola, "MIMO Radar with Frequency Diversity," *IEEE International Waveform Diversity & Design Conference*, Orlando, FL, Feb. 8–13, 2009.
[25] P. Gonzalez-Blano, E.D. Diego, E. Millan, B. Errasti, I. Montiel, "Stepped-Frequency Waveform in Radar Demonstrator and Its Jamming," *IEEE International Waveform Diversity & Design Conference*, Orlando, FL, Feb. 8–13, 2009.
[26] P.F. Sammartino, C. Baker, "The Frequency Diverse Bistatic System," *IEEE International Waveform Diversity & Design Conference*, Orlando, FL, Feb. 8–13, 2009.
[27] T. Sturman, "Enabling Chaotic Waveform Diversity," *IEEE International Waveform Diversity & Design Conference*, Orlando, FL, Feb. 8–13, 2009.
[28] J. Li, J.R. Guerci, L. Xu, "Signal Waveform's Optimal-Under-Restriction Design for Active Sensing," *IEEE Signal Process. Lett.*, vol. 13, no. 9, pp. 565–568, Sep. 2006.
[29] N. Bharadwaj, V. Chandrasekar, "Distributed Waveforms for Networked Meteorological Radar," *IEEE International Waveform Diversity & Design Conference*, Orlando, FL, Feb. 8–13, 2009.
[30] T. Higgins, S. Blunt, "Analysis of Range-Angle Coupled Beam forming with Frequency-Diverse Chirps," *IEEE International Waveform Diversity & Design Conference*, Orlando, FL, Feb. 8–13, 2009.
[31] H.B. Manjunath, J. Zhang, A. Papandreou-Suppappola, D. Morrell, "Waveform-Agile Sensing for Range and DoA Estimation in MIMO Radars," *IEEE International Waveform Diversity & Design Conference*, Orlando, FL, Feb. 8–13, 2009.
[32] M.C. Wicks, "History of Waveform Diversity and Future Benefits to Military Systems," *NATO Lecture Series on Waveform Diversity (SET119)* held in UK, Italy, Germany, and France, pp. 8–19, Sep. 2008.
[33] G. Genello, "Information Processing for Waveform Diversity Control," *IEEE International Waveform Diversity & Design Conference*, Orlando, FL, Feb. 8–13, 2009.
[34] I. Bradaric, G. Capraro, M. Wicks "Multistatic Ambiguity Function—A Tool for Waveform Selection in Distributed Radar Systems," *IEEE International Waveform Diversity & Design Conference*, Orlando, FL, Feb. 8–13, 2009.
[35] D.D. Weiner, M.C. Wicks, G.T. Capraro, "Waveform Diversity and Sensors as Robots in Advanced Military Systems," *1st International Waveform Diversity and Design Conference*, Edinburgh, UK, Nov. 2004.
[36] I. Bradaric, G.T. Capraro, P. Zulch, "Signal Processing and Waveform Selection Strategies in Multistatic Radar Systems," *International Waveform Diversity and Design Conference*, Pisa, Italy, June 2007, invited paper.
[37] R. Bonneau, "A Coherence Gradient Method in Waveform Design for Apertures," *IEEE International Waveform Diversity & Design Conference*, Orlando, FL, Feb. 8–13, 2009.
[38] G. Dingley, C. Alabaster, "Radar Based Automatic Target System," *IEEE International Waveform Diversity & Design Conference*, Orlando, FL, Feb. 8–13, 2009.

[39] R. Romero, N. Goodman, "Improved Waveform Design for Target Recognition with Multiple Transmissions," *IEEE International Waveform Diversity & Design Conference*, Orlando, FL, Feb. 8–13, 2009.
[40] M.R. Bell, "Information Theory and Radar Waveform Design," *IEEE Trans. Inform. Theory*, vol. 39, no. 5, pp. 1578–1597, Sep. 1993.
[41] S.U. Pillai, H.S. Oh, D.C. Youla, J.R. Guerci, "Optimum Transmit Receiver Design in the Presence of Signal-Dependent Interference and Channel Noise," *IEEE Trans. Inform. Theory*, vol. 46, no. 2, pp. 577–584, Mar. 2000.
[42] S. Kay, "Optimal Signal Design for Detection of Gaussian Point Targets in Stationary Gaussian Clutter/Reverberation," *IEEE J. Sel. Topics in Sig. Proc. Mag.*, vol. 1, no. 1, pp. 31–41, Jun. 2007.
[43] D.A. Garren, M.K. Osborn, A.C. Odom, J.S. Goldstein, S.U. Pillai, J.R. Guerci, "Enhanced Target Detection and Identification via Optimized Radar Transmission Pulse Shape," *Proc. IEEE*, vol. 148, no. 3, pp. 130–138, Jun. 2001.
[44] D.T. Gjessing, *Target Adaptive Matched Illumination Radar: Principles and Applications*, London, England: Peter Peregrinus, 1986.
[45] F. Ahmad, M. Amin, H. Estephan, "Multistatic Waveform Design for Seeing Through the Wall," *IEEE International Waveform Diversity & Design Conference*, Orlando, FL, Feb. 8–13, 2009.
[46] J. Stiles, J. Jenshak, "Sparse Array Construction using Marginal Fisher's Information," *IEEE International Waveform Diversity & Design Conference*, Orlando, FL, Feb. 8–13, 2009.
[47] T. Ch. Tang, C.C. Ho, "EMI Suppression for High-Speed PCB," *IEEE International Waveform Diversity & Design Conference*, Orlando, FL, Feb. 8–13, 2009.
[48] L.L. Monte, R. Ansari, D. Erricolo, M. Wicks., "The Use of Geometric Diversity for Spectral Dominance in Underground Imaging," *IEEE International Waveform Diversity & Design Conference*, Orlando, FL, Feb. 8–13, 2009.
[49] M. Wicks, "RF Sensors & Spectrum Allocation," *IEEE International Waveform Diversity & Design Conference*, Orlando, FL, Feb. 8–13, 2009.
[50] S. Ciccarelli, V. Amuso, "Simulation of the Front-End of a MEMS Based Ultra Narrow Band Tomographic Imaging System," *IEEE International Waveform Diversity & Design Conference*, Orlando, FL, Feb. 8–13, 2009.
[51] H. Deng, "Polyphase Coding Signal Design for Orthogonal Netted Radar Systems," *IEEE Transactions on Signal Processing*, vol. 52, no. 11, Nov. 2004.
[52] Y. Yang, R. Blum, "Waveform Design for MIMO Radar Based on Mutual Information and Minimum Mean-Square Error Estimation,"*Proc. 40th Annual Conference on Information Sciences and Systems*, pp. 111–116, March 2006.
[53] S.U. Pillai, H.S. Oh, J.R. Guerci, "Multichannel Matched Transmit Receiver Design in Presence of Signal-Dependent Interference and Noise," *IEEE Sensor Array and Multichannel Signal Processing Workshop*, pp. 385–389, March 2000.
[54] S.U. Pillai, "Multichannel Waveform Diversity for Improved Target Identification," *Second Tri-Service Waveform Diversity Workshop*, Verona, NY, Feb. 2–3, 2004.
[55] S. Sen, M. Hurtado, A. Nehorai, "Adaptive OFDM Radar for Detecting a Moving Target in Urban Scenarios," *IEEE International Waveform Diversity & Design Conference*, Orlando, FL, Feb. 8–13, 2009.
[56] M. Inggs, M. Brooker, "Extensible Simulator for Waveform Diversity Testing," *IEEE International Waveform Diversity & Design Conference*, Orlando, FL, Feb. 8–13, 2009.
[57] Y. Paichard, M. Brooker, M. Inggs, "A Signal Level Simulator for Netted Radar Waveforms Evaluation," *IEEE International Waveform Diversity & Design Conference*, Orlando, FL, Feb. 8–13, 2009.
[58] I. Shirokov, I. Serdyuk, "The Investigation of Phase Synchronization of Reference Oscillators Through Atmospheric Channel," *IEEE International Waveform Diversity & Design Conference*, Orlando, FL, Feb. 8–13, 2009.

[59] I. Akduman, A. Yapar, F. Akleman, H. Sahinturk, "Electromagnetic Imaging of Objects Buried under a Dielectric Rough Surface," *IEEE International Waveform Diversity & Design Conference*, Orlando, FL, Feb. 8–13, 2009.

[60] M. Inggs, "Passive Coherent Location as Cognitive Radar," *IEEE International Waveform Diversity & Design Conference*, Orlando, FL, Feb. 8–13, 2009.

[61] Y. Paichard, "Orghogonal Multicarrier Phased Coded Signal for netted Radar Systems," *IEEE International Waveform Diversity & Design Conference*, Orlando, FL, Feb. 8–13, 2009.

CHAPTER 2

Waveform Design and Matched Filtering

Waveform diversity in the transmitter-receiver design context refers to all aspects of waveform design for improving the overall performance. The general waveform diversity problem is reviewed in Sec. 2.1, where the various design components are outlined along with a criterion for measuring performance. Since the goal is to enhance the desired signal output and suppress the undesired interference and noise components, the signal to interference plus noise ratio (*SINR*) or the signal to noise ratio (*SNR*) can be used as the optimization criterion. The simple receiver design problem is considered in Sec. 2.2.1, where a known signal is received in white noise. The classic matched filter receiver is the optimum solution in this case. This is followed by the colored noise case in Sec. 2.2.2, where both the causal and noncausal receiver solutions are derived [4–8].

Matched filters lead to pulse compression and the general pulse compression problem is examined in Sec. 2.3 where the superiority of the chirp pulse in this regard is demonstrated. Section 2.4 examines the general transmitter-receiver design problem in the presence of white noise. Since the classic matched filter is the optimum receiver here, the resulting output *SNR* is a (quadratic) function of the transmit signal as well as the target characteristics. The natural step is to maximize the output *SNR* over the transmit signal subject to an energy constraint. This leads to an eigenfunction solution whose kernel involves the target response function. An interesting aspect of this approach is that the optimum transmit signal turns out to be minimum phase.

As an application, Sec. 2.5 addresses the time-bandwidth optimization problem and various special cases are examined.

2.1 Introduction

Waveform diversity refers to the use of various waveforms (signals) in both transmitter (Tx) and receiver (Rx) designs for improving the overall performance such as detection and/or identification of targets in interference and noise. Waveform diversity can be exploited spatially using a multiple set of sensors for both transmission and/or reception, and in the time-frequency domain using distinct waveforms of different durations over different spectral bands. In addition, other features such as polarization, energy distribution of various transmit signals can be used for further diversity [9–11].

Spatial diversity can be realized from a single platform or multiple platforms for both transmission and reception. For example, in monostatic mode the same set of sensors is used for transmission and reception, whereas in the bistatic mode different sensors are used for these two functions (Fig. 2.1). Thus in the bistatic case, a dedicated sensor is used for transmission and multiple sets of sensor outputs can be used as the receiver. In both cases, the platform involved may be stationary

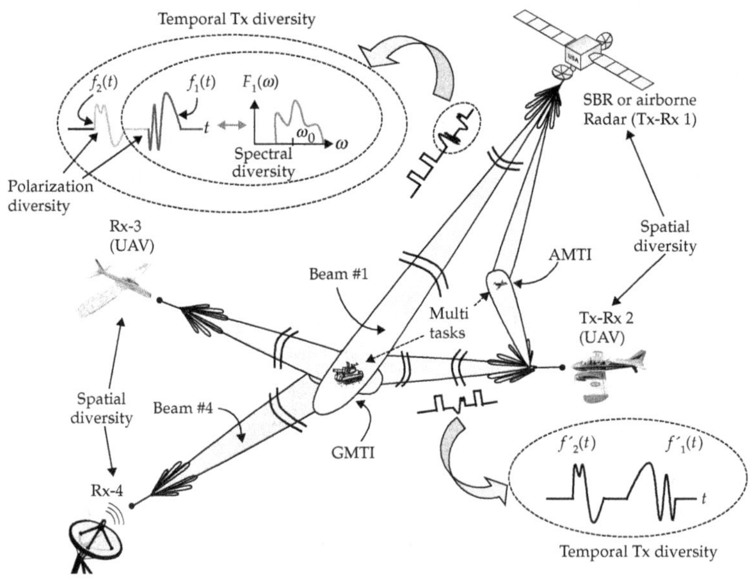

FIGURE 2.1 Multichannel waveform diversity.

or in motion. In a typical airborne or space-based radar situation, the platform is moving and both transmitter and receiver may be time-sharing the same set of antennas in different configurations giving rise to a monostatic situation. If we envision a set of unmanned aerial vehicles (UAVs) in this situation purely in the receiver mode, then we have a multisensor bistatic situation at hand [12–14].

The objective in such situations also may comprise multiple tasks such as ground moving target indication (GMTI), airborne moving target indication (AMTI), and synthetic aperture radar (SAR). In the case of GMTI, the task is to improve the overall system design so that slowly moving targets can be detected and identified in the presence of clutter, jamming signals, and noise. Once the transmitter and receiver array configurations and their relative positions are known, the problem is to select various transmitter and receiver waveforms for target return enhancement and simultaneous suppression of interference and noise. The target scene may consist of single or multiple targets and which may be stationary or moving under various conditions. The system goal is to detect and identify all of them, and estimate their relevant parameters such as their range, Doppler, elevation, and azimuth angles, etc. [15–19]. Usually target return is buried in competing interference which consists of stochastic return signals from the environment and noise. These random returns are transmit signal-dependent, and hence they directly compete with the target returns and usually dominating them. One task of waveform diversity is to distinguish the interference spectral characteristics and design waveforms accordingly to minimize the competing signal-dependent interferences.

In the spectral domain, it is easy to state the waveform design goal. The transmit waveform should radiate energy at those spectral regions where target characteristics are dominant, and minimize transmission at those regions where interference and noise are dominant. However, this is to be attained using finite duration waveforms in some optimal manner, and the problem becomes all the more challenging when target, interference, and noise have overlapping spectral regions.

The problem considered here is equally applicable in the communication scene as well, where the channel response plays the role of the target. In that case, the transmitted signal passes through the channel to generate an output signal that also gets modified by interference and noise. Once again the goal is to enhance the signal part of the channel output and minimize the interference and noise components through proper joint transmitter-receiver design. Clearly, the target or channel characteristics must play a role in the transmitter as well as the receiver design—to enhance the corresponding final desired output. At the same time, the design should suppress the interference and noise output components. In this context, the receiver output signal

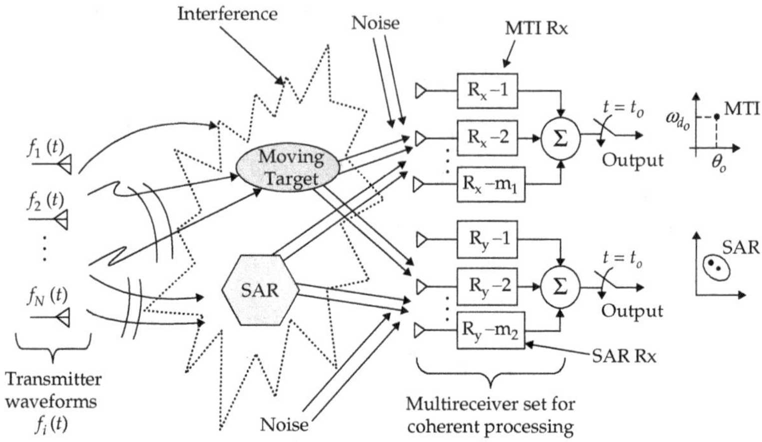

FIGURE 2.2 Transmitter-receiver design.

to interference plus noise ratio (*SINR*) may be used as an optimization goal. This is illustrated in Fig. 2.2 where the receiver outputs are combined to make a decision at $t = t_o$. Thus

$$SINR = \frac{\text{Receiver Output Signal (Target) Power at } t = t_o}{\text{Average Interference plus Noise Output Power}}. \tag{2.1}$$

In the absence of interference, (2.1) reduces to the familiar *SNR* criterion, where

$$SINR = \frac{\text{Receiver Output Signal Power at } t = t_o}{\text{Average Noise Power}}. \tag{2.2}$$

To maximize the output *SINR*, the following design components must be simultaneously taken into account:

Transmitter: Set of finite duration waveforms $f_{ij}(t) \leftrightarrow F_{ij}(\omega)$ (i : spatial and j : temporal) with an energy constraint on the whole set or subsets of waveforms. Known transmitter output filters $P_{ij}(\omega)$ can be used to control the transmitter bandwidth B_o. The spatial set of waveforms can be realized by partitioning the transmit array, and the temporal set by utilizing a pulse train of waveforms.

Target (Channel): Impulse response $q(t) \leftrightarrow Q(\omega)$. Simplest model is the point (impulse) target ($Q(\omega) \equiv 1$, flat spectrum). In general, $Q(\omega)$ is arbitrary. In the multi-input multi-output (multichannel) case, $Q(\omega)$ represents the target transform matrix whose $(i, j)^{th}$ entry $Q_{ij}(\omega)$ is the transfer function between the ith transmitter and the jth receiver.

Noise: Stationary stochastic process $w_n(t)$ with power spectrum $G_n(\omega) = \sigma_n^2$ that corresponds to the white (flat) noise case. In general, $G_n(\omega)$ is an arbitrary spectrum. [High interference (clutter dominant) situation may be represented as $G_n(\omega) \approx 0$.] In the multichannel case, $G_n(\omega)$ is represented by a positive definite spectral matrix.

Interference (Clutter): Interference can be modeled as a stationary stochastic process $w_c(t)$ with power spectrum $G_c(\omega) = \sigma_c^2$ that corresponds to flat interference. In general $G_c(\omega)$ is an arbitrary spectrum with specific bandwidth. [Noise dominant situation may be represented as $G_c(\omega) \approx 0$.] In the multichannel case, $G_c(\omega)$ represents a positive-definite spectral matrix.

Receiver Filter: A multiple set of waveforms $h_i(t) \leftrightarrow H_i(\omega)$ that may be causal or noncausal.

Decision Instant t_o: The overall goal is to maximize the receiver output *SINR* at the decision instant.

The simplest situation is to consider a single transmitter and a single receiver with a target (channel) and no interference signal. To start with assume the transmitter signal and the target to be known. Hence the goal is to design the optimum receiver. The receiver in this case is presented with the target output (fixed signal) that is corrupted by noise. The simplest noise situation is white noise (equal strength at all frequencies). Thus the simplest waveform design problem is to design a receiver to maximize the receiver output signal to noise ratio (*SNR*) for a given incoming signal that is buried in white noise. The solution is the well-known "matched filter" receiver [20–28].

2.2 Matched Filter Receivers

Consider the problem of designing a receiver to detect an incoming deterministic signal $s(t)$ that is buried in noise. This situation occurs all the time—in radar, sonar, and communication scenarios [1]. In a communication scene, a transmitted signal $f(t)$ goes through a channel with impulse response $q(t)$. The channel output signal $s(t)$ is further corrupted by noise $n(t)$, so that the observed signal $r(t)$ as shown in Fig. 2.3(a) is given by

$$r(t) = s(t) + n(t), \tag{2.3}$$

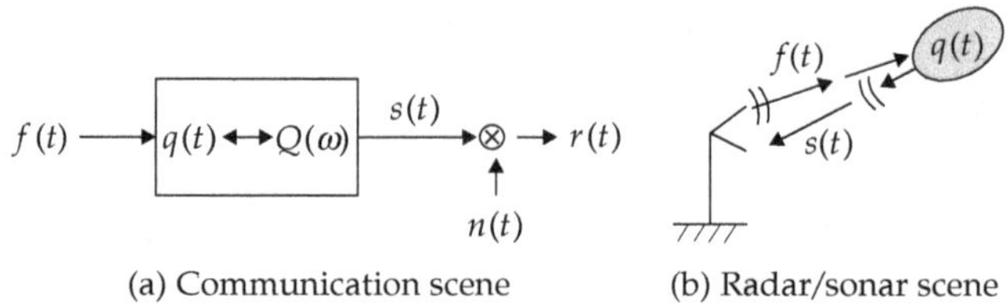

FIGURE 2.3 Noisy channel output/radar return.

where

$$s(t) = f(t) * q(t) \triangleq \int_{-\infty}^{+\infty} f(\tau)q(t-\tau)d\tau. \tag{2.4}$$

In radar and sonar, the same model represents the target return in noise, where $s(t)$ represents the target return signal (if any), and $q(t)$ the target impulse response [Fig. 2.3(b)].

In all these cases, the received signal $r(t)$ is passed through a receiver with transfer function $H(\omega)$ to minimize the effect of noise as shown in Fig. 2.4. The problem is how to design a good receiver.

Toward this, let $y(t)$ represent the receiver output. Then

$$y(t) = \hat{s}(t) + w(t) \tag{2.5}$$

where $\hat{s}(t)$ represents the signal part of the output and $w(t)$ the output noise. Clearly

$$\hat{s}(t) = s(t) * h(t) \tag{2.6}$$

and

$$w(t) = n(t) * h(t), \tag{2.7}$$

where

$$h(t) \leftrightarrow H(\omega) \tag{2.8}$$

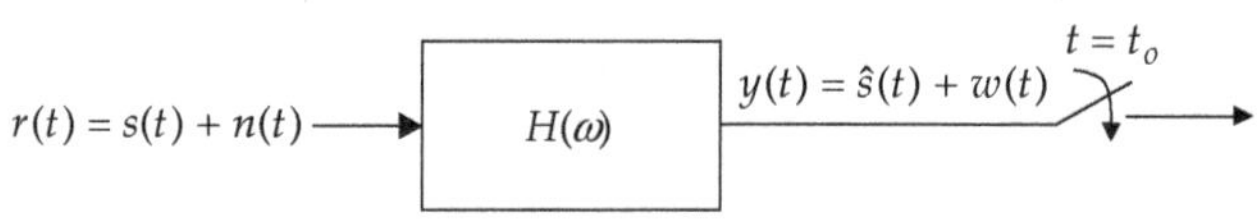

FIGURE 2.4 Receiver.

represents the unknown receiver characteristics. Obviously the receiver transfer function $H(\omega)$ should be selected so as to maximize the signal and minimize the output noise.

2.2.1 Matched Filter Receivers in White Noise

Consider the detection problem where the receiver output must be used to decide in favor of the target being present or absent in the data. Toward this, the output SNR of the received signal evaluated at some desired time instant $t = t_o$ can be used to select the optimum receiver so as to maximize the receiver output SNR at $t = t_o$. Note that the output SNR at $t = t_o$ is given by the instantaneous signal power at $t = t_o$ divided by the average noise power [1, 2].[1] Thus [see (2.1)]

$$SNR|_{t=t_o} \triangleq \frac{|\hat{s}(t_o)|^2}{E\{|w(t)|^2\}}. \tag{2.9}$$

From (2.6)

$$\hat{s}(t) \leftrightarrow \hat{S}(\omega) = S(\omega)H(\omega) \tag{2.10}$$

so that

$$\hat{s}(t) = \frac{1}{2\pi}\int_{-\infty}^{+\infty} \hat{S}(\omega)e^{j\omega t}d\omega, \tag{2.11}$$

and using (2.11), we obtain

$$\hat{s}(t_o) = \frac{1}{2\pi}\int_{-\infty}^{+\infty} S(\omega)H(\omega)e^{j\omega t_o}d\omega. \tag{2.12}$$

Let $G_n(\omega)$ and $G_w(\omega)$ represent the noise power spectral densities of $n(t)$ and $w(t)$ respectively. Hence

$$E\{|w(t)|^2\} = \frac{1}{2\pi}\int_{-\infty}^{+\infty} G_w(\omega)d\omega. \tag{2.13}$$

[1] The definition in (2.9) uses the *instantaneous* output signal power versus average output noise power. The numerator here is *not* an average value of the signal power as in the case of the denominator. This makes sense since the incoming signal is assumed to be deterministic, as opposed to the noise which is modeled as a stationary stochastic process.

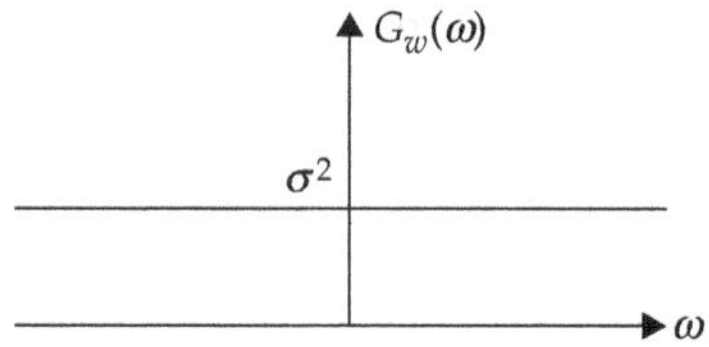

FIGURE 2.5 White noise.

But from (2.7)

$$G_w(\omega) = G_n(\omega)|H(\omega)|^2 \tag{2.14}$$

and hence

$$E\{|w(t)|^2\} = \frac{1}{2\pi}\int_{-\infty}^{+\infty} G_n(\omega)|H(\omega)|^2 d\omega. \tag{2.15}$$

If we assume $G_n(\omega)$ to be white noise as shown in Fig. 2.5, we have

$$G_n(\omega) = \sigma^2, \tag{2.16}$$

and using this in (2.15), we get

$$E\{|w(t)|^2\} = \frac{\sigma^2}{2\pi}\int_{-\infty}^{+\infty} |H(\omega)|^2 d\omega. \tag{2.17}$$

Substituting (2.12) and (2.17) into (2.9), we get the output SNR at $t = t_o$ to be

$$SNR_o = \frac{\left|\int_{-\infty}^{+\infty} S(\omega)H(\omega)e^{j\omega t_o} d\omega\right|^2}{2\pi\sigma^2 \int_{-\infty}^{+\infty} |H(\omega)|^2 d\omega}. \tag{2.18}$$

Obviously, the unknown receiver $H(\omega)$ in (2.18) should be chosen so as to maximize the output SNR at the observation time $t = t_o$. We can use

Schwarz' inequality to simplify (2.18). Note that Schwarz' inequality states that

$$\left| \int A(\omega) B(\omega) d\omega \right|^2 \leq \int |A(\omega)|^2 d\omega \int |B(\omega)|^2 d\omega \tag{2.19}$$

and equality in (2.19) is achieved if and only if

$$B(\omega) = \mu A^*(\omega), \tag{2.20}$$

where μ is a constant. To make use of (2.19) in the present problem, let

$$A(\omega) = S(\omega) e^{j\omega t_o} \tag{2.21}$$

and

$$B(\omega) = H(\omega). \tag{2.22}$$

We can choose the constant multiplier μ in (2.20) to be unity, since it gets canceled out from the numerator and the denominator of (2.18). Using (2.21) and (2.22) in (2.18), from (2.19) we obtain

$$\left| \int_{-\infty}^{+\infty} S(\omega) H(\omega) e^{j\omega t_o} d\omega \right|^2 \leq \int_{-\infty}^{+\infty} |S(\omega)|^2 d\omega \int_{-\infty}^{+\infty} |H(\omega)|^2 d\omega. \tag{2.23}$$

With (2.23) in (2.18), we get

$$SNR_o \leq \frac{1}{2\pi\sigma^2} \int_{-\infty}^{+\infty} |S(\omega)|^2 d\omega. \tag{2.24}$$

Thus the maximum value of the output *SNR* is given by

$$SNR_{\max} = \frac{\frac{1}{2\pi} \int_{-\infty}^{+\infty} |S(\omega)|^2 d\omega}{\sigma^2} = \frac{\int_{-\infty}^{+\infty} |s(t)|^2 dt}{\sigma^2} = \frac{E}{\sigma^2}, \tag{2.25}$$

a quantity *independent* of $H(\omega)$, and from (2.22) this maximum value is achieved if and only if

$$H(\omega) = A^*(\omega) = \left(S(\omega) e^{j\omega t_o} \right)^* = S^*(\omega) e^{-j\omega t_o}, \tag{2.26}$$

or, in the time domain we obtain

$$h(t) = s^*(t_o - t). \tag{2.27}$$

If $s(t)$ is real, the matched filter solution reduces to the classic form

$$h(t) = s(t_o - t) \tag{2.28}$$

which is a time-reversed and shifted version of $s(t)$. Thus, under additive white noise, the optimum receiver that maximizes the output *SNR* at $t = t_o$ is given by (2.26) and (2.27). Notice that $h(t)$ depends only on the receiver input signal waveform $s(t)$, or the receiver is *matched* to its input signal $s(t)$. To summarize, if $s(t) + n(t)$ is received and the noise $n(t)$ is white, then the optimum receiver is given by the classic matched filter solution as in Fig. 2.6.

In Fig. 2.6, the time index t_o represents the time-instant at which the output *SNR* is maximized and hence that instant ($t = t_o$) must be used to make the decision. Note that the optimum filter in (2.27) need not represent a causal solution. However if $s(t)$ is a finite duration signal, then by suitably selecting t_o, the receiver $h(t)$ can be made causal.

From (2.4), we have

$$S(\omega) = F(\omega)Q(\omega) \tag{2.29}$$

and hence the optimum receiver

$$H(\omega) = S^*(\omega)e^{-j\omega t_o} = F^*(\omega)Q^*(\omega)e^{-j\omega t_o}. \tag{2.30}$$

From (2.30), the optimum filter is matched to the input signal and channel characteristics. From (2.25), the maximum value of output

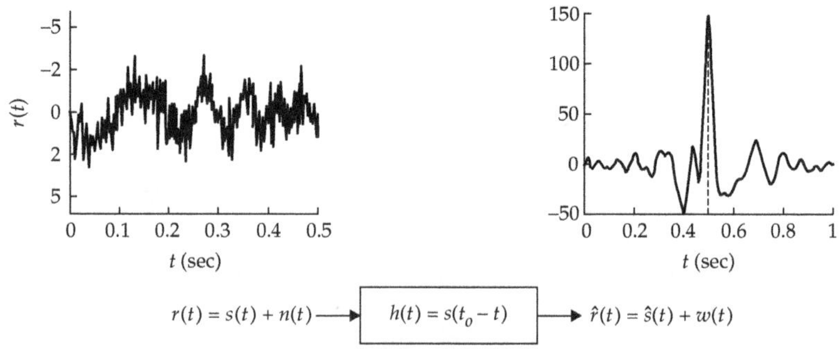

FIGURE 2.6 Optimum receiver (matched filter).

SNR is also given by

$$SNR_{\max} = \frac{\int_{-\infty}^{+\infty} |s(t)|^2\, dt}{\sigma^2} = \frac{\frac{1}{2\pi}\int_{-\infty}^{+\infty} |S(\omega)|^2\, d\omega}{\sigma^2} = \frac{\frac{1}{2\pi}\int_{-\infty}^{+\infty} |F(\omega)Q(\omega)|^2\, d\omega}{\sigma^2}. \tag{2.31}$$

The same conclusions can also be reached by making use of a time domain analysis. To see this, referring back to (2.6), (2.7), and (2.9), we have

$$SNR_t = \frac{\left|\int_{-\infty}^{+\infty} h(\tau)s(t-\tau)d\tau\right|^2}{E\left\{\left|\int_{-\infty}^{+\infty} h(\tau)n(t-\tau)d\tau\right|^2\right\}} = \frac{\left|\int_{-\infty}^{+\infty} h(\tau)s(t-\tau)d\tau\right|^2}{\int_{-\infty}^{+\infty}\int_{-\infty}^{+\infty} h(\tau_1)h^*(\tau_2)E\{n(t-\tau_1)n^*(t-\tau_2)\}d\tau_1 d\tau_2}. \tag{2.32}$$

If we assume $n(t)$ to be stationary white noise, then

$$E\left\{n(t-\tau_1)n^*(t-\tau_2)\right\} = R_{nn}(\tau_1-\tau_2) = \sigma^2\delta(\tau_1-\tau_2), \tag{2.33}$$

and (2.32) becomes

$$SNR_{t=t_o} = \frac{\left|\int_{-\infty}^{+\infty} h(\tau)s(t_o-\tau)d\tau\right|^2}{\int_{-\infty}^{+\infty}\int_{-\infty}^{+\infty} h(\tau_1)h^*(\tau_2)\sigma^2\delta(\tau_1-\tau_2)d\tau_1 d\tau_2} = \frac{\left|\int_{-\infty}^{+\infty} h(\tau)s(t_o-\tau)d\tau\right|^2}{\sigma^2\int_{-\infty}^{+\infty} |h(\tau)|^2\, d\tau}. \tag{2.34}$$

Once again, using Schwarz' inequality, the numerator in the above expression gives

$$\left|\int_{-\infty}^{+\infty} h(\tau)s(t_o-\tau)d\tau\right|^2 \leq \int_{-\infty}^{+\infty} |h(\tau)|^2\,d\tau \int_{-\infty}^{+\infty} |s(t_o-\tau)|^2\,d\tau \tag{2.35}$$

and (2.34) reduces to

$$SNR_{\max}|_{t=t_o} \leq \frac{1}{\sigma^2}\int_{-\infty}^{+\infty} |s(t_o-\tau)|^2\,d\tau = \frac{1}{\sigma^2}\int_{-\infty}^{+\infty} |s(t)|^2\,dt = \frac{E}{\sigma^2} \tag{2.36}$$

with equality if and only

$$h(t) = s^*(t_o - t), \tag{2.37}$$

which agrees with (2.27).

Example 2.1 A triangular pulse $f(t)$ as shown in Fig. 2.7 is transmitted through an ideal channel with transfer function

$$Q(\omega) = e^{-j\omega t_c}. \tag{2.38}$$

If white noise corrupts the input signal, find the optimum receiver.

Solution

Since

$$r(t) = s(t) + n(t) \tag{2.39}$$

and $n(t)$ is white noise, the optimum receiver is the matched filter given by (2.26) and (2.27), that is,

$$h(t) = s(t_o - t). \tag{2.40}$$

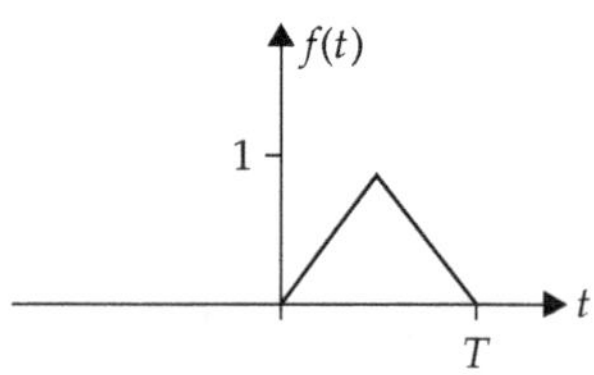

FIGURE 2.7 Triangular pulse.

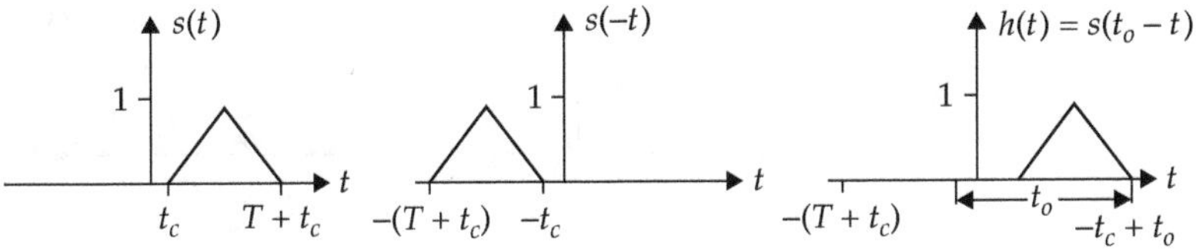

(a) $s(t)$ (b) $s(-t)$ (c) $h(t)$

FIGURE 2.8 Optimum receiver at decision instant $t = t_o$.

Here

$$S(\omega) = F(\omega)Q(\omega) = F(\omega)e^{-j\omega t_c} \tag{2.41}$$

so that

$$s(t) = f(t - t_c) \tag{2.42}$$

and this is illustrated in Fig. 2.8.

In particular, if $t_o = t_c$, then $h_c(t) = s(t_c - t)$ and it is shown in Fig. 2.9(a). On the other hand, if $t_o = t_c + T$, then we get $h(t)$ as shown in Fig. 2.9(b). In Fig. 2.9(a), $h(t)$ is noncausal, whereas in Fig. 2.9(b) it represents a causal filter.

From (2.25) and (2.31), the maximum output SNR in this case at $t = t_o$ is given by

$$SNR_{\max} = \frac{\int_{-\infty}^{+\infty} |s(t)|^2\, dt}{\sigma^2} = \frac{2\int_0^{T/2} (2t/T)^2\, dt}{\sigma^2} = \frac{8}{3T^2\sigma^2}\left(\frac{T}{2}\right)^3 = \frac{T}{3\sigma^2}. \tag{2.43}$$

To summarize, if $r(t) = s(t) + n(t)$, where $s(t)$ has the form in Fig. 2.10(a), then if $n(t)$ is white noise, the optimum receiver impulse response is given by

$$h(t) = s(t_o - t), \tag{2.44}$$

and this is illustrated in Fig. 2.10(a)–(c) for some arbitrary t_o.

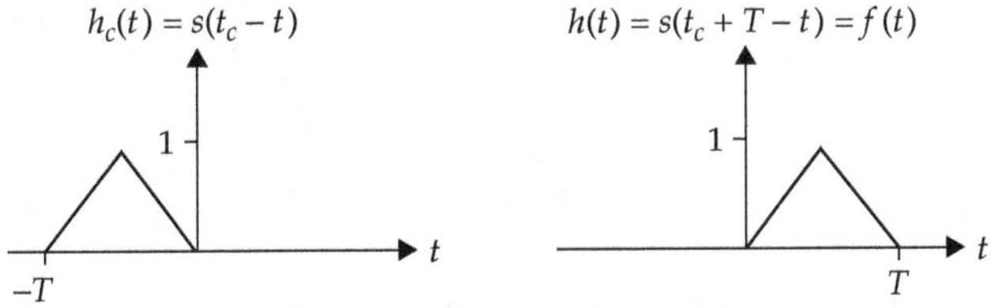

(a) Optimum receiver at decision instant $t_o = t_c$.

(b) Optimum receiver at decision instant $t_o = t_c + T$.

FIGURE 2.9 Optimum receiver at decision instant $t_o = t_c$.

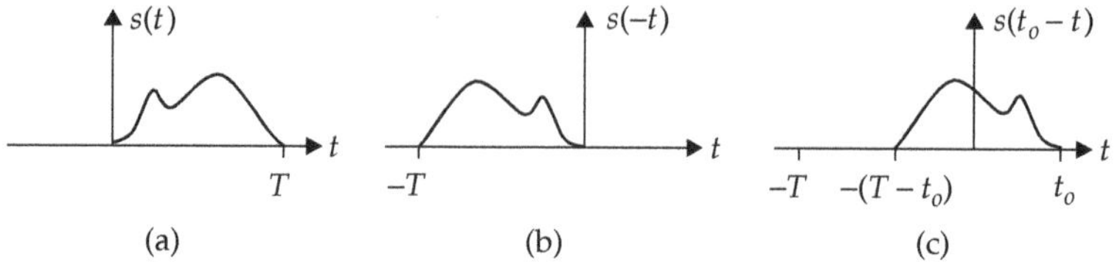

FIGURE 2.10 Optimum receiver for some arbitrary t_o.

Matched Filter as a Correlator Receiver

Suppose the received data $r(t)$ is available in the interval $(0, t)$. In that case using (2.44), the matched filter output is given by

$$\hat{r}(t) = r(t) * h(t) = \int_0^t r(\tau)h(t-\tau)d\tau = \int_0^t r(\tau)s(t_o - t + \tau)d\tau, \quad (2.45)$$

since $h(t-\tau) = s(t_o - (t-\tau)) = s(t_o - t + \tau)$. The matched filter maximizes its output SNR at $t = t_o$, and hence the output $\hat{r}(t)$ must be used at that instant to make further decision. Thus from (2.45), we have

$$\hat{r}(t_o) = \int_0^{t_o} r(\tau)s(\tau)d\tau = \int_0^{t_o} r(t)s(t)dt. \quad (2.46)$$

Notice that (2.46) can be interpreted as a correlator receiver and it can be represented as in Fig. 2.11.

From Fig. 2.11, the received waveform is correlated with $s(t)$ and integrated until $t = t_o$ to generate the output $\hat{r}(t_o)$. Since the correlator receiver in (2.46) follows from the matched filter receiver in (2.44), we conclude that, in the case of additive white noise, both matched

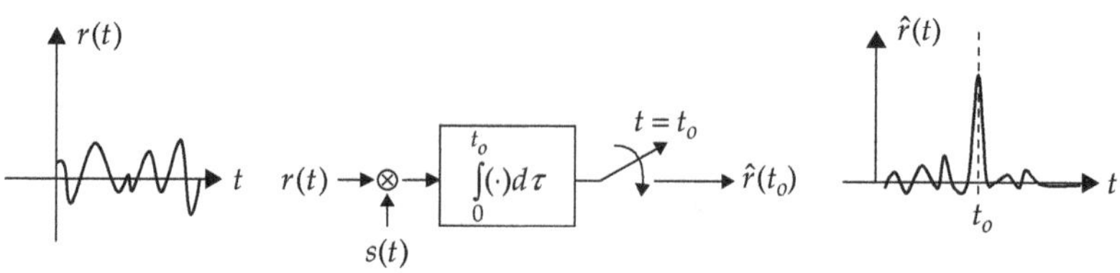

FIGURE 2.11 Correlator receiver.

filter receiver and the correlator receiver are equivalent realizations [29–32].

2.2.2 Matched Filter Receivers in Colored Noise

What if the receiver noise is of nonwhite nature? [3] How does one design the optimum causal filter in the colored noise case? Interestingly, by increasing t_o alone the optimum solution cannot be made causal in this case. To see this, let $G_n(\omega)$ represent the input noise spectrum so that the receiver output noise power is given by

$$E\left\{w^2(t)\right\} = \frac{1}{2\pi}\int_{-\infty}^{+\infty} G_n(\omega)\,|H(\omega)|^2 d\omega. \tag{2.47}$$

Every spectrum $G(\omega)$ that is nonnegative and satisfying the Paley-Wiener condition given by [33]

$$\int_{-\infty}^{+\infty} \frac{|\ln G(\omega)|}{1+\omega^2} d\omega < \infty, \tag{2.48}$$

can be factorized in terms of its Wiener factor $L(s)$ as

$$G(\omega) = L(j\omega)L^*(j\omega), \tag{2.49}$$

where $L(s)$ is *analytic* together with its *inverse* $L^{-1}(s)$ in the right-half plane of the complex $s = \sigma + j\omega$ plane (minimum phase system). Let $L_n(s)$ represent the Wiener factor associated with $G_n(\omega)$ so that

$$G_n(\omega) = |L_n(j\omega)|^2 \tag{2.50}$$

and

$$SNR|_{t=t_o} = \frac{\left|\frac{1}{2\pi}\int_{-\infty}^{+\infty} H(\omega)S(\omega)e^{j\omega t_o}d\omega\right|^2}{\frac{1}{2\pi}\int_{-\infty}^{+\infty} G_n(\omega)\,|H(\omega)|^2\,d\omega} = \frac{\left|\frac{1}{2\pi}\int_{-\infty}^{+\infty} H(\omega)S(\omega)e^{j\omega t_o}d\omega\right|^2}{\frac{1}{2\pi}\int_{-\infty}^{+\infty} |L_n(\omega)H(\omega)|^2\,d\omega}. \tag{2.51}$$

To accommodate the new denominator, let us rewrite the term $H(\omega)S(\omega)$ in the numerator as

$$\{L_n(j\omega)H(\omega)\}\left\{L_n^{-1}(j\omega)S(\omega)\right\}. \tag{2.52}$$

Substituting this into (2.51) and a straightforward application of Schwarz' inequality gives

$$\rho(t_o) \le \frac{1}{2\pi}\int_{-\infty}^{+\infty}\left|L_n^{-1}(j\omega)S(\omega)\right|^2 d\omega = \frac{1}{2\pi}\int_{-\infty}^{+\infty}\frac{|S(\omega)|^2}{G_n(\omega)}d\omega, \tag{2.53}$$

with equality if and only if

$$L_n(j\omega)H(\omega) = \left(L_n^{-1}(j\omega)\right)^* S^*(\omega)e^{-j\omega t_o} \tag{2.54}$$

or

$$H(\omega) = \frac{S^*(\omega)e^{-j\omega t_o}}{G_n(\omega)}. \tag{2.55}$$

Thus in the time domain if

$$l_{inv}(t) \leftrightarrow L_n^{-1}(j\omega), \tag{2.56}$$

then the optimum receiver impulse response is given by

$$h(t) = l_{inv}(t) * l_{inv}^*(-t) * s(t_o - t). \tag{2.57}$$

Clearly (2.57) represents a noncausal waveform, and a simple finite shift by t_o alone (however large) may not be enough to maintain the causal nature of the receiver [2, 3].

Optimum Causal Matched Filter in Colored Noise

To obtain the optimum causal receiver, it is necessary to proceed differently. Toward this, let [34]

$$v(t) \leftrightarrow H(\omega)L_n(j\omega) \tag{2.58}$$

and let $g(t)$ represent the inverse transform of $L_n^{-1}(j\omega)S(\omega)$

$$g(t) \leftrightarrow L_n^{-1}(j\omega)S(\omega) = L_n^{-1}(j\omega)Q(\omega)F(\omega) \tag{2.59}$$

so that $g(t)$ is causal and

$$g^*(-t) \leftrightarrow \left(L_n^{-1}(j\omega)S(\omega)\right)^* \tag{2.60}$$

that gives

$$g^*(t_o - t) \leftrightarrow \left(L_n^{-1}(j\omega)\right)^* S^*(\omega)e^{-j\omega t_o}. \tag{2.61}$$

Since $L_n(s)$ and $L_n^{-1}(s)$ are analytic in Re $s > 0$, we have $v(t)$ and $g(t)$ are causal waveforms and using Parseval's theorem

$$\begin{aligned}\int_{-\infty}^{+\infty} v(t)g(t_o - t)dt &= \int_{-\infty}^{+\infty} v(t)\left\{g^*(t_o - t)\right\}^* dt \\ &= \frac{1}{2\pi}\int_{-\infty}^{+\infty} \{H(\omega)L_n(j\omega)\}\left\{L_n^{-1}(j\omega)S(\omega)e^{j\omega t_o}\right\} d\omega \\ &= \frac{1}{2\pi}\int_{-\infty}^{+\infty} H(\omega)S(\omega)e^{j\omega t_o} d\omega \triangleq Q,\end{aligned} \tag{2.62}$$

same as the numerator factor in (2.51). But from (2.62)

$$Q = \int_{-\infty}^{+\infty} v(t)g(t_o - t)dt = \int_{0}^{+\infty} v(t)g(t_o - t)dt, \tag{2.63}$$

because of the causal nature of $v(t)$. However, by making use of the unit step function $u(t)$, we can rewrite the later integral in (2.63) as follows

$$Q = \int_{0}^{+\infty} v(t)g(t_o - t)dt = \int_{-\infty}^{+\infty} v(t)g(t_o - t)u(t)dt. \tag{2.64}$$

Let $K(\omega)$ represent the transform of the causal pulse $g^*(t_o - t)u(t)$ as shown in Fig. 2.12. Thus

$$g^*(t_o - t)u(t) \leftrightarrow K(\omega), \tag{2.65}$$

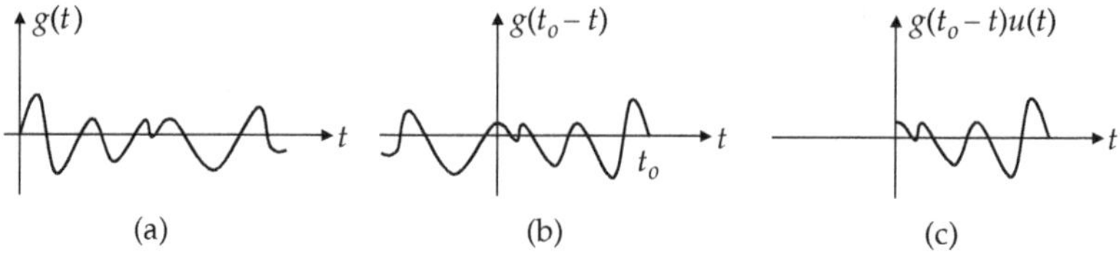

FIGURE 2.12 Causal part of time-shifted waveform.

and once again Parseval's theorem applied to (1.64) gives

$$\begin{aligned} Q &= \int_{-\infty}^{+\infty} \{v(t)\}\left\{g^*(t_o - t)u(t)\right\}^* dt \\ &= \frac{1}{2\pi}\int_{-\infty}^{+\infty} \{H(\omega)L_n(j\omega)\}\left\{K^*(\omega)\right\} d\omega. \end{aligned} \tag{2.66}$$

Using this in (2.51), we get

$$\begin{aligned} \rho(t_o) \triangleq SNR|_{t=t_o} &= \frac{\left|\frac{1}{2\pi}\int_{-\infty}^{+\infty} H(\omega)L_n(j\omega)K^*(\omega)d\omega\right|^2}{\frac{1}{2\pi}\int_{-\infty}^{+\infty} |L_n(\omega)H(\omega)|^2\, d\omega} \\ &\leq \frac{1}{2\pi}\int_{-\infty}^{+\infty} |K(\omega)|^2\, d\omega, \end{aligned} \tag{2.67}$$

with equality if and only if

$$H(\omega)L_n(j\omega) = K(\omega) \tag{2.68}$$

or equivalently

$$H(\omega) = L_n^{-1}(j\omega)K(\omega) \leftrightarrow l_{inv}(t) * g^*(t_o - t)u(t). \tag{2.69}$$

Note that the impulse response $h(t)$ of this matched filter is necessarily causal in this case. Thus

$$h_c(t) = l_{inv}(t) * g^*(t_o - t)u(t). \tag{2.70}$$

In addition using (2.65) in (2.67) we obtain

$$\rho_{\max}(t_o) = \frac{1}{2\pi}\int_{-\infty}^{+\infty} |K(\omega)|^2\,d\omega = \int_0^{t_o} |g(t)|^2\,dt. \tag{2.71}$$

Equations (2.68)–(2.71) represent the optimum causal solution, where $K(\omega)$ and $g(t)$ are as defined in (2.65) and (2.59) respectively.

Notice that the casual solution in (2.69) is neither a simple shifted version of the noncausal solution in (2.57), nor can it be obtained directly from (2.57). Using (2.55), (2.56), and (2.61), we can rewrite (2.57) as

$$h_{nc}(t) = l_{inv}(t) * g^*(t_o - t) \tag{2.72}$$

and on comparing (2.72) with (2.69), we notice that the causal solution in (2.69) and (2.70) drops the noncausal portion of $g^*(t_o - t)$ *prior to* its convolution with $l_{inv}(t)$ as in Fig. 2.13. We also have

$$h_{nc}(t) = h_c(t) + l_{inv}(t) * g^*(t_o - t)u(-t). \tag{2.73}$$

Since $l_{inv}(t) * g^*(t_o - t)u(-t)$ has a causal part, it will be impossible to obtain the optimum causal filter $h_c(t)$ from the noncausal one at the final state.

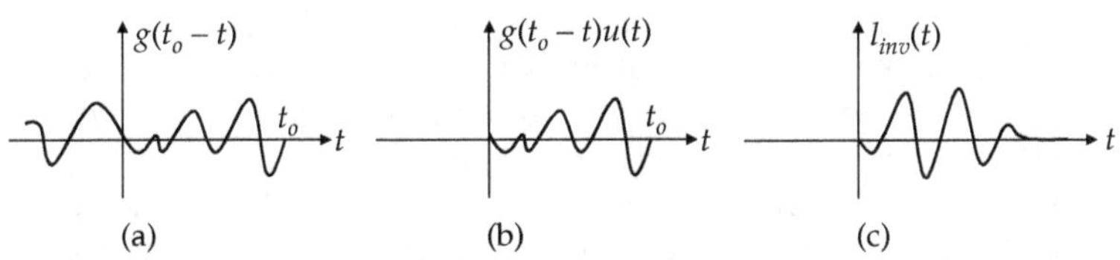

FIGURE 2.13 Causal solution of the matched filter.

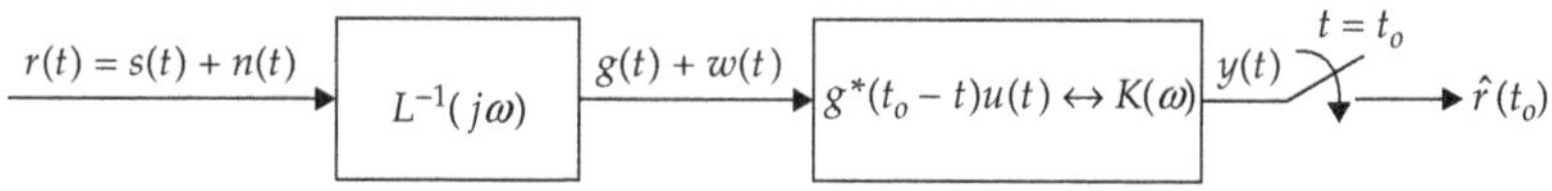

FIGURE 2.14 Matched filter in colored noise.

The causal matched filter receiver in (2.69) that corresponds to the colored noise case can be given an interesting interpretation as shown in Fig. 2.14.

From Fig. 2.14, the input noise $n(t)$ is first "whitened" using the filter $L^{-1}(j\omega)$ to generate white noise $w(t)$. During this transformation, the input signal $s(t)$ becomes $g(t)$ that is given by (2.59). Since $g(t)$ is contaminated by white noise, from (2.27), $g^*(t_o - t)$ represents the optimum *noncausal* matched filter receiver for the input $g(t) + w(t)$. However from (2.69) and (2.70), the causal matched filter receiver for the input $g(t) + w(t)$ is given by $g^*(t_o - t)u(t) \leftrightarrow K(\omega)$. Hence taken together the concatenated overall receiver has the form in (2.69), where the input is first whitened followed by matched filtering to generate the desired output.

Finally, in the white noise case $G_n(\omega) = \sigma^2$ so that from (2.59), $g(t) = s(t)/\sigma$ and substituting this into (2.71), we obtain

$$SNR_{\max} = \rho(t_o) = \frac{1}{\sigma^2} \int_0^{t_o} |s(t)|^2 \, dt. \tag{2.74}$$

As a result, the causal matched filter in the white noise case is given by (compared with the noncausal receiver in (2.27))

$$h_c(t) = s^*(t_o - t)u(t). \tag{2.75}$$

From (2.71) and (2.74), in the causal case the signal to noise ratio at the matched filter output at $t = t_o$ is proportional to the energy in the input signal *up to that instant*. This relation is *unlike* (2.36), where the *SNR* is proportional to the *total* energy. Thus in the case of a causal receiver, the maximum *SNR* output is a monotonically nondecreasing function of the observation instant t_o, and as we shall see later, this property has certain interesting implications.

Since the matched filter can generate outputs with large peaks, they contain the potential for generating time-compressed waveforms that may lead to accurate target range estimation. Next, we investigate this possibility and show that the chirp waveform indeed possesses the desired time (or range) compression property.

2.3 Chirp and Pulse Compression

A receiver matched to the incoming signal produces large peaks at the output (Fig. 2.15) and hence it has the potential for time (or range) compression [35]. For example, the matched filter receiver corresponding to the incoming signal $s(t)$ is given by $s^*(t_o - t)$, where t_o represents the output observation instant and for finite duration input signals, t_o is usually chosen to be the signal duration T itself, so that the matched filter represents a causal filter (see Fig. 2.9). In that case, from (2.45) the matched filter output $y(t)$ equals

$$y(t) = \int_0^t s(\tau)s^*(t_o - t + \tau)d\tau \tag{2.76}$$

which peaks at $t = t_o$ and the peak value is given by

$$y(t_o) = \int_0^{t_o} |s(t)|^2\, dt. \tag{2.77}$$

It follows that if the input waveform is delayed by T_o due to target echo, the matched filter output also is delayed by the same amount, and hence the method can be used for range estimation as well, provided the output is time-compressed enough to give the desired range resolution [36–39].

To analyze this concept further, consider a finite energy signal $s(t)$ with corresponding transform $S(\omega)$. Then $s(t)$ is square-integrable and hence

$$E_s = \int_{-\infty}^{+\infty} |s(t)|^2\, dt = \frac{1}{2\pi}\int_{-\infty}^{+\infty} |S(\omega)|^2\, d\omega < \infty. \tag{2.78}$$

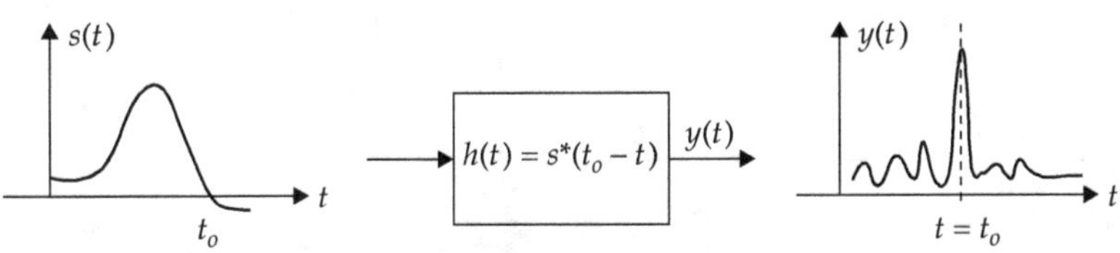

FIGURE 2.15 Matched filter and time compression.

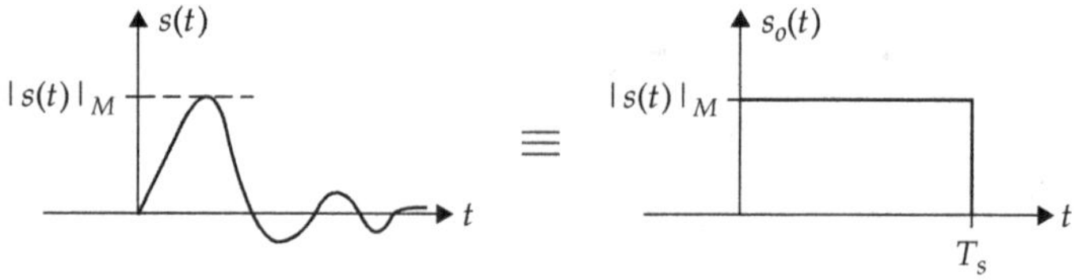

FIGURE 2.16 An arbitrary signal and its equal energy equivalent rectangular pulse.

Assume that $|s(t)|_M$, the maximum value of $s(t)$ is finite. Then we may define the effective duration of $s(t)$ to be [35]

$$T_s = \frac{E_s}{|s(t)|_M^2}, \tag{2.79}$$

and similarly, its effective bandwidth equals

$$B_s = \frac{E_s}{|S(\omega)|_M^2}, \tag{2.80}$$

where $|S(\omega)|_M$ represents the maximum value of its transform $S(\omega)$. From (2.79), a rectangular pulse with amplitude $|s(t)|_M$ and width T_s has the same energy as $s(t)$, as shown in Figs. 2.16 and 2.17.

A filter matched to $s(t)$ has transform [Fig. 2.15 or (2.26) with $t_o = 0$]

$$H(\omega) = S^*(\omega), \tag{2.81}$$

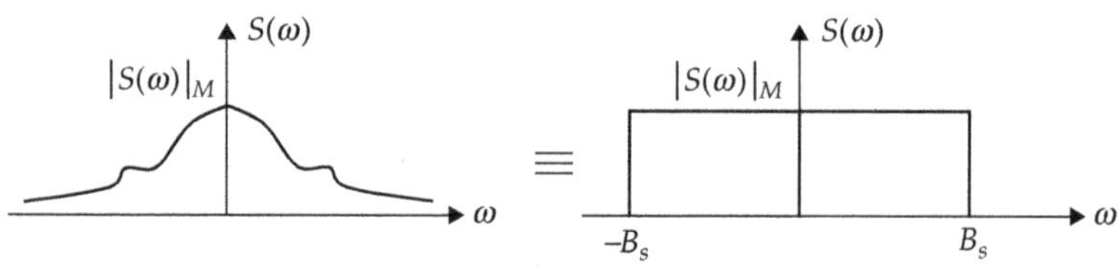

FIGURE 2.17 An arbitrary spectrum and its equal energy equivalent rectangular spectrum.

and hence its output signal transform is given by $H(\omega)S(\omega) = |S(\omega)|^2$. This gives the filter output signal to be

$$y(t) = \frac{1}{2\pi}\int_{-\infty}^{+\infty} |S(\omega)|^2 e^{j\omega t} d\omega, \tag{2.82}$$

whose peak value is obtained at $t = 0$. Hence, its peak value is given by

$$|y(t)|_M = y(0) = \frac{1}{2\pi}\int_{-\infty}^{+\infty} |S(\omega)|^2 d\omega = E_s. \tag{2.83}$$

The output signal energy E_y equals

$$\begin{aligned} E_y &= \int_{-\infty}^{+\infty} |y(t)|^2 dt = \frac{1}{2\pi}\int_{-\infty}^{+\infty} |S(\omega)|^4 d\omega \\ &\le |S(\omega)|_M^2 \frac{1}{2\pi}\int_{-\infty}^{+\infty} |S(\omega)|^2 d\omega = |S(\omega)|_M^2 E_s. \end{aligned} \tag{2.84}$$

Hence

$$E_y = \frac{|S(\omega)|_M^2}{\eta} E_s \tag{2.85}$$

where η is an unknown constant that is *greater than* unity. Thus we have

$$\eta = \frac{|S(\omega)|_M^2 \int_{-\infty}^{+\infty} |S(\omega)|^2 d\omega}{\int_{-\infty}^{+\infty} |S(\omega)|^4 d\omega} > 1. \tag{2.86}$$

Using (2.79) and (2.83)–(2.85), the effective duration of the output signal $y(t)$ is given by

$$T_y = \frac{E_y}{|y(t)|_M^2} = \frac{|S(\omega)|_M^2}{\eta}\frac{E_s}{E_s^2} = \frac{|S(\omega)|_M^2}{\eta}\frac{1}{E_s} = \frac{1}{\eta B_s} \tag{2.87}$$

where B_s represents the effective bandwidth of $s(t)$ as in (2.80). (See Fig. 2.18).

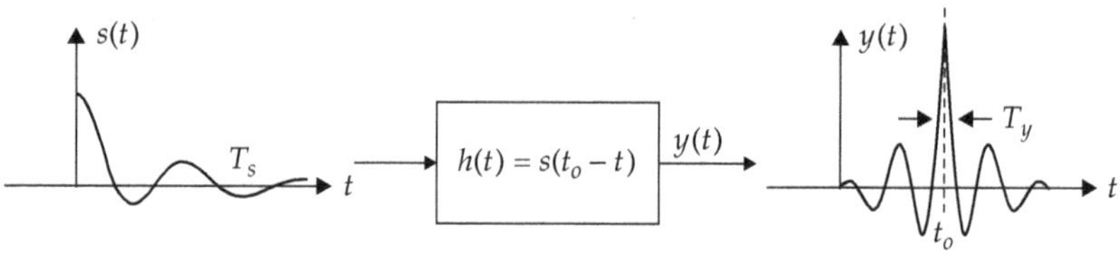

FIGURE 2.18 Effective duration of the matched filter output $y(t)$.

To determine the pulse compression realized by the matched filter output compared to its input, we can examine the input to output effective pulse length ratio T_s/T_y. From (2.87) [35],

$$\frac{T_{input}}{T_{output}} = \frac{T_s}{T_y} = T_s \eta B_s = \eta T_s B_s \geq T_s B_s \tag{2.88}$$

since $\eta > 1$. Thus large time compression ratios can be realized at the matched filter output by *any* input signal with large time-bandwidth product $T_s B_s$. In this case, the matched filter output will appear as a narrow pulse whose peak corresponds to the input delay [40–42].

As Fig. 2.19 shows, the bandwidth of a rectangular pulse is inversely proportional to its pulse width T, and hence its time-bandwidth product is essentially a constant. As a result, a rectangular pulse has poor time-compression property. In general, the time-bandwidth product of any pulse is a constant since the bandwidth is usually inversely proportional to its duration. In those cases, output pulse compression must result from large values for the factor η in (2.86). Hence an alternate way to realize large pulse compression ratio is to transmit a waveform whose η factor is large.

Interestingly, as Sec. 2.6 shows, the chirped pulse signal (both modulated as well as centered at baseband) has the capacity to realize

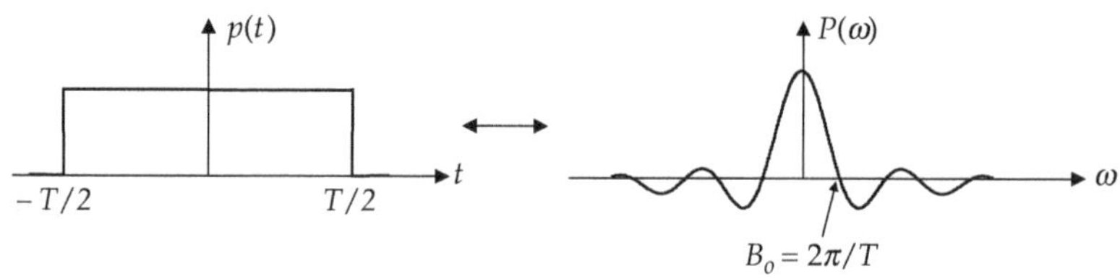

FIGURE 2.19 A rectangular pulse and its transform.

large time-bandwidth products. For example, from (2.202)–(2.204) the transform of the causal chirped signal

$$s(t) = \begin{cases} \cos\left(\omega_o t + \beta t^2\right), & 0 < t < T \\ 0, & \text{otherwise,} \end{cases} \tag{2.89}$$

is given by [33]

$$|S(\omega)| \approx \begin{cases} \frac{1}{2}\sqrt{\pi/\beta}, & \omega_o < |\omega| < \omega_o + 2\beta T \\ 0, & \text{otherwise,} \end{cases} \tag{2.90}$$

provided $\beta T \gg 1$.

From (2.90), the bandwidth B_s of a chirped signal equals $2\beta T$ and the transform is flat over that region. Notice that the bandwidth can be increased by increasing the duration of the chirp signal. Finally, substituting these values into (2.88) we get

$$\frac{T_s}{T_y} = T\eta 2\beta T = 2\eta\beta T^2 \geq 1 \tag{2.91}$$

since $\beta T \gg 1$. From (2.84), (2.85), and (2.90),

$$\eta = \frac{|S(\omega)|_M^2}{\frac{1}{2\pi}\int_{-\infty}^{+\infty} |S(\omega)|^4 \, d\omega} E_s \approx \frac{(\pi/\beta)^2}{(\pi/\beta)^2} = 1. \tag{2.92}$$

From (2.91) by increasing the pulse duration of the chirp signal, it is possible to realize output pulse compression to any desired degree as shown in Fig. 2.20. This remarkable property of the chirp signal is exploited in radar for accurate target range estimation. Together with its ease in generation, the chirp signal has become indispensable in radar [43–46].

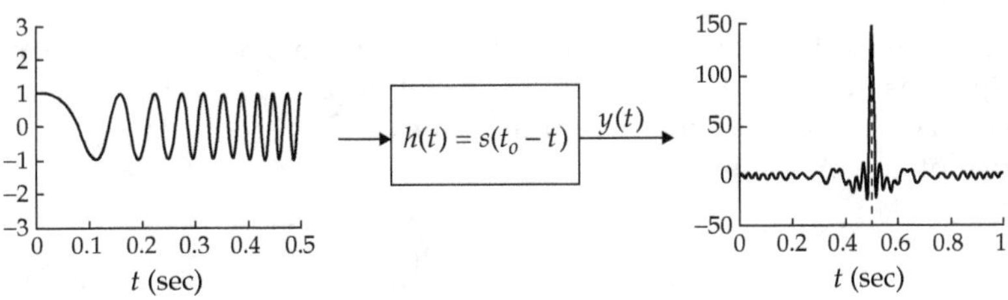

FIGURE 2.20 Chirp signal and its matched filter output $y(t)$.

The above analysis assumes the channel is ideal so that the transmit signal $s(t)$ is returned with an unknown delay τ_o as $s(t-\tau_o)$. In that case in the presence of additive noise, the matched filter achieves excellent pulse compression when $s(t)$ is a chirp signal. However, if the target (channel) interaction is significant enough to change the shape of the return signal, the chirp waveform may not be the optimum transmit signal. Clearly in that case the designer must pay attention to the target (channel) characteristics as well [47].

Before addressing this general design problem, we show that if the target/channel effect is simply to modulate the chirp signal with a baseband waveform $a(t)$ that is narrowband so that

$$b(t) = a(t)e^{-j(\omega_o t+\beta t^2)}, \quad 0 < t < T \tag{2.93}$$

is received at the receiver input, then remarkably enough the original matched filter matched to the chirp signal still retains the desirable pulse compression property.

The target interaction can result in either modulating the transmit chirp signal by a time-varying gain function $a(t)$ as in (2.93), or it can result in convolving the chirp by a target impulse response function $a_o(t)$. Interestingly, in the later case also the output $b_o(t)$ has the same form as in (2.93), since

$$\begin{aligned} b_o(t) = a_o(t) * e^{-j\beta t^2} &= \int_0^T a_o(\tau)e^{-j\beta(t-\tau)^2}d\tau \\ &= e^{-j\beta t^2}\int_0^T a_o(\tau)e^{-j\beta\tau(\tau-2t)}d\tau = A_o(t)e^{-j\beta t^2}. \end{aligned} \tag{2.94}$$

Thus target interaction either in the form of convolution or multiplicative gain can be modeled as in (2.93) and as we show below in both cases it is still possible to attain pulse compression property [48].

Modulated Chirp Signal

Suppose a suitable baseband signal $a(t)$ is used to modulate the chirp signal as in (2.197) to generate the output in (2.93). Clearly, $a(t)$ and $b(t)$ have the same duration but the bandwidth of $b(t)$ is larger than that of $a(t)$. Also, since

$$|a(t)| = |b(t)| \tag{2.95}$$

the energy of both signals are the same. Consider the receiver $h_o(t)$ that is matched to only the chirp signal as shown in Fig. 2.21.

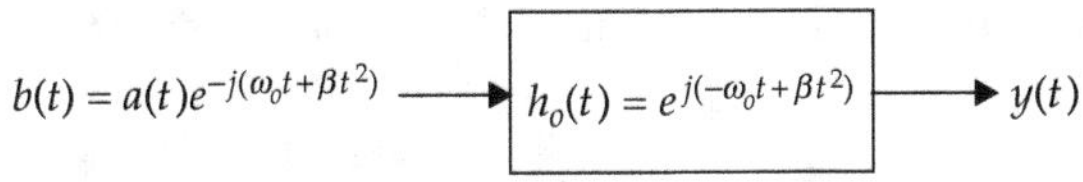

FIGURE 2.21 Modulated chirp signal and matched filter output.

(use (2.27) with $t_o = 0$), that is,

$$h_o(t) = s^*(-t) = e^{j(-\omega_0 t+\beta t^2)} \tag{2.96}$$

so that the matched filter output $y(t)$ equals

$$\begin{aligned} y(t) &= \int_{-\infty}^{+\infty} b(\tau)h_o(t-\tau)d\tau \\ &= \int_{-\infty}^{+\infty} a(\tau)e^{-j(\omega_0\tau+\beta\tau^2)}e^{j\left(-\omega_0(t-\tau)+\beta(t-\tau)^2\right)}d\tau \\ &= e^{j(-\omega_0 t+\beta t^2)}\int_{-\infty}^{+\infty} a(\tau)e^{-j2\beta t\tau}d\tau = e^{j(-\omega_0 t+\beta t^2)}A(2\beta t). \end{aligned} \tag{2.97}$$

Thus the output pulse width T_y depends only on the width of $A(\omega)$ and β, and it is given by

$$T_y = \frac{\text{width of } A(\omega)}{2\beta}. \tag{2.98}$$

As a result by selecting the chirp modulation parameter β large enough, once again the output can be made as narrow as possible, even when an unknown carrier modulates the transmit chirp signal. This property is very desirable when the channel has time-dependent amplification gains. Equation (2.93) can be given another interpretation as well. From a signal design viewpoint, the freedom present in selecting $a(t) \leftrightarrow A(\omega)$ can be used for enhancing the target features and suppressing the interference.

2.4 Joint Transmitter-Receiver Design in Noise

Additive White Noise Case

Consider the problem of transmitting a signal $f(t)$ over a channel with impulse response $q(t)$. The channel output $s(t)$ gets corrupted by additive white noise $n(t)$, and the received signal is given by (see Fig. 2.3)

$$r(t) = s(t) + n(t). \tag{2.99}$$

If maximization of the output *SNR* at the decision instant $t = t_o$ is chosen to be the criterion for receiver design, then as we have seen in the previous section, for a given $s(t)$ the matched filter is the optimum solution, and the optimum causal receiver $h(t)$ is given by (2.75). Thus

$$h(t) = s^*(t_o - t)u(t). \tag{2.100}$$

In that case, since $s(t) = f(t) * q(t)$, the maximum *SNR* given by (2.36) or (2.74) is a function of $f(t)$ and it makes sense to select $f(t)$ optimally so as to further maximize the output *SNR* at $t = t_o$.

In the causal case, since $f(t)$ and $q(t)$ are causal, $s(t)$ is also causal, and from the relation

$$s(t) = f(t) * q(t) = \int_{-\infty}^{+\infty} f(\tau)q(t-\tau)d\tau = \int_{0}^{t} f(\tau)q(t-\tau)d\tau, \tag{2.101}$$

where $s(t_o)$ depends only on the segment of $q(t)$ up to $t = t_o$. Since the maximum *SNR* also depends only on $s(t)$ up to $t = t_o$, the signal design problem reduces to designing the optimum pulse $f(t)$, for $0 < t < t_o$ that maximizes (2.74). Notice that the *SNR* in (2.74) can be obviously increased by simply scaling $f(t)$, and hence it makes sense to restrict the energy in $f(t)$ to a prescribed constant for this optimization problem. Thus the problem reduces to finding an $f(t)$ for $0 < t < t_o$, such that

$$\int_{0}^{t_0} |f(t)|^2 \, dt = 1, \tag{2.102}$$

and

$$\rho(t_o) \triangleq \int_{0}^{t_0} |s(t)|^2 \, dt \tag{2.103}$$

is maximized where $s(t)$ is given by (2.101).

This formulation is meaningful in many communication problems as well where $f(t)$, for example, can represent the baseband carrier used to modulate the information bearing symbols. Usually a rectangular pulse is used as the baseband carrier. That formulation assumes that we use the same rectangular pulse, *irrespective* of the actual channel characteristics. But the rectangular pulse may not be the best

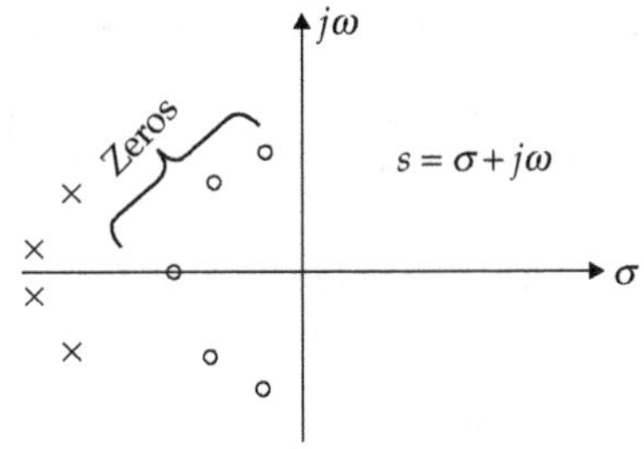

FIGURE 2.22 Zeros and poles of a minimum phase signal transform.

pulse shape for all channels. Thus, given some additional information about the channel, the problem is to find a meaningful way to incorporate that information in selecting the best baseband pulse. Notice that since information symbols arrive at every T_o seconds in this model, at the output, decisions must be made also at every T_o seconds, and hence t_o could be chosen equal to T_o in this case.

Minimum Phase Character of the Optimum Solution

Interestingly, it can be shown that the input pulse $f(t)$ that maximizes $\rho(t_o)$ in (2.103) must be a *minimum phase* signal [34]. A signal $f(t)$ is said to be minimum phase, if its Laplace transform $F(s)$ is free of zeros and poles in the open right-half plane ($\text{Re}\, s > 0$) as in Fig. 2.22. To establish the minimum phase condition, assume that the finite causal pulse $f(t)$ is not minimum phase. Then

$$F(s) = \int_0^{t_o} f(t)e^{-st}dt \tag{2.104}$$

has at least one zero in $\text{Re}\, s > 0$, say $s = s_o$. Rewrite $F(s)$ as

$$F(s) = F_o(s)\left(\frac{s - s_o}{s + s_o^*}\right) \triangleq F_o(s)A(s), \;\; \sigma_o = \text{Re}\, s > 0, \tag{2.105}$$

where $A(s) = (s - s_o)/(s + s_o^*)$ is a regular all-pass function.[2] Let $f_o(t)$ represent the inverse Laplace transform corresponding to $F_o(s)$. By extracting an all-pass, the remaining portion $F_o(s)$ is guaranteed to have the same energy as the original pulse $f(t)$. However, for $f_o(t)$ to act as a possible candidate in this optimization problem, it should

[2]A regular all-pass function has all its poles in the open left-half plane, and represents a causal system.

also be a finite pulse of duration t_o. To examine this, notice that

$$F_o(s) = F(s)\left(\frac{s+s_o^*}{s-s_o}\right) = F(s) + 2\sigma_o \frac{F(s)}{s-s_o} \tag{2.106}$$

so that, its inverse transform in given by

$$f_o(t) = f(t) + 2\sigma_o f(t) * e^{s_o t}u(t) = f(t) + 2\sigma_o \int_0^\infty f(\tau)e^{s_o(t-\tau)}u(t-\tau)d\tau$$

$$= \begin{cases} f(t) + 2\sigma_o e^{s_o t}\displaystyle\int_0^t f(\tau)e^{-s_o\tau}d\tau, & t < t_o \\ f(t) + 2\sigma_o e^{s_o t}\displaystyle\int_0^{t_o} f(\tau)e^{-s_o\tau}d\tau, & t \geq t_o \end{cases}$$

$$= \begin{cases} f(t) + 2\sigma_o e^{s_o t}\displaystyle\int_0^t f(\tau)e^{-s_o\tau}d\tau, & t < t_o \\ 2\sigma_o e^{s_o t}F(s_o) = 0, & t \geq t_o \end{cases}. \tag{2.107}$$

Thus $f_o(t)$ is also of finite duration t_o, and it possesses the same energy as $f(t)$, establishing the fact that extraction of an all-pass factor from a finite pulse transform still retains its finite pulse shape characteristics.

Toward establishing the minimum phase character of the optimum pulse $f(t)$ that maximizes (2.103), let $s(t)$ and $s_o(t)$ represent the channel output due to $f(t)$ and $f_o(t)$, respectively as in Fig. 2.23. Recall that $f_o(t)$ is also a finite pulse with the same energy, but its transform contains one less zero in the right-half plane compared to that of $f(t)$. Thus

$$s(t) = f(t) * q(t); \quad s_o(t) = f_o(t) * q(t). \tag{2.108}$$

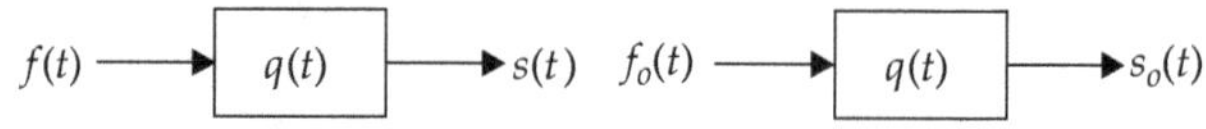

FIGURE 2.23 Transmit waveform $f(t)$ and $f_o(t)$.

We will show that the *running energy* up to $t = t_o$ in $s_o(t)$ is greater than that in $s(t)$; that is,

$$\int_0^{t_o} |s_o(t)|^2\, dt \geq \int_0^{t_o} |s(t)|^2\, dt, \tag{2.109}$$

indicating that extraction of a right-half plane zero, or more precisely a regular all-pass function, from the input excitation pulse will result in greater running energy for the system output. Clearly, by continuing this process of regular all-pass extraction, maximum running energy at the output will be realized by a pulse whose transform is free of zeros in the right-half plane; that is, among all pulses with the same input energy, the *minimum phase* pulse generates maximum running energy at the output of *any* system.

We can make use of (2.105) to complete the minimum phase proof. Together with (2.108), this gives (Fig. 2.24)

$$s(t) = a(t) * s_o(t) = \int_0^{t} s_o(\tau)a(t-\tau)d\tau, \tag{2.110}$$

that is, $s(t)$ is the output of the all-pass filter $a(t) \leftrightarrow A(s)$ due to the input $s_o(t)$.

By making use of the regular all-pass nature of $A(s)$ and Parseval's relation, we have

$$\int_0^{\infty} |s_o(t)|^2\, dt = \int_0^{\infty} |s(t)|^2\, dt. \tag{2.111}$$

Let $x(t)$ represent the output of the same all-pass filter $A(s)$ due to the truncated input $s_1(t)$ as shown in Fig. 2.25. Thus

$$s_1(t) = \begin{cases} s_o(t), & 0 < t < t_o \\ 0, & \text{otherwise.} \end{cases} \tag{2.112}$$

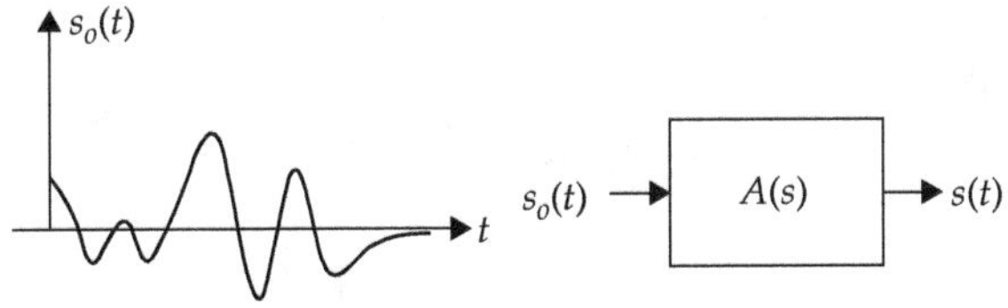

FIGURE 2.24 Minimum phase proof.

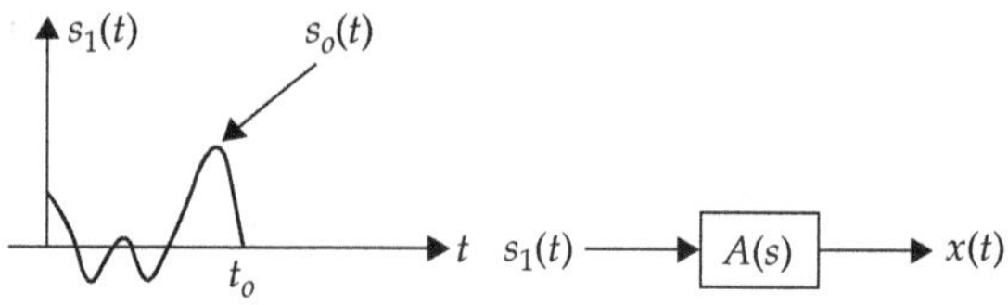

FIGURE 2.25 Truncated input.

Then once again using Parseval's relation, by equating the total input and output energy, we get

$$\int_0^\infty |s_1(t)|^2\,dt = \int_0^{t_o} |s_o(t)|^2\,dt = \int_0^\infty |x(t)|^2\,dt. \tag{2.113}$$

But, because of the *causal* nature of $A(s)$, we have

$$x(t) = \int_0^t s_o(\tau)a(t-\tau)d\tau = s(t),\ 0 < t < t_o, \tag{2.114}$$

(i.e., the outputs in Figs. 2.24 and Fig. 2.25 agree up to $t < t_o$) and substituting this into the later half of (2.113), we get

$$\int_0^{t_o} |s_o(t)|^2\,dt = \int_0^{t_o} |s(t)|^2\,dt + \int_{t_o}^\infty |x(t)|^2\,dt, \tag{2.115}$$

or

$$\int_0^{t_o} |s_o(t)|^2\,dt \geq \int_0^{t_o} |s(t)|^2\,dt, \tag{2.116}$$

establishing our original claim. Thus to maximize the running energy of any system output, the input pulse $f(t)$ must be necessarily minimum phase.

Optimum Solution

Returning to the original optimization problem in (2.103), to make further progress, we can substitute (2.101) into (2.103). This gives

$$\begin{aligned}\rho(t_o) &= \int_0^{t_o} \left| \int_0^{\infty} f(\tau) q(t-\tau) d\tau \right|^2 dt \\ &= \int_0^{t_o} \left(\int_0^{\infty}\int_0^{\infty} f^*(\tau_1) q^*(t-\tau_1) f(\tau_2) q(t-\tau_2) d\tau_1 d\tau_2 \right) dt \\ &= \int_0^{\infty} f^*(\tau_1) d\tau_1 \int_0^{\infty} \underbrace{\left(\int_0^{t_o} q^*(t-\tau_1) q(t-\tau_2) dt \right)}_{K(\tau_1, \tau_2)} f(\tau_2) d\tau_2.\end{aligned} \tag{2.117}$$

Define

$$K(\tau_1, \tau_2) \triangleq \int_0^{t_o} q^*(t-\tau_1) q(t-\tau_2) dt \tag{2.118}$$

to represent the positive-definite kernel associated with the channel impulse response $q(t)$. Note that only the channel response up to $t = t_o$ takes part in defining the above kernel. Further, $q(t)$ causal implies that

$$K(\tau_1, \tau_2) = 0, \text{ if } |\tau_1| > t_o, \text{ or } \quad |\tau_2| > t_o. \tag{2.119}$$

Using (2.118) in (2.117), it simplifies to

$$\begin{aligned}\rho(t_o) &= \int_0^{\infty} f^*(\tau_1) d\tau_1 \int_0^{\infty} K(\tau_1, \tau_2) f(\tau_2) d\tau_2 \\ &= \int_0^{t_o} f^*(\tau_1) d\tau_1 \int_0^{t_o} K(\tau_1, \tau_2) f(\tau_2) d\tau_2,\end{aligned} \tag{2.120}$$

where the later limits in the above integral follow from (2.119). The above integral defines an integral operator $T(f)$ given by

$$T(f) = \int_0^{t_o} K(\tau_1, \tau_2) f(\tau_2) d\tau_2 \triangleq \psi(\tau_1), \tag{2.121}$$

and using this, (2.120) can be expressed also as an inner product given by

$$\rho(t_o) = \int_0^{t_o} f^*(\tau_1)\psi(\tau_1)d\tau_1. \tag{2.122}$$

However, using Schwarz' inequality, the above expression yields

$$\rho(t_o) \leq \left(\int_0^{t_o} |f(\tau_1)|^2 \, d\tau_1 \int_0^{t_o} |\psi(\tau_1)|^2 \, d\tau_1 \right)^{1/2} = \left(\int_0^{t_o} |\psi(t)|^2 \, dt \right)^{1/2} \tag{2.123}$$

and (2.122) is maximized if and only if $\psi(\tau_1)$ is chosen to be proportional to $f(\tau_1)$ in $0 < \tau_1 < t_o$ that is, for maximum *SNR* $f(\tau_1)$ must be a solution of the integral equation[3]

$$\psi(\tau_1) = \int_0^{t_o} K(\tau_1, \tau_2) f(\tau_2) d\tau_2 = \lambda f(\tau_1), \;\; 0 < \tau_1 < t_o. \tag{2.124}$$

Substituting (2.124) into (2.122), we get

$$\rho(t_o) = \lambda \int_0^{t_o} |f(\tau_1)|^2 \, d\tau_1 = \lambda \leq \lambda_1, \tag{2.125}$$

where λ_1 is the largest (positive) eigenvalue of the above integral equation. Notice that the right-hand side of (2.125) is independent of $f(t)$ and hence to maximize $\rho(t_o)$, λ in (2.125) must be chosen as the largest eigenvalue of the above integral equation. In other words, $\rho(t_o)$ is maximized by selecting $f(t)$ to correspond to the eigenfunction associated with the largest eigenvalue λ of the integral equation (2.124), and the maximum possible *SNR* is given by

$$\rho(t_o) = \lambda_{\max}(t_o), \tag{2.126}$$

where $\lambda_{\max}(t_o)$ is the largest value of λ that satisfies (2.124). From the previous argument, the eigenfunction $f(t)$ corresponding to the

[3]Every eigenvalue that results as a solution of the integral equation in (2.124) will be positive, because of the symmetric positive-definite nature of the kernel $K(\tau_1, \tau_2)$ in (2.118). Note that $K(x, y)$ is said to be symmetric if $K(x, y) = K^*(y, x)$, and positive definite if $\iint K(x, y) f^*(x) f(y) dx\, dy > 0$ for any $f(x)$ as in the case here [see (2.120)].

largest eigenvalue of the kernel in (2.124) also represents a minimum phase function [49].

Optimum Input and the Matched Filter Receiver

It is easy to show that in the white noise case, the optimum solution for $f(t)$ that satisfies the integral equation (2.124) is identical to the causal matched filter receiver solution. In fact, more generally if $f(t)$ satisfies (2.124) for some eigenvalue λ, then

$$h(t) = f(t), \quad 0 < t < t_o, \tag{2.127}$$

where $h(t)$ is the optimum causal receiver defined in (2.100). To see this, from (2.100) and (2.101), we get

$$h(t) = s^*(t_o - t)u(t) = \int_0^{t_o} q^*(t_o - t - \tau) f^*(\tau)d\tau, \tag{2.128}$$

and consider the integral

$$\int_0^{t_o} K(\tau_1, \tau_2)h(\tau_2)d\tau_2, \tag{2.129}$$

obtained by replacing $f(\tau_2)$ in (2.124) with $h(\tau_2)$. Substituting (2.128) into (2.129), and making use of (2.118), we obtain

$$\begin{aligned}
&\int_0^{t_o} K(\tau_1, \tau_2) \int_0^{t_o} q^*(t_o - \tau_2 - \tau) f^*(\tau) d\tau_2\, d\tau \\
&= \int_0^{t_o}\int_0^{t_o} q^*(t - \tau_1) q(t - \tau_2) \int_0^{t_o} q^*(t_o - \tau_2 - \tau) f^*(\tau) d\tau_2\, d\tau\, dt \qquad (2.130)\\
&= \int_0^{t_o} q^*(t - \tau_1) \int_0^{t_o}\int_0^{t_o} q(t - \tau_2) q^*(t_o - \tau_2 - \tau) f^*(\tau) d\tau_2\, d\tau\, dt.
\end{aligned}$$

Let $t_o - \tau_2 = u$, so $\tau_2 = t_o - u$, and the above integral becomes

$$\int_0^{t_o} q^*(t - \tau_1) \int_0^{t_o}\int_0^{t_o} q(t - t_o + u) q^*(u - \tau) f^*(\tau) du\, d\tau\, dt. \tag{2.131}$$

But by definition

$$\int_0^{t_o} q\,(u-(t_o-t))\,q^*(u-\tau)du = K^*(t_o-t,\tau). \tag{2.132}$$

Thus (2.131) becomes

$$\begin{aligned}\int_0^{t_o} q^*(t-\tau_1)\left(\int_0^{t_o} K(t_o-t,\tau)f(\tau)d\tau\right)^* dt \\ = \int_0^{t_o} q^*(t-\tau_1)\left(\lambda f^*(t_o-t)\right)dt,\end{aligned} \tag{2.133}$$

where we have made use of (2.124). Finally, let $t_o - t = v$, then $t = t_o - v$, and hence

$$\lambda \int_0^{t_o} q^*(t-\tau_1)f^*(t_o-t)dt = \lambda \int_0^{t_o} q^*(t_o-\tau_1-v)f^*(v)dv = \lambda h(\tau_1), \tag{2.134}$$

where we have made use of (2.128). Thus from (2.129)–(2.134) we have shown that

$$\int_0^{t_o} K(\tau_1,\tau_2)h(\tau_2)d\tau_2 = \lambda h(\tau_1),\ \ 0<\tau_1<t_o, \tag{2.135}$$

where $f(t)$ in (2.128)–(2.133) is the actual eigenfunction in (2.124) corresponding to eigenvalue λ there. Obviously, (2.135) is true even when $f(t)$ corresponds to the largest eigenvalue $\lambda_{\max}$, and in that case $h_o(t)$ is the desired matched filter.

Figures 2.26–2.28 show the optimal input waveforms for various channel characteristics with details as indicated there. In Fig. 2.26, the channel is assumed to be the usual rectangular pulse of width T_o, and the optimum input waveform $f(t)$ for $t_o = T_o$ obtained by solving (2.124) and the corresponding matched filter responses are shown in Fig. 2.26(b)–(d). Figure 2.26(e) shows the optimum input pulse shape $f(t)$, for different values of t_o.[4] Similarly, Figs. 2.27 and 2.28 exhibit

[4]For the decision instant t_o, since $f(t)$ exists only in the interval $0 < t < t_o$, it should be easy to identify the corresponding input pulse in Fig. 2.26(e) for various values of t_o.

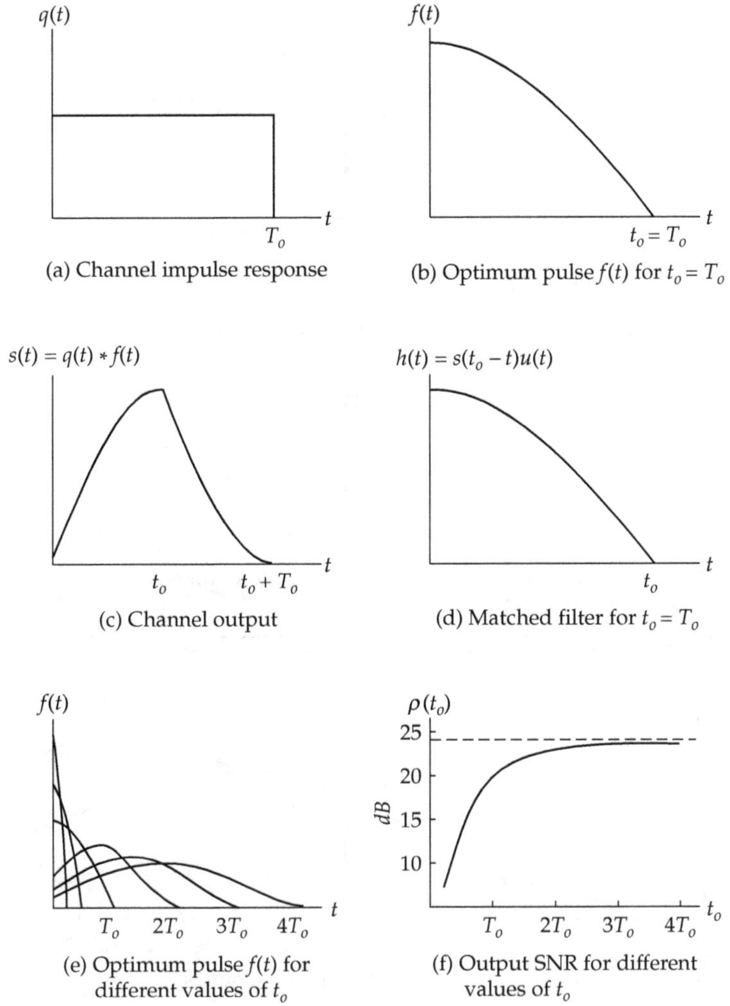

FIGURE 2.26 Optimum pulse and matched filter for a rectangular channel response.

the desired transmit pulse $f(t)$ as well as the corresponding matched filters for various channel characteristics.

This leads us to an interesting question: for a given channel, is there a best value for t_o at which decision should be made? If we use the output *SNR* $\rho(t_o)$ as the optimality criterion, this question can be answered by examining $\rho(t_o)$ versus t_o.

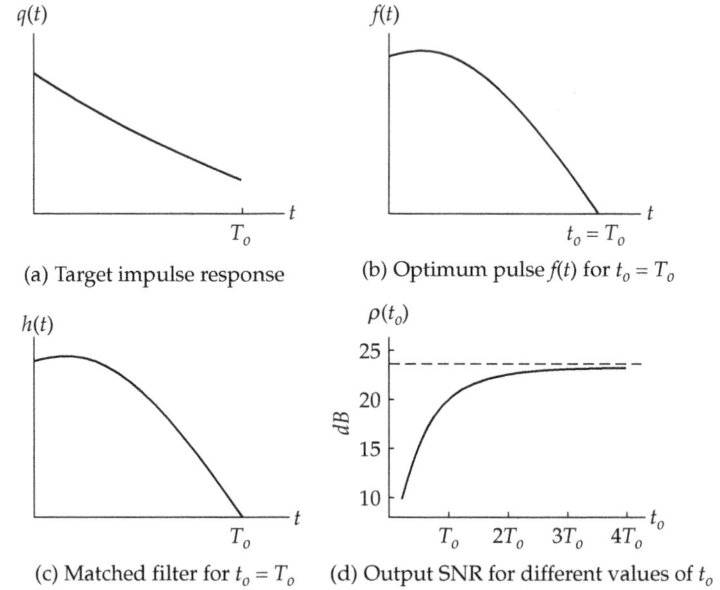

FIGURE 2.27 Optimum pulse and matched filter example 2.

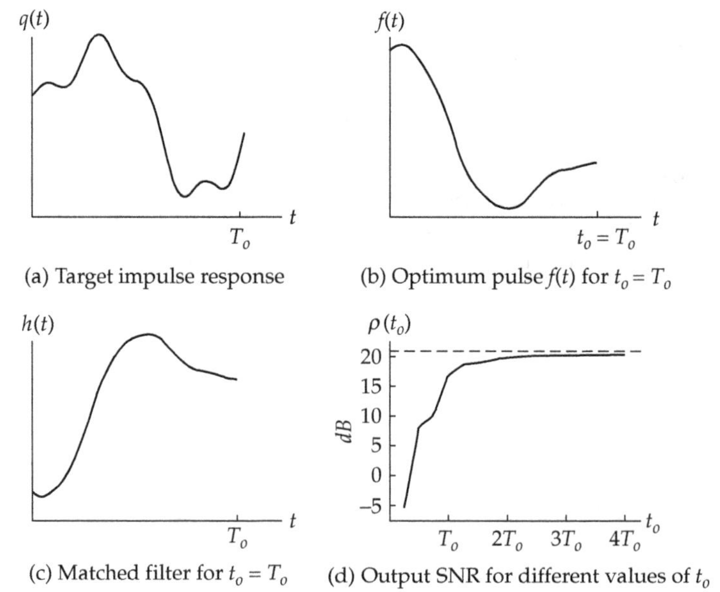

FIGURE 2.28 Optimum pulse and matched filter example 3.

It is easy to see that $\rho(t_o)$ is in fact a monotonically nondecreasing function of t_o, that is,

$$\rho(t_1) \geq \rho(t_o), \text{ if } t_1 > t_o. \tag{2.136}$$

To prove this, let us assume the contrary, and suppose that for some $t_1 > t_o$, the optimum eigenvector $f_1(t)$ associated with t_1 leads to a smaller SNR compared to the optimum eigenvector $f_o(t)$ associated with $t = t_o$. In that case, since $s_o(t) = f_o(t) * q(t)$, $s_1(t) = f_1(t) * q(t)$, and if we have

$$\rho(t_o) = \int_0^{t_o} |s_o(t)|^2\, dt > \rho(t_1) = \int_0^{t_1} |s_1(t)|^2\, dt, \tag{2.137}$$

by redefining

$$f_1(t) = \begin{cases} f_o(t), & 0 < t \leq t_o, \\ 0, & t_o < t \leq t_1, \end{cases} \tag{2.138}$$

the inequality in (2.136) can be satisfied with equality. Thus (2.136) is always true, indicating that under the joint optimality condition, delaying the decision instant t_o cannot make it worse in terms of discriminating the signal from the noise. Figures 2.26(f), 2.27(d), and 2.28(d) show the monotonic nondecreasing nature of $\rho(t_o)$ as a function of t_o. These figures indicate that $\rho(t_o) \to \rho(\infty)$, the maximum possible value of the output SNR obtained by letting $t_o \to \infty$, and the $\rho(t_o)$ curve can be used to determine the best value for t_o [50].

To summarize, for a given pulse $f(t)$, the channel output results in a specific waveform $s(t)$ and maximization of the output SNR at $t = t_o$ leads to the matched filter receiver. Since the output SNR depends on $f(t)$, further improvement in the output SNR can be obtained by exploiting the freedom present in selecting $f(t)$ at the transmitter. Thus for a given channel, by combining the matched filter design together with the input pulse design problem, we have been able to achieve joint optimal transmitter-receiver design. The optimum input pulse will be always minimum phase, and moreover in the white noise case the matched filter is identical to the optimum input pulse.

The key idea in this section is to make use of the kernel of the channel characteristics to define an optimal input waveform, and as we show in the next section, this approach has been used to solve another interesting problem in communication theory.

2.5 Joint Time-Bandwidth Optimization

Suppose a pulse $e(t) \leftrightarrow E(\omega)$ of duration t_o needs to be transmitted through a channel with impulse response $q(t) \leftrightarrow Q(\omega)$. Let $r(t) \leftrightarrow R(\omega)$ denote the channel output.

Then

$$r(t) = e(t) * q(t) = \int_0^{t_o} e(\tau)q(t-\tau)d\tau, \tag{2.139}$$

or, in terms of their transforms

$$R(\omega) = E(\omega)Q(\omega). \tag{2.140}$$

Suppose $-B_o < \omega < B_o$ represents the desired spectral band at the receiver as shown in Fig. 2.29(b). We would like to choose the input signal $e(t)$ such that the output signal $r(t)$ has maximum energy concentrated in the above band [51–53].

This formulation makes sense in communication problems for minimizing co-channel interference, since the above design criterion makes sure that most of the energy of the transmitted signal will be concentrated in the desired band $(-B_o, B_o)$, and hence very little energy will spill over to the adjacent bands. In a different context, we may be interested in finding the best time-limited signal that also has maximum energy in a given band. Since time-limited signals cannot be band limited, the next best thing that one could hope for in terms of band-limiting a time-limited signal is to maximize the energy within a given bandwidth [54,55].

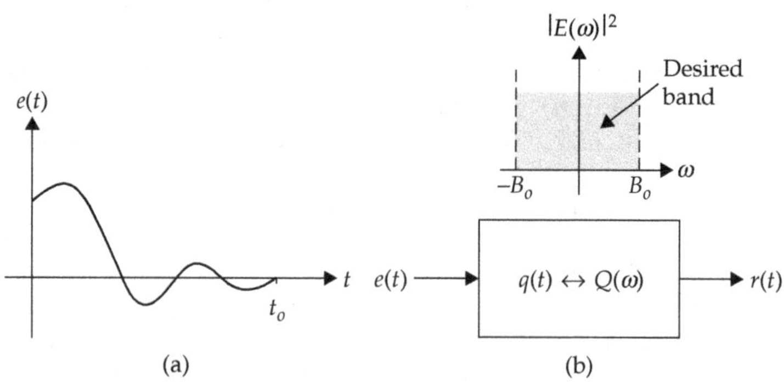

FIGURE 2.29 Joint time-bandwidth optimization, a pulse goes through the channel.

Returning to the original problem, the objective is to select the best pulse $e(t)$ such that the output energy

$$E = \frac{1}{2\pi}\int_{-B_o}^{B_o} |R(\omega)|^2\,d\omega = \frac{1}{2\pi}\int_{-B_o}^{B_o} |Q(\omega)|^2\,|E(\omega)|^2\,d\omega \tag{2.141}$$

in the spectral band $(-B_o, B_o)$ is maximized. Since simple scaling of $e(t)$ can increase the value of the above E, it is necessary to normalize $e(t)$ by requiring it to possess unit (or, constant) energy. Thus our problem reduces to maximizing (2.141) by selecting a suitable time-limited signal $e(t)$, $0 < t < t_o$ under the energy constraint

$$\int_0^{t_o} |e(t)|^2\,dt = 1. \tag{2.142}$$

To solve this problem, once again we can make use of the concept of the channel kernel discussed in the previous section. However, it is necessary to bring in the desired bandwidth information into this kernel in some suitable manner.

To express this problem in the time domain, we can make use of Parseval's theorem. If we define

$$\Psi(\omega) \triangleq \begin{cases} |Q(\omega)|^2\,E(\omega), & -B_o < \omega < B_o \\ 0, & \text{otherwise,} \end{cases} \tag{2.143}$$

then using Parseval's theorem in (2.141) we obtain

$$E = \frac{1}{2\pi}\int_{-\infty}^{+\infty} \Psi(\omega)E^*(\omega)d\omega = \int_{-\infty}^{+\infty} \psi(t)e^*(t)dt = \int_0^{t_o} \psi(t)e^*(t)dt, \tag{2.144}$$

where $\psi(t)$ represents the inverse transform of $\Psi(\omega)$ defined above. To simplify $\psi(t)$, define $K(\tau)$ to represent the inverse Fourier transform of the channel transfer function gain $|Q(\omega)|^2$ *restricted* to the frequency band $(-B_o < \omega < B_o)$ as in (2.143). Thus (Fig. 2.30)

$$K(\tau) \triangleq \frac{1}{2\pi}\int_{-B_o}^{B_o} |Q(\omega)|^2\,e^{j\omega\tau}d\omega = K^*(-\tau). \tag{2.145}$$

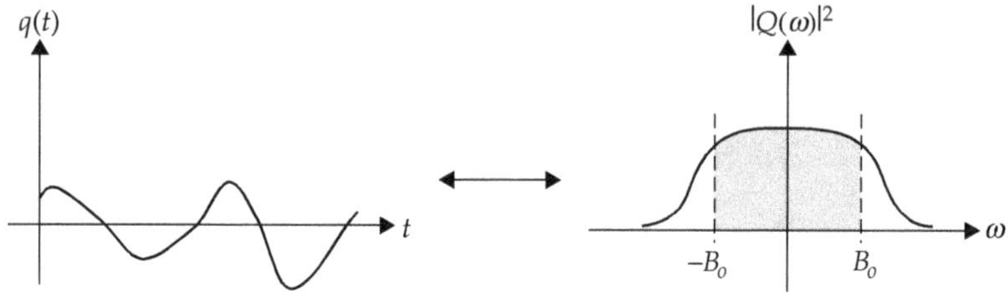

FIGURE 2.30 Inverse Fourier transform of the channel restricted to $(-B_o < \omega < B_o)$.

The function $K(\tau)$ acts as the kernel associated with the channel transfer function gain $|Q(\omega)|^2$ over the desired bandwidth $(-B_o, B_o)$. Note that $K(\tau)$ represents a stationary kernel (autocorrelation function), since it is a function of only one argument.

Using (2.143)–(2.145) and $\psi(t) \leftrightarrow \Psi(\omega)$, we get

$$\psi(t) = K(t) * e(t) = \int_{-\infty}^{+\infty} K(t-\tau)e(\tau)d\tau = \int_{0}^{t_o} K(t-\tau)e(\tau)d\tau, \quad (2.146)$$

since $e(\tau)$ is non-zero only in the interval $0 < \tau < t_o$. Note that $\psi(t)$ in (2.146) defines an integral operator $T(e)$ through the above kernel, and using this, the positive quantity E in (2.144) can be represented as an inner product. Thus

$$E = \int_{0}^{t_o} \psi(t)e^*(t)dt, \quad (2.147)$$

and once again using Schwarz' inequality, the above expression gives

$$E \leq \left(\int_{0}^{t_o} |\psi(t)|^2 dt \int_{0}^{t_o} |e(t)|^2 dt \right)^{1/2} = \left(\int_{0}^{t_o} |\psi(t)|^2 dt \right)^{1/2}, \quad (2.148)$$

and it is maximized if and only if we choose

$$\psi(t) = \lambda e(t), 0 \leq t \leq t_o, \quad (2.149)$$

that is, if and only if $e(t)$ satisfies the integral equation

$$\int_0^{t_o} K(t-\tau)e(\tau)d\tau = \lambda e(\tau), \quad 0 \le t \le t_o. \tag{2.150}$$

In this case

$$E = \lambda \int_0^{t_o} |e(t)|^2 \, dt = \lambda. \tag{2.151}$$

Since the right side of the above equation is independent of $e(t)$, to maximize E, the eigenvalue λ in (2.150) must be chosen as the largest possible value. Thus the optimal input waveform $e(t)$ that maximizes E is the solution of the integral equation in (2.150) corresponding to the largest eigenvalue of λ, where the positive definite kernel $K(t-\tau)$ is as defined in (2.145). From (2.148),

$$E_{\max} \triangleq \sqrt{\int_0^{t_o} |\psi_{\max}(t)|^2 \, dt} = \lambda_{\max} \tag{2.152}$$

represents the maximum value for the energy, where $\psi_{\max}(t)$ is as defined (2.146) with $e(t)$ there representing the eigenvector associated with the largest eigenvalue in (2.150). Since this situation is identical to that in (2.103), using the arguments developed in that section it follows that to maximize $E_{\max}$, the input waveform $e(t)$ must be necessarily minimum phase. Thus the above optimum solution given by the eigenfunction associated with the largest eigenvalue of (2.150) must be necessarily minimum phase. In addition, it is easy to show that if $e(t)$ satisfies the integral equation (2.150), then so does

$$f(t) \triangleq e^*(t_o - t), \; 0 \le t \le t_o. \tag{2.153}$$

This follows easily, since

$$\begin{aligned} \int_0^{t_o} K(t-\tau) f(\tau)d\tau &= \int_0^{t_o} K(t-\tau)e^*(t_o-\tau)d\tau \\ &= \left(\int_0^{t_o} K(t_o - t - x)e(x)dx \right)^* \\ &= \lambda e^*(t_o - t) = \lambda f(t), \quad 0 < t < t_o, \end{aligned} \tag{2.154}$$

where we have used the substitution $t_o - \tau = x$, and made use of the symmetric nature of the kernel given in (2.145). Thus if λ represents a distinct eigenvalue, then these two solutions must equal. This gives

$$e(t) = e^*(t_o - t), \ \ 0 < t < t_o, \tag{2.155}$$

that is, $e(t)$ represents a symmetric solution in the sense that it agrees with its matched filter with respect to the interval $(0, t_o)$. Alternatively, in terms of their transforms

$$E(s) = e^{-t_o s} E^*(-s^*). \tag{2.156}$$

Thus $E(s)$ must possess zeros symmetrically in both half planes. However, because of its minimum phase character, the eigenvector corresponding to the largest eigenvalue has no zeros in the strict right-half plane. Consequently it is also free of zeros in the strict left-half plane. Thus all of its zeros must lie on the $j\omega -$ axis.

To summarize, the best input pulse $e(t)$ of finite duration t_o that also maximizes the channel output energy over the band $(-B_o, B_o)$ is given by the eigenfunction associated with the largest eigenvalue $\lambda_{\max}(t_o)$ of the integral equation in (2.150). It represents a minimum phase symmetric function with all the zeros of its Laplace transform $E(s)$ distributed along the $j\omega -$ axis. The maximum possible value of the energy in $(-B_o, B_o)$ is given by $\lambda_{\max}(t_o)$. Note that the channel characteristics comes into play in the kernel through $|Q(\omega)|^2$ along with the desired frequency band $(-B_o, B_o)$.

Once again, it is easy to show that $\lambda_{\max}(t_o)$ is a monotone nondecreasing function of t_o. To see this, using (2.148), we have

$$E_{\max} = \lambda_{\max}(t_o) = \sqrt{\int_0^{t_o} |\psi(t)|^2 \, dt}, \tag{2.157}$$

where $\psi(t)$ is given by (2.146). With $t_1 > t_o$, let $e_1(t)$ and $e(t)$ represent the solution of (2.150) for t_1 and t_o, respectively. Thus

$$\lambda_{\max}(t_1) = \sqrt{\int_0^{t_1} |\psi_1(t)|^2 \, dt} \tag{2.158}$$

where $\psi_1(t) = \int_0^{t_1} K(t-\tau)e_1(\tau)d\tau$, and if $\lambda_{\max}(t_1) < \lambda_{\max}(t_o)$, then it is easy to satisfy this inequality with an equality by redefining $e_1(t)$ as

$$e_1(t) = \begin{cases} e(t), & 0 < t < t_o \\ 0, & \text{otherwise.} \end{cases} \tag{2.159}$$

Thus if $t_1 > t_o$ then $\lambda_{\max}(t_1) \geq \lambda_{\max}(t_o)$, which establishes the monotone nondecreasing nature of $\lambda_{\max}(t_o)$.

To illustrated the scheme presented here, Figs. 2.31 and 2.32 show the optimum waveform $e(t)$ for a typical low-pass channel gain function $|Q(\omega)|^2$ given by

$$|Q(\omega)|^2 = \frac{1}{1+\omega^2} \tag{2.160}$$

for different values of t_o and B_o. In this case, from (2.145) we have

$$K(\tau) = \frac{1}{\pi}\int_0^{B_o} \frac{\cos\omega\tau}{1+\omega^2}d\omega, \tag{2.161}$$

and together with (2.150) it can be used to solve for the desired $e(t)$. Figure 2.31 shows the optimum $e(t)$ for various values of t_o with B_o fixed equal to the normalized bandwidth of 0.5. Similarly, Fig. 2.32 shows the corresponding $e(t)$ for different values of B_o with t_o fixed equal to unity. Notice the symmetric nature of these waveforms. As

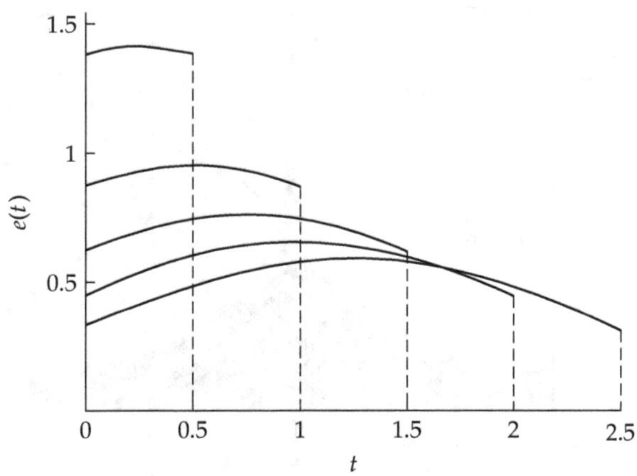

FIGURE 2.31 Optimum $e(t)$ for $t_o = 0.5, 1, 1.5, 2,$ and 2.5 with B_o fixed to 0.5 normalized bandwidth.

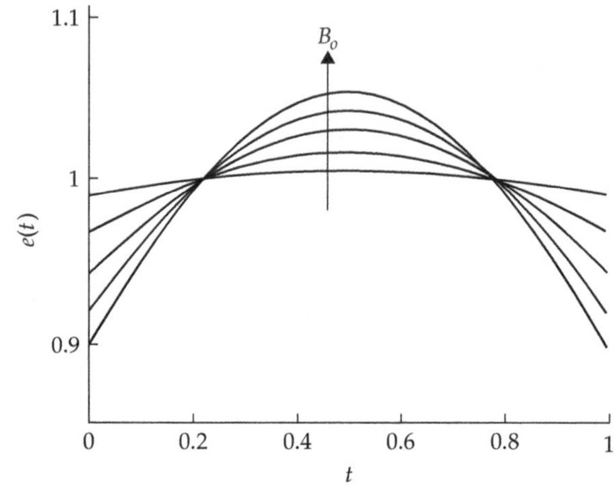

FIGURE 2.32 Optimum $e(t)$ for $B_o = 0.2, 0.4, 0.6, 0.8,$ and 1 with t_o fixed equal to 1.

noted earlier, in all these cases the transforms of these waveforms have all their zeros on the $j\omega$ – axis.

Figure 2.33 shows the square magnitude of the transforms of the optimum waveforms in Fig. 2.32 as a function of the normalized frequency and for comparison purposes, they are plotted along with $|Q(\omega)|^2$.

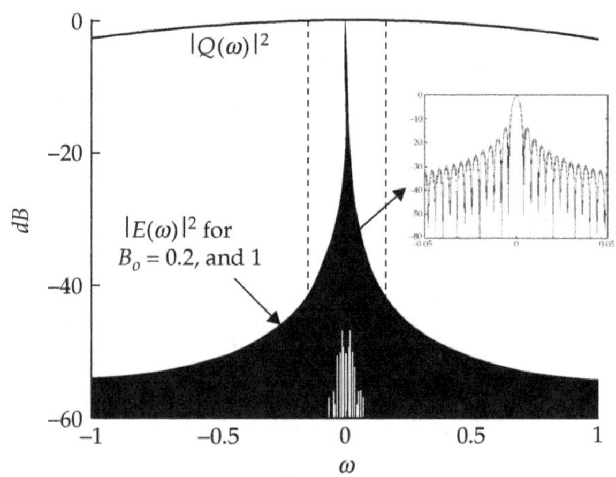

FIGURE 2.33 Transforms of optimum waveform for $B_o = 0.2, 0.4, 0.6, 0.8,$ and 1 with $t_o = 1$ versus $|Q(\omega)|^2$.

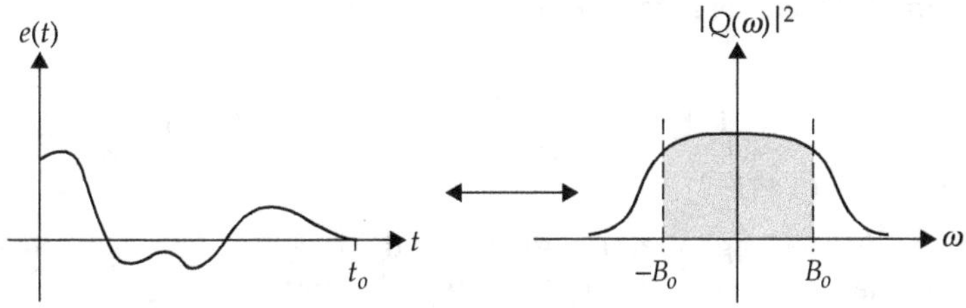

FIGURE 2.34 Best time-limited signal of unit energy that has maximum energy within a given bandwidth.

Prolate Spheroidal Functions

As a special case, suppose we need to determine the best time-limited signal that is also maximally band-limited. In other words, this is equivalent to finding the best time-limited signal of unit energy that has maximum energy within a given bandwidth B_o as in Fig. 2.34.

Thus, the problem is to maximize

$$E = \frac{1}{2\pi} \int_{-B_o}^{B_o} |E(\omega)|^2 \, d\omega \tag{2.162}$$

subject to

$$\int_0^{t_o} |e(t)|^2 \, dt = \frac{1}{2\pi} \int_{-\infty}^{+\infty} |E(\omega)|^2 \, d\omega = 1. \tag{2.163}$$

Notice that E in (2.162) represents the energy compaction ratio inside the frequency band $(-B_o, B_o)$ for $e(t)$, and has the same form as (2.141), with $|Q(\omega)|^2$ as shown in Fig. 2.35.

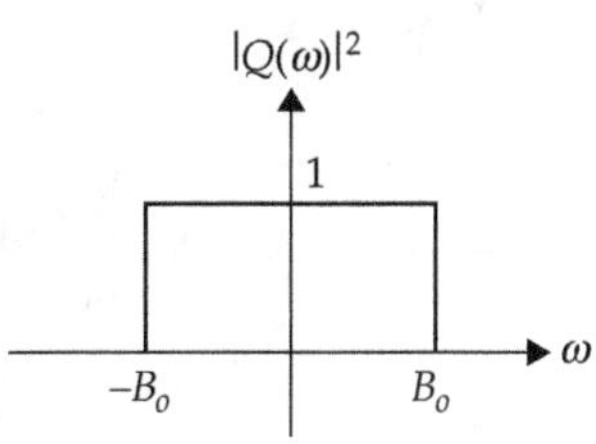

FIGURE 2.35 Flat spectrum for $|Q(\omega)|^2$.

From (2.145), this gives the corresponding kernel to be

$$K(\tau) = \frac{1}{2\pi}\int_{-B_o}^{B_o} e^{j\omega\tau} d\tau = \frac{1}{\pi}\frac{\sin B_o\tau}{\tau} = \frac{B_o}{\pi}\text{sinc}\,(B_o\tau), \tag{2.164}$$

and the desired signal $e(t)$ is given by the eigenfunction associated with the largest eigenvalue of the integral equation

$$\int_0^{t_o} \frac{1}{\pi}\frac{\sin B_o(t_1 - t_2)}{t_1 - t_2} e(t_2)dt_2 = \lambda e(t_1),\ 0 < t_1 < t_0. \tag{2.165}$$

By a simple change of variables ($t = B_o t_1,\ \tau = B_o t_2$), we can rewrite the above equation as

$$\frac{1}{\pi}\int_0^{t_o B_o} \frac{\sin(t-\tau)}{t-\tau} e(\tau)d\tau = \lambda e(\tau),\ 0 < t < t_0 B_o, \tag{2.166}$$

which shows that the time-bandwidth product $\eta \triangleq t_o B_o$ is the *only* independent parameter in this problem. Solutions of the integral equation

$$\frac{1}{\pi}\int_0^{\eta} \frac{\sin(t-\tau)}{t-\tau} e(\tau)d\tau = \lambda(\eta)e(\tau),\ 0 < t < \eta \tag{2.167}$$

for various values of the time-bandwidth product η represent the well-known prolate spheroidal functions [56]. These functions are plotted in Fig. 2.36 for various values of η. As pointed out earlier, the

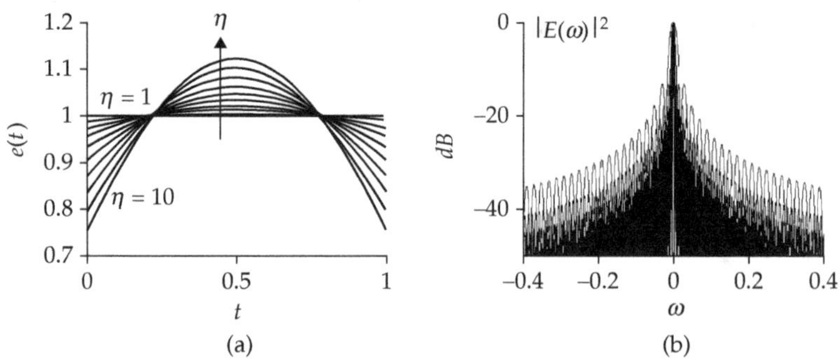

FIGURE 2.36 Energy compaction ratio for prolate spheroidal functions.

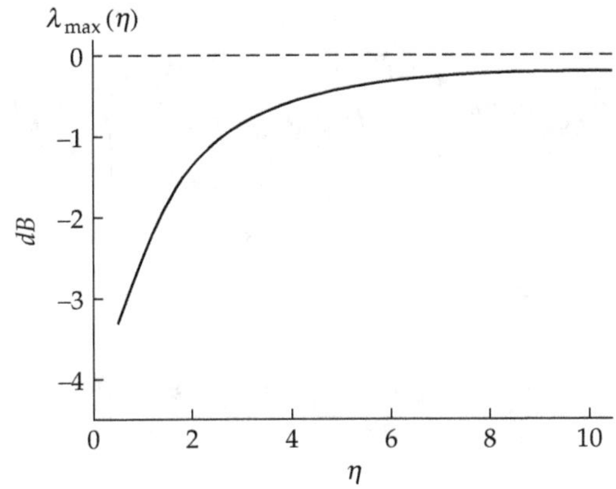

FIGURE 2.37 $\lambda_{max}(\eta)$ as a function of η with t_o fixed equal to 1.

Laplace transform $E(s)$ will have all of its zeros on the $j\omega$ – axis. Figure 2.37 shows $\lambda_{max}(\eta)$ as a function of η with t_o fixed equal to unity.

As $\eta \to \infty$, $\lambda_{max}(\eta) \to 1$, since this is equivalent to fixing $t = t_o$ and letting $B_o \to \infty$. In this case, E in (2.162) tends to unity as well.

2.6 Transform of a Chirp Signal

Consider the chirped rectangular symmetric pulse (noncausal)

$$f_o(t) = \begin{cases} e^{j\beta t^2}, & -T/2 < t < T/2, \\ 0, & \text{otherwise,} \end{cases} \tag{2.168}$$

whose transform is given by

$$F_o(\omega) = \int_{-T/2}^{T/2} e^{j\beta t^2} e^{-j\omega t} dt = e^{-j\omega^2/4\beta} \int_{-T/2}^{T/2} e^{j\beta\left(t-\frac{\omega}{2\beta}\right)^2} dt. \tag{2.169}$$

Let $\sqrt{\beta}\left(t - \dfrac{\omega}{2\beta}\right) = \tau$. Then

$$F_o(\omega) = \frac{1}{\sqrt{\beta}} e^{-j\omega^2/4\beta} \int_{-(\beta T+\omega)/2\sqrt{\beta}}^{(\beta T-\omega)/2\sqrt{\beta}} e^{j\tau^2} d\tau, \tag{2.170}$$

or

$$F_o(\omega) = \frac{1}{\sqrt{\beta}} e^{-j\omega^2/4\beta} \left(\int_0^{(\beta T-\omega)/2\sqrt{\beta}} e^{j\tau^2} d\tau + \int_0^{(\beta T+\omega)/2\sqrt{\beta}} e^{j\tau^2} d\tau \right). \tag{2.171}$$

Define the Fresnel integral

$$K(\omega) = \sqrt{\frac{2}{\pi}} \int_0^{\omega} e^{jt^2} dt \tag{2.172}$$

so that

$$\begin{aligned} K(\infty) &= \sqrt{\frac{2}{\pi}} \int_0^{\infty} e^{jt^2} dt = \frac{1}{2}\sqrt{\frac{2}{\pi}} \int_{-\infty}^{\infty} e^{jt^2} dt \\ &= \frac{1}{\sqrt{2\pi}} \sqrt{j\pi} = \sqrt{\frac{j}{2}} = \frac{1}{\sqrt{2}} e^{j\pi/4}, \end{aligned} \tag{2.173}$$

and hence

$$\int_0^{\infty} \cos t^2 dt = \int_0^{\infty} \sin t^2 dt = \frac{1}{2}\sqrt{\pi/2}. \tag{2.174}$$

In terms of (2.172), (2.171) becomes

$$F_o(\omega) = e^{-j\omega^2/4\beta} \sqrt{\frac{\pi}{2\beta}} \left\{ K\left(\frac{\beta T + \omega}{2\sqrt{\beta}}\right) + K\left(\frac{\beta T - \omega}{2\sqrt{\beta}}\right) \right\} \tag{2.175}$$

which represents the transform of the pulsed symmetric chirp signal in (2.168). Notice that near the baseband region, (i.e., for $\sqrt{\beta}T \gg 1$)

$$K\left(\frac{\beta T + \omega}{2\sqrt{\beta}}\right) \approx K\left(\frac{\sqrt{\beta}T}{2}\right) \approx K(\infty) = \sqrt{j/2}. \tag{2.176}$$

Hence for large values of $\sqrt{\beta}T$, in the baseband region [see (2.173)]

$$F_o(\omega) \approx \sqrt{\frac{\pi}{2\beta}} \left(\sqrt{\frac{j}{2}} + \sqrt{\frac{j}{2}} \right) \tag{2.177}$$

or

$$|F_o(\omega)| \approx \sqrt{\frac{\pi}{\beta}}, \quad |\omega| < \beta T \tag{2.178}$$

and for $\omega > \beta T$ [use $K(-\omega) = -K(\omega)$]

$$K\left(\frac{\beta T - \omega}{2\sqrt{\beta}}\right) = -K\left(\frac{\omega - \beta T}{2\sqrt{\beta}}\right) \approx -K(\omega) \tag{2.179}$$

and hence

$$F_o(\omega) = 0, \quad \omega > \beta T. \tag{2.180}$$

Hence

$$|F_o(\omega)| = \begin{cases} \sqrt{\pi/\beta}, & |\omega| < \beta T, \\ 0, & |\omega| > \beta T. \end{cases} \tag{2.181}$$

Thus the chirp signal in (2.168) has constant magnitude in the baseband region, and it is essentially zero outside that region as shown in Fig. 2.38. Notice that even in the passband the magnitude goes down as $1/\sqrt{\beta}$ for the chirp.

Next, consider the modulated chirped pulse signal

$$f(t) = e^{j(\omega_o t + \beta t^2)}, \quad -\frac{T}{2} < t < \frac{T}{2}. \tag{2.182}$$

Then

$$F(\omega) = F_o(\omega - \omega_o). \tag{2.183}$$

Hence we get

$$F(\omega) = e^{-j(\omega-\omega_o)^2/4\beta}\sqrt{\frac{\pi}{2\beta}}\left[K\left(\frac{\omega - \omega_1}{2\sqrt{\beta}}\right) + K\left(\frac{\omega_2 - \omega}{2\sqrt{\beta}}\right)\right] \tag{2.184}$$

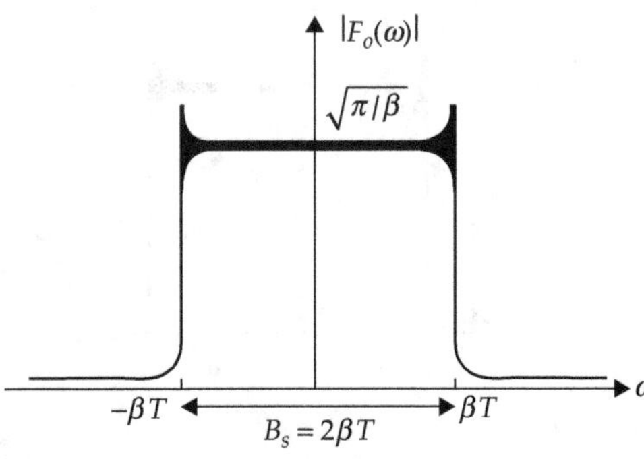

FIGURE 2.38 Transform of a chirp signal $f_o(t) = e^{j\beta t^2}$, $-T/2 < t < T/2$.

where

$$\omega_1 = \omega_o - \beta T, \quad \omega_2 = \omega_o + \beta T. \tag{2.185}$$

Once again for large values of $\sqrt{\beta}T$ compared to ω, we have

$$|F(\omega)| \approx 2\sqrt{\frac{\pi}{2\beta}}\left|K\left(\frac{\sqrt{\beta}T}{2}\right)\right| \approx \sqrt{\frac{\pi}{2\beta}}2\left|\sqrt{\frac{j}{2}}\right| = \sqrt{\frac{\pi}{\beta}}, \tag{2.186}$$

and for $|\omega - \omega_o| \gg \beta T$, we have

$$F(\omega) \approx \sqrt{\frac{\pi}{2\beta}}\{K(\omega) + K(-\omega)\} \approx 0, \text{ for } \omega_o - \beta T > \omega > \omega_o + \beta T. \tag{2.187}$$

Hence (Fig. 2.39)

$$|F(\omega)| = \begin{cases} \sqrt{\pi/\beta}, & \omega_o - \beta T < \omega < \omega_o + \beta T, \\ 0, & \text{otherwise.} \end{cases} \tag{2.188}$$

To determine the transform of a causal chirp, define using (2.182)

$$\begin{aligned} f_1(t) &= f(t - T/2) = e^{j\{\omega_o(t-T/2)+\beta(t-T/2)^2\}} \\ &= e^{-j(2\omega_o T - \beta T^2)/4}\, e^{j(\omega_o - \beta T)t + \beta t^2}, \quad 0 < t < T. \end{aligned} \tag{2.189}$$

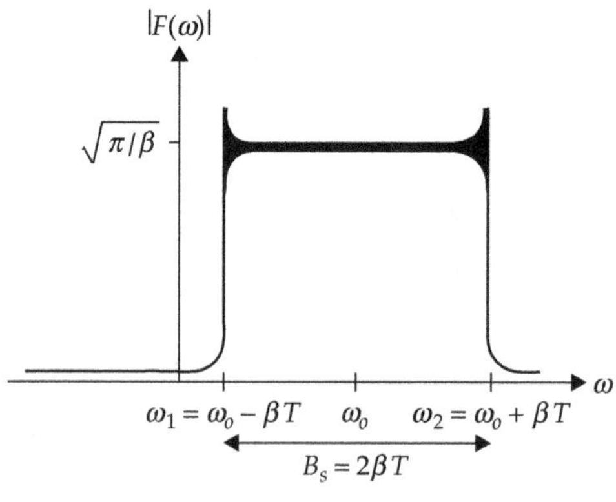

FIGURE 2.39 Transform of a chirp signal $f(t) = e^{j(\omega_o t + \beta t^2)}$, $-\frac{T}{2} < t < \frac{T}{2}$.

Notice that $f_1(t)$ has a new carrier at

$$\Omega_o = \omega_o - \beta T \tag{2.190}$$

rather than at ω_o. Hence with $\phi_o = (2\omega_o T - \beta T^2)/4$, we have

$$f_1(t) = e^{-j\phi_o} e^{j(\Omega_o t + \beta t^2)}, \; 0 < t < T \tag{2.191}$$

has a transform equal to $F(\omega)e^{-j\omega T/2}$. Hence

$$F_1(\omega) = F(\omega)e^{-j\omega T/2}. \tag{2.192}$$

Let

$$f_2(t) = e^{j(\Omega_o t + \beta t^2)}, \; 0 < t < T. \tag{2.193}$$

From (2.191) we have

$$f_2(t) = e^{j\phi_o} f_1(t), \; 0 < t < T, \tag{2.194}$$

and hence

$$\begin{aligned} F_2(\omega) &= e^{j\phi_o} F_1(\omega) = e^{j\phi_o} e^{-j\omega T/2} F(\omega) \\ &= e^{j(2\omega_o T - \beta T^2)/4} e^{-j\omega T/2} e^{-j(\omega-\omega_o)^2/4\beta} G(\omega) \\ &= e^{-j\left(\frac{(\omega-\omega_0)^2}{4\beta} + \frac{\beta T^2}{4} + \frac{(\omega-\omega_0)T}{2}\right)} G(\omega) = e^{-j\left(\frac{\omega-\omega_0}{2\sqrt{\beta}} + \frac{\sqrt{\beta}T}{2}\right)^2} G(\omega) \\ &= e^{-j\left(\frac{\omega-(\omega_0-\beta T)}{2\sqrt{\beta}}\right)^2} G(\omega) = e^{-j\left(\frac{\omega-\Omega_0}{2\sqrt{\beta}}\right)^2} G(\omega), \end{aligned} \tag{2.195}$$

where from (2.184)

$$G(\omega) = \sqrt{\frac{\pi}{2\beta}} \left\{ K\left(\frac{\omega - \Omega_o}{2\sqrt{\beta}}\right) + K\left(\frac{\Omega_o + 2\beta T - \omega}{2\sqrt{\beta}}\right) \right\}, \tag{2.196}$$

since $\omega_o = \Omega_o + \beta T$. Notice that the transform in (2.195) and (2.196) is nonzero in the interval $\Omega_o < \omega < \Omega_o + 2\beta T$ for $\beta T \gg 1$ and zero elsewhere.

As shown in Fig. 2.40, when a carrier modulated chirp as in (2.193) is transmitted in $0 < t < T$ (causal waveform), its nonzero transform region is in $(\Omega_o, \; \Omega_o + 2\beta T)$ where Ω_o represents the carrier frequency. Thus, in standard form

$$f_3(t) = \begin{cases} e^{j(\omega_o t + \beta t^2)}, & 0 < t < T, \\ 0, & \text{otherwise} \end{cases} \tag{2.197}$$

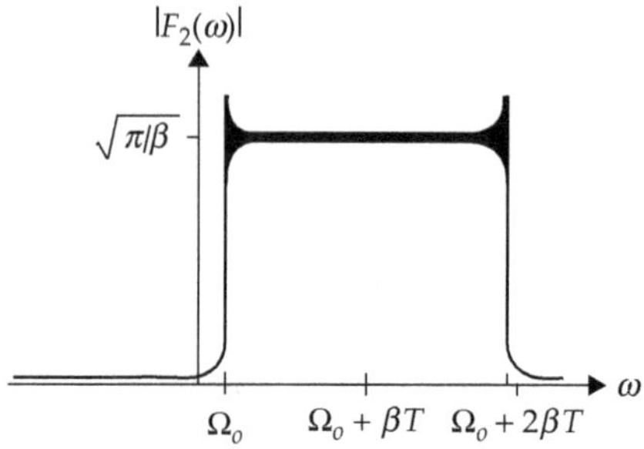

FIGURE 2.40 Transform of a chirp signal $f_2(t) = e^{j(\Omega_o t + \beta t^2)}, 0 < t < T$.

has the transform (see (2.197), (2.198), and Fig. 2.41),

$$F_3(\omega) = e^{-j\left(\frac{\omega-\omega_0}{2\sqrt{\beta}}\right)^2} \sqrt{\frac{\pi}{2\beta}} \left[K\left(\frac{\omega-\omega_0}{2\sqrt{\beta}}\right) + K\left(\frac{\omega_0 + 2\beta T - \omega}{2\sqrt{\beta}}\right) \right]. \tag{2.198}$$

To down convert $F_3(\omega)$ to baseband, one must translate it by the center frequency $\omega_o + \beta T$, and *not* by the carrier frequency ω_o alone. Equation (2.198) may be also rewritten as

$$F_3(\omega) = e^{-j\left(\frac{\omega-\omega_0}{2\sqrt{\beta}}\right)^2} \sqrt{\frac{1}{\beta}} \int_{x-\sqrt{\beta}T}^{x} e^{jt^2} dt \tag{2.199}$$

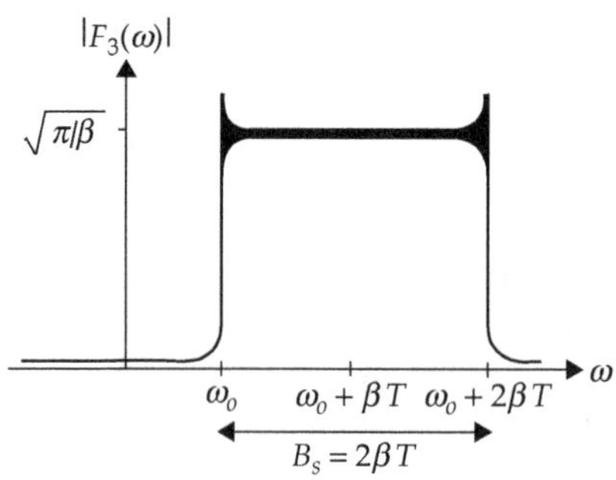

FIGURE 2.41 Transform of a chirp signal $f_3(t) = e^{j(\omega_o t + \beta t^2)}, 0 < t < T$.

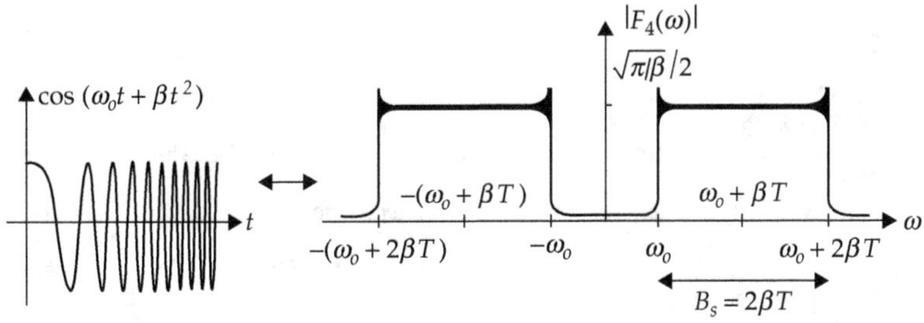

FIGURE 2.42 Transform of a real and causal chirp signal $f_4(t) = \cos(\omega_o t + \beta t^2), 0 < t < T$.

where

$$x = \frac{\omega - \omega_o}{2\sqrt{\beta}}. \tag{2.200}$$

Finally since

$$\cos(\omega_o t + \beta t^2) = \frac{f_3(t) + f_3^*(t)}{2}, \tag{2.201}$$

from (2.197)–(2.201) we have the transform pair

$$f_4(t) = \cos(\omega_o t + \beta t^2),\ 0 < t < T \ \leftrightarrow \ F_4(\omega) = \frac{F_3(\omega) + F_3^*(\omega)}{2}. \tag{2.202}$$

From (2.199)

$$F_3^*(-\omega) = e^{-j\left(\frac{\omega+\omega_o}{2\sqrt{\beta}}\right)^2} \sqrt{\frac{1}{\beta}} \int_y^{y+\sqrt{\beta}T} e^{-jt^2} dt \tag{2.203}$$

where $y = \frac{\omega+\omega_o}{2\sqrt{\beta}}$. Using (2.197)–(2.201) we get (Fig. 2.42)

$$|F_4(\omega)| = \begin{cases} \frac{1}{2}\sqrt{\pi/\beta}, & \omega_o < |\omega| < \omega_o + 2\beta T, \\ 0, & \text{otherwise.} \end{cases} \tag{2.204}$$

References

[1] D. Middleton, *Introduction to Statistical Communication Theory*, California, Peninsula Pub, 1960.

[2] J. V. Difranco, W. L. Rubin, *Radar Detection*, New York, SciTech Publishing, Jan., 2004.

[3] H. L. V. Trees, *Detection, Estimation, and Modulation Theory, Part I*. New York, John Wiley and Sons, 1968.

[4] P. Stoica, J. Li, X. Zhu, "Waveform Synthesis for Diversity-Based Transmit Beampattern Design," *IEEE Transactions on Signal Processing*, vol. 56, no. 6, June 2008.

[5] K.W. Forsythe, D.W. Bliss, "Waveform Correlation and Optimization Issues for MIMO Radar," *Proc. 39th Asilomar Conf. Signals, Systems, Computer*, Pacific Grove, CA, pp. 1306–1310, Nov. 2005.

[6] B. Friedlander, "Waveform Design for MIMO Radars," *IEEE Transactions on Aerospace and Electronic Systems*, vol. 43, no. 3, pp. 1227–1238, Oct. 2007.

[7] Y. Chi, A. Pezeshki, R. Calderbank, S. Howard, "Range Sidelobe Suppression in a Desired Doppler Interval," *IEEE International Waveform Diversity & Design Conference*, Orlando, FL, Feb. 8–13, 2009.

[8] Y. Li, W. Moran, S. Sira, A. Papandreou-Suppappola, "Adaptive Waveform Design in Rapidly Varying Radar Scenes," *IEEE International Waveform Diversity & Design Conference*, Orlando, FL, Feb. 8–13, 2009.

[9] J. Goodman, B. Miller, M. Herman, M. Wai, P. Monticciolo, "Extending the Dynamic Range of RF Receiver with Nonlinear Equalization," *IEEE International Waveform Diversity & Design Conference*, Orlando, FL, Feb. 8–13, 2009.

[10] J. Zhang, A. Papandreou-Suppappola, "MIMO Radar with Frequency Diversity," *IEEE International Waveform Diversity & Design Conference*, Orlando, FL, Feb. 8–13, 2009.

[11] P. Gonzalez-Blano, E.D. Diego, E. Millan, B. Errasti, I. Montiel, "Stepped-Frequency Waveform in Radar Demonstrator and its Jamming," *IEEE International Waveform Diversity & Design Conference*, Orlando, FL, Feb. 8–13, 2009.

[12] HI. Demirkiran, D. Weiner, A. Drozd, I. Kasperovich, "Equivalent Based Analysis for Narrow Weakly Nonlinear Bandpass Systems for Waveform Diversity Applications," *IEEE International Waveform Diversity & Design Conference*, Orlando, FL, Feb. 8–13, 2009.

[13] M. Stuff, B. Thelen, N. Subotic, J. Parker, J. Browning, "Optimization and Waveforms for Compressive Sensing Applications in the Presence of Interference," *IEEE International Waveform Diversity & Design Conference*, Orlando, FL, Feb. 8–13, 2009.

[14] H. Mendelson, P. Zulch, "Mitigation of Radar Range Ambiguities via Waveform Diversity," *IEEE International Waveform Diversity & Design Conference*, Orlando, FL, Feb. 8–13, 2009.

[15] S. Sen, M. Hurtado, A. Nehorai, "Adaptive OFDM Radar for Detecting a Moving Target in Urban Scenarios," *IEEE International Waveform Diversity & Design Conference*, Orlando, FL, Feb. 8–13, 2009.

[16] M. Inggs, M. Brooker, "Extensible simulator for Waveform Diversity Testing," *IEEE International Waveform Diversity & Design Conference*, Orlando, FL, Feb. 8–13, 2009.

[17] Y. Paichard, M. Brooker, M. Inggs, "A Signal Level Simulator for Netted Radar Waveforms Evaluation," *IEEE International Waveform Diversity & Design Conference*, Orlando, FL, Feb. 8–13, 2009.

[18] I. Shirokov, I. Serdyuk, "The Investigation of Phase Synchronization of Reference Oscillators Through Atmospheric Channel," *IEEE International Waveform Diversity & Design Conference*, Orlando, FL, Feb. 8–13, 2009.

[19] I. Akduman, A. Yapar, F. Akleman, H. Sahinturk, "Electromagnetic Imaging of Objects Buried under a Dielectric Rough Surface," *IEEE International Waveform Diversity & Design Conference*, Orlando, FL, Feb. 8–13, 2009.
[20] P.F. Sammartino, C. Baker, "The Frequency Diverse Bistatic System," *IEEE International Waveform Diversity & Design Conference*, Orlando, FL, Feb. 8–13, 2009.
[21] T. Sturman, "Enabling Chaotic Waveform Diversity," *IEEE International Waveform Diversity & Design Conference*, Orlando, FL, Feb. 8–13, 2009.
[22] J. Li, J.R. Guerci, L. Xu, "Signal Waveform's Optimal-Under-Restriction Design for Active Sensing," *IEEE Signal Process. Lett.*, vol. 13, no. 9, pp. 565–568, Sep. 2006.
[23] N. Bharadwaj, V. Chandrasekar, "Distributed Waveforms for Networked Meteorological Radar," *IEEE International Waveform Diversity & Design Conference*, Orlando, FL, Feb. 8–13, 2009.
[24] T. Higgins, S. Blunt, "Analysis of Range-Angle Coupled Beamforming with Frequency-Diverse Chirps," *IEEE International Waveform Diversity & Design Conference*, Orlando, FL, Feb. 8–13, 2009.
[25] H.B. Manjunath, J. Zhang, A. Papandreou-Suppappola, D. Morrell, "Waveform-Agile Sensing for Range and DoA Estimation in MIMO Radars," *IEEE International Waveform Diversity & Design Conference*, Orlando, FL, Feb. 8–13, 2009.
[26] M. Inggs, "Passive Coherent Location as Cognitive Radar," *IEEE International Waveform Diversity & Design Conference*, Orlando, FL, Feb. 8–13, 2009.
[27] Y. Paichard, "Orghogonal Multicarrier Phased Coded Signal for Netted Radar Systems," *IEEE International Waveform Diversity & Design Conference*, Orlando, FL, Feb. 8–13, 2009.
[28] J. Stiles, J. Jenshak, "Sparse Array Construction Using Marginal Fisher's Information," *IEEE International Waveform Diversity & Design Conference*, Orlando, FL, Feb. 8–13, 2009.
[29] M.C. Wicks, "History of Waveform Diversity and Future Benefits to Military Systems," *NATO Lecture Series on Waveform Diversity (SET119)* held in UK, Italy, Germany,and France, Sep. 8–19, 2008.
[30] G. Genello, "Information Processing for Waveform Diversity Control," *IEEE International Waveform Diversity & Design Conference*, Orlando, FL, Feb. 8–13, 2009.
[31] I. Bradaric, G. Capraro, M.C. Wicks "Multistatic Ambiguity Function—A Tool for Waveform Selection in Distributed Radar Systems," *IEEE International Waveform Diversity & Design Conference*, Orlando, FL, Feb. 8–13, 2009.
[32] D.D. Weiner, M.C. Wicks, G.T. Capraro, "Waveform Diversity and Sensors as Robots in Advanced Military Systems," *1st International Waveform Diversity and Design Conference*, Edinburgh, UK, Nov. 2004.
[33] A. Papoulis, *Signal Analysis*, New York: McGraw-Hill, 1977.
[34] S.U. Pillai, H.S. Oh, D.C. Youla, J.R. Guerci, "Optimum Transmit-Receiver Design in the Presence of Signal-Dependent Interference and Channel Noise," *IEEE Transactions on Information Theory*, vol. 46, no. 2, pp. 577–584, March 2000.
[35] D.C. Youla, "Chirp and Pulse Compression," Private communications, Polytechnic University, Melville, NY, 2004.
[36] I. Bradaric, G.T. Capraro, P. Zulch, "Signal Processing and Waveform Selection Strategies in Multistatic Radar Systems," *International Waveform Diversity and Design Conference*, Pisa, Italy, June 2007, invited paper.
[37] R. Bonneau, "A Coherence Gradient Method in Waveform Design for Apertures," *IEEE International Waveform Diversity & Design Conference*, Orlando, FL, Feb. 8–13, 2009.
[38] G. Dingley, C. Alabaster, "Radar Based Automatic Target System," *IEEE International Waveform Diversity & Design Conference*, Orlando, FL, Feb. 8–13, 2009.

[39] R. Romero, N. Goodman, "Improved Waveform Design for Target Recognition with Multiple Transmissions," *IEEE International Waveform Diversity & Design Conference*, Orlando, FL, Feb. 8–13, 2009.

[40] T. Ch. Tang, C.C. Ho, "EMI Suppression for High-Speed PCB," *IEEE International Waveform Diversity & Design Conference*, Orlando, FL, Feb. 8–13, 2009.

[41] L.L. Monte, R. Ansari, D. Erricolo, M. wicks., "The Use of Geometric Diversity for Spectral Dominance in Underground Imaging," *IEEE International Waveform Diversity & Design Conference*, Orlando, FL, Feb. 8–13, 2009.

[42] M. Wicks, "RF Sensors & Spectrum Allocation," *IEEE International Waveform Diversity & Design Conference*, Orlando, FL, Feb. 8–13, 2009.

[43] M.R. Bell, "Information Theory and Radar Waveform Design," *IEEE Trans. Inform. Theory*, vol. 39, no. 5, pp. 1578–1597, Sep. 1993.

[44] S.U. Pillai, H.S. Oh, D.C. Youla, J.R. Guerci, "Optimum Transmit Receiver Design in the Presence of Signal-Dependent Interference and Channel Noise," *IEEE Trans. Inform. Theory*, vol. 46, no. 2, pp. 577–584, March 2000.

[45] S. Kay, "Optimal Signal Design for Detection of Gaussian Point Targets in Stationary Gaussian Clutter/Reverberation ," *IEEE J. Sel. Topics in Sig. Proc. Mag.*, vol. 1, no. 1, pp. 31–41, June 2007.

[46] D.A. Garren, M.K. Osborn, A.C. Odom, J.S. Goldstein, S.U. Pillai, J.R. Guerci, "Enhanced Target Detection and Identification via Optimized Radar Transmission Pulse Shape," *Proc. IEEE*, vol. 148, no. 3, pp. 130–138, June 2001.

[47] S. Ciccarelli, V. Amuso, "Simulation of the Front-End of a MEMS Based Ultra Narrow Band Tomographic Imaging system," *IEEE International Waveform Diversity & Design Conference*, Orlando, FL, Feb. 8–13, 2009.

[48] D.T. Gjessing, *Target Adaptive Matched Illumination Radar: Principles and Applications*, London, England: Peter Peregrinus, 1986.

[49] Y. Yang and R. Blum, "Waveform Design for MIMO Radar Based on Mutual Information and Minimum Mean-Square Error Estimation," *Proc. 40th Annual Conference on Information Sciences and Systems*, pp. 111–116, March 2006.

[50] F. Ahmad, M. Amin, H. Estephan, "Multistatic Waveform Design for Seeing Through the Wall," *IEEE International Waveform Diversity & Design Conference*, Orlando, FL, Feb. 8–13, 2009.

[51] Y. Yang, R. Blum, "Waveform Design for MIMO Radar Based on Mutual Information and Minimum Mean-Square Error Estimation," *Proc. 40th Annual Conference on Information Sciences and Systems*, pp. 111–116, March 2006.

[52] S.U. Pillai, H.S. Oh, J.R. Guerci, "Multichannel Matched Transmit Receiver Design in Presence of Signal-Dependent Interference and Noise," *IEEE Sensor Array and Multichannel Signal Processing Workshop*, pp. 385–389, March 2000.

[53] S.U. Pillai, "Multichannel Waveform Diversity for Improved Target Identification," *Second Tri-Service Waveform Diversity Workshop*, Verona, NY, Feb. 2–3, 2004.

[54] S. Ciccarelli, V. Amuso, "Simulation of the Front-End of a MEMS Based Ultra Narrow Band Tomographic Imaging System," *IEEE International Waveform Diversity & Design Conference*, Orlando, FL, Feb. 8–13, 2009.

[55] H. Deng, "Polyphase Coding Signal Design for Orthogonal Netted Radar Systems," *IEEE Transactions on Signal Processing*, vol. 52, no. 11, Nov. 2004.

[56] D. Slepian, H.O. Pollak, "Prolate Spheroidal Wave Functions, Fourier Analysis and Uncertainty—I," *Bell Syst. Tech. J. 40*, pp. 43–63, Jan. 1961.

CHAPTER 3

New Methods for Optimum Transmitter and Receiver Design

3.1 Introduction

In the general waveform design problem, the desired target is buried in both interference and noise. The transmit signal excites the desired target and the interference simultaneously. The interference can be foliage returns in the form of clutter for radar, scattered returns of the transmit signal from the sea-bottom and different ocean-layers in the case of sonar, or multipath returns in the communication scene. In all these cases, like the target return, the interference returns are also transmit signal-dependent, and hence it puts conflicting demands on the receiver. In general, the receiver input consists of target returns, interferences, and the ever-present noise. The goal of the receiver is to enhance the target returns and simultaneously suppress both the interference and noise signals. In a detection environment, the decision regarding the presence or absence of the target is made at some specified instant $t = t_o$ using the receiver output data, and hence to maximize detection, the signal power to average interference plus noise ratio (*SINR*) at the receiver output can be used as an optimization goal. This scheme is illustrated in Fig. 3.1 [1–3].

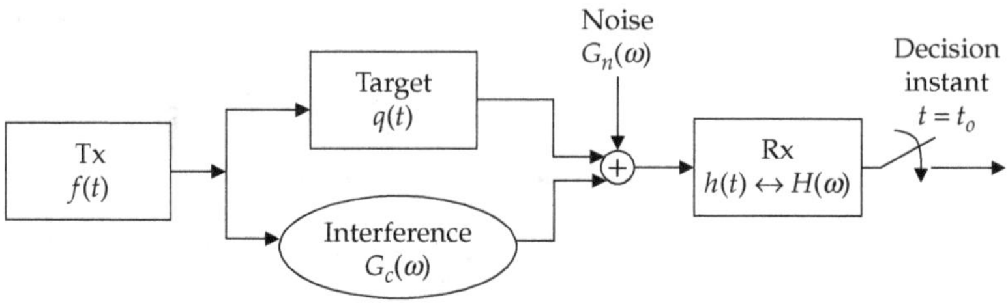

FIGURE 3.1 Transmitter and receiver in presence of interference and noise.

The transmitter output bandwidth can be controlled using a known transmitter output filter $P_1(\omega)$ (Fig. 3.2). A similar filter with transform characteristics $P_2(\omega)$ can be used at the receiver input to control the processing bandwidth as well.

The transmit waveform set can have spatial and temporal components to it each designated for a specific goal. The simplest situation is that shown in Fig. 3.2 where a finite duration waveform $f(t)$ of energy E is to be designed. Thus

$$\int_0^{T_0} |f(t)|^2 dt = E. \tag{3.1}$$

Usually, transmitter output filter characteristics $p_1(t) \leftrightarrow P_1(\omega)$ are known, and for design purposes, it is best to incorporate the transmitter output filter and the receiver input filter along with the target and clutter spectral characteristics. This is shown in Fig. 3.3(b) where

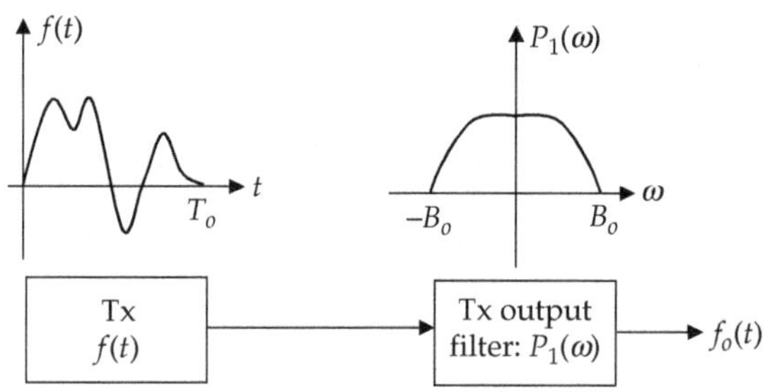

FIGURE 3.2 Transmitter output filter.

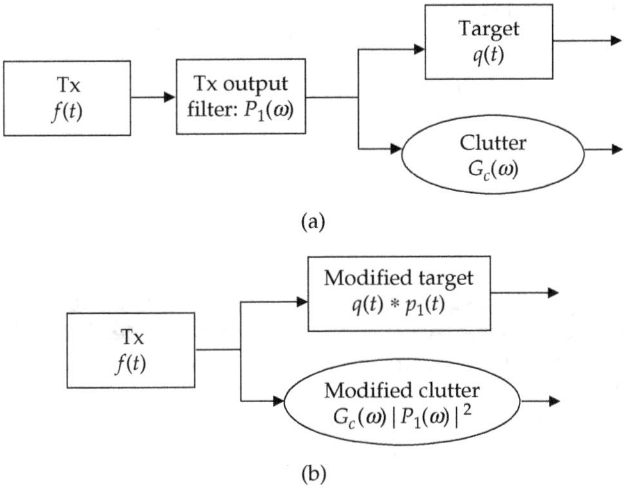

FIGURE 3.3 Two equivalent representations for the transmitter output filter.

the original target response $q(t)$ gets modified as $q(t) * p_1(t)$, and the clutter power spectrum $G_c(\omega)$ gets modified as $G_c(\omega)|P_1(\omega)|^2$ [4].

In what follows the design is first carried out *without* considering the receiver input filter, and in Sec. 5.1 the effect of introducing the receiver input filter is investigated.

Let $q(t) \leftrightarrow Q(\omega)$ represent the target impulse response and its transform. In general $q(t)$ can be any arbitrary waveform. In a linear domain setup, the transmit signal $f(t)$ interacts with the target $q(t)$ to generate the output

$$s(t) = f(t) * q(t) = \int_0^{T_0} f(\tau)q(t-\tau)d\tau \tag{3.2}$$

that represents the desired signal.

The interference returns are usually due to the random scattered returns of the transmit signal from the environment, and hence can be modeled as a stochastic signal $w_c(t)$ that is excited by the transmit signal $f(t)$. If the environment returns are stationary, then the interference can be represented by its power spectrum $G_c(\omega)$. This gives the average interference power to be $G_c(\omega)|F(\omega)|^2$. Finally, let $n(t)$ represent the receiver input noise with power spectral density $G_n(\omega)$. Thus the receiver input signal equals

$$r(t) = s(t) + w_c(t) * f(t) + n(t), \tag{3.3}$$

and the input interference plus noise power equals

$$G_I(\omega) = G_c(\omega)|F(\omega)|^2 + G_n(\omega). \tag{3.4}$$

The received signal is presented to the receiver with impulse response $h(t)$. The simplest receiver is of the noncausal type. Causality puts additional restrictions on the receiver and will be treated separately (Sec. 5.2).

With no restrictions on the receiver, its output signal and interference noise components are given by

$$y_s(t) = s(t) * h(t) = \frac{1}{2\pi}\int_{-\infty}^{+\infty} S(\omega)H(\omega)e^{j\omega t}d\omega \tag{3.5}$$

and

$$y_n(t) = \{w_c(t) * f(t) + n(t)\} * h(t). \tag{3.6}$$

The output $y_n(t)$ represents a second order stationary stochastic process with power spectrum [5]

$$G_o(\omega) = \left(G_c(\omega)|F(\omega)|^2 + G_n(\omega)\right)|H(\omega)|^2 \tag{3.7}$$

and hence the total output interference plus noise power is given by

$$\begin{aligned} \sigma_{I+N}^2 &= \frac{1}{2\pi}\int_{-\infty}^{+\infty} G_o(\omega)d\omega \\ &= \frac{1}{2\pi}\int_{-\infty}^{+\infty} \left(G_c(\omega)|F(\omega)|^2 + G_n(\omega)\right)|H(\omega)|^2\, d\omega. \end{aligned} \tag{3.8}$$

Referring back to Fig. 3.1, the signal component $y_s(t)$ in (3.5) at the receiver output needs to be maximized at the decision instant t_o in presence of the above interference and noise. Hence the instantaneous output signal power at $t = t_o$ is given by

$$P_o = |y_s(t_o)|^2 = \left| \frac{1}{2\pi} \int_{-\infty}^{+\infty} S(\omega) H(\omega) e^{j\omega t_o} d\omega \right|^2. \tag{3.9}$$

This gives the receiver output *SINR* at $t = t_o$ to be

$$SINR = \frac{P_o}{\sigma_{I+N}^2} = \frac{\left| \frac{1}{2\pi} \int_{-\infty}^{+\infty} S(\omega) H(\omega) e^{j\omega t_o} d\omega \right|^2}{\frac{1}{2\pi} \int_{-\infty}^{+\infty} G_I(\omega) \, |H(\omega)|^2 \, d\omega}. \tag{3.10}$$

We can apply Cauchy-Schwarz inequality in (3.10) to eliminate $H(\omega)$. This gives [see (2.53)–(2.55)]

$$\begin{aligned} SINR &\leq \frac{1}{2\pi} \int_{-\infty}^{+\infty} \frac{|S(\omega)|^2}{G_I(\omega)} d\omega \\ &= \frac{1}{2\pi} \int_{-\infty}^{+\infty} \frac{|Q(\omega)|^2 |F(\omega)|^2}{G_c(\omega)|F(\omega)|^2 + G_n(\omega)} d\omega = SINR_{\text{max}}. \end{aligned} \tag{3.11}$$

Thus the maximum obtainable *SINR* is given by (3.11), and using (2.57), this is achieved if and only if

$$\begin{aligned} H_{opt}(\omega) &= \frac{S^*(\omega)}{G_c(\omega)|F(\omega)|^2 + G_n(\omega)} e^{-j\omega t_o} \\ &= \frac{Q^*(\omega) F^*(\omega)}{G_c(\omega)|F(\omega)|^2 + G_n(\omega)} e^{-j\omega t_o}. \end{aligned} \tag{3.12}$$

In (3.12), the phase shift $e^{-j\omega t_o}$ can be retained to approximate causality for the receiver waveform. Interestingly, even with a point target ($Q(\omega) \equiv 1$), flat noise ($G_n(\omega) = \sigma_n^2$), and flat clutter ($G_c(\omega) = \sigma_c^2$), the optimum receiver is *not* conjugately matched to the transmit signal, since in that case from (3.12) we have

$$H_{opt}(\omega) = \frac{F^*(\omega)}{\sigma_c^2 |F(\omega)|^2 + \sigma_n^2} e^{-j\omega t_o} \neq F^*(\omega) \, e^{-j\omega t_o}. \tag{3.13}$$

3.2 Transmitter Waveform Design

When the receiver design satisfies (3.12), the output *SINR* is given by the right side of (3.11), where the free parameter $|F(\omega)|^2$ can be chosen to further maximize the output *SINR*, subject to the transmit energy constraint in (3.1). Thus the transmit signal design reduces to the following optimization problem:

Maximize

$$SINR_{\max} = \frac{1}{2\pi}\int_{-\infty}^{+\infty} \frac{|Q(\omega)|^2|F(\omega)|^2}{G_c(\omega)|F(\omega)|^2 + G_n(\omega)}\, d\omega, \tag{3.14}$$

subject to the energy constraint

$$\int_0^{T_o} |f(t)|^2 dt = \frac{1}{2\pi}\int_{-\infty}^{+\infty} |F(\omega)|^2 d\omega = E. \tag{3.15}$$

To solve this new constrained optimization problem, combine (3.14) and (3.15) to define the modified Lagrange optimization function

$$\Lambda_o = \int_{-\infty}^{+\infty} \underbrace{\left\{ \frac{|Q(\omega)|^2 y^2(\omega)}{G_c(\omega)y^2(\omega) + G_n(\omega)} - \frac{1}{\lambda^2} y^2(\omega) \right\}}_{\Lambda(\omega)} d\omega \tag{3.16}$$

where

$$y(\omega) = |F(\omega)| \tag{3.17}$$

is the free design parameter. As Sec. 3.5 shows, the maximum of (3.16) is given by $\frac{\partial \Lambda(\omega)}{\partial y} = 0$ which gives

$$\frac{\partial \Lambda(\omega)}{\partial y} = 2y(\omega)\left\{ \frac{G_n(\omega)|Q(\omega)|^2}{\{G_c(\omega)y^2(\omega) + G_n(\omega)\}^2} - \frac{1}{\lambda^2} \right\} = 0. \tag{3.18}$$

From (3.18), either

$$y(\omega) = 0 \tag{3.19}$$

or

$$\frac{G_n(\omega)|Q(\omega)|^2}{\{G_c(\omega)y^2(\omega) + G_n(\omega)\}^2} - \frac{1}{\lambda^2} = 0, \tag{3.20}$$

which gives

$$y^2(\omega) = \frac{\sqrt{G_n(\omega)}\left(\lambda|Q(\omega)| - \sqrt{G_n(\omega)}\right)}{G_c(\omega)} \tag{3.21}$$

provided $y^2(\omega) > 0$.

To see whether (3.21) indeed maximizes (3.16), following (3.124) we get

$$\begin{aligned} \frac{\partial^2 \Lambda(\omega)}{\partial y^2} &= 2\left\{\frac{G_n(\omega)|Q(\omega)|^2}{\left\{G_c(\omega)y^2(\omega) + G_n(\omega)\right\}^2} - \frac{1}{\lambda^2}\right\} \\ &\quad - 8\frac{G_n(\omega)|Q(\omega)|^2 G_c(\omega)}{\left\{G_c(\omega)y^2(\omega) + G_n(\omega)\right\}^3} y^2(\omega) \end{aligned} \tag{3.22}$$

which is clearly negative when evaluated at the solution given in (3.21) since in that case the first term above goes to zero. Hence for the $y(\omega)$ given in (3.21), we have

$$\left.\frac{\partial^2 \Lambda(\omega)}{\partial y^2}\right|_{y=y(\omega)} < 0 \tag{3.23}$$

which shows that the optimum solution of $y(\omega)$ indeed maximizes the *SINR* output.

Let Ω_+ represent the frequencies over which $y^2(\omega)$ in (3.21) is strictly positive, and let Ω_o represent the complement of Ω_+. Observe that the set Ω_+ is a function of the noise and target spectral characteristics as well as the constraint constant λ. In terms of Ω_+, we have

$$|F(\omega)|^2 = \begin{cases} y^2(\omega), & \omega \in \Omega_+, \\ 0, & \omega \in \Omega_o. \end{cases} \tag{3.24}$$

From (3.21), $y^2(\omega) > 0$ over Ω_+ gives the necessary condition

$$\lambda \geq \max_{\omega \in \Omega_+} \frac{\sqrt{G_n(\omega)}}{|Q(\omega)|} \tag{3.25}$$

and the energy constraint in (3.15) when applied to (3.21)–(3.24) gives

$$E = \frac{1}{2\pi}\int_{\Omega_+} y^2(\omega)d\omega = \frac{\lambda}{2\pi}\int_{\Omega_+} \frac{\sqrt{G_n(\omega)}|Q(\omega)|}{G_c(\omega)}d\omega - \frac{1}{2\pi}\int_{\Omega_+} \frac{G_n(\omega)}{G_c(\omega)}d\omega \tag{3.26}$$

or, for a given value of E, we have

$$\lambda = \frac{E + \frac{1}{2\pi}\int\limits_{\Omega_+} \frac{G_n(\omega)}{G_c(\omega)} d\omega}{\frac{1}{2\pi}\int\limits_{\Omega_+} \frac{\sqrt{G_n(\omega)}|Q(\omega)|}{G_c(\omega)} d\omega} \stackrel{\Delta}{=} \lambda(E). \tag{3.27}$$

Clearly, $\lambda(E)$ in (3.27) must satisfy the inequality in (3.25) as well. This gives rise to the concept of transmitter energy threshold that is characteristic to this design approach.

3.2.1 Transmitter Threshold Energy[1]

From (3.25)–(3.27), the transmit energy E must be such that $\lambda(E)$ obtained from (3.27) should satisfy (3.25). If not, E must be increased to accommodate it, and hence it follows that there exists a *minimum threshold* value for the transmit energy below which it will not be possible to maintain $|F(\omega)|^2 > 0$. This threshold value is given by

$$E_{\min} = \left(\max_{\omega \in \Omega_+} \frac{\sqrt{G_n(\omega)}}{|Q(\omega)|} \right) \frac{1}{2\pi} \int\limits_{\Omega_+} \frac{\sqrt{G_n(\omega)}|Q(\omega)|}{G_c(\omega)} d\omega - \frac{1}{2\pi} \int\limits_{\Omega_+} \frac{G_n(\omega)}{G_c(\omega)} d\omega \tag{3.28}$$

and for any operating condition, the transmit energy E must exceed $E_{\min}$. Clearly, the minimum threshold energy depends on the target, clutter, and noise characteristics as well as the bandwidth under consideration. With $E > E_{\min}$, substituting (3.20) and (3.21) into the $SINR_{\max}$ in (3.14) we get

$$\begin{aligned} SINR_{\max} &= \frac{1}{2\pi} \int\limits_{\Omega_+} \frac{|Q(\omega)|^2 y^2(\omega)}{\lambda \sqrt{G_n(\omega)}|Q(\omega)|} d\omega \\ &= \frac{1}{2\pi} \int\limits_{\Omega_+} \frac{|Q(\omega)|}{\lambda(E)\sqrt{G_n(\omega)}} \frac{\sqrt{G_n(\omega)}\left(\lambda(E)|Q(\omega)| - \sqrt{G_n(\omega)}\right)}{G_c(\omega)} d\omega \\ &= \frac{1}{2\pi} \int\limits_{\Omega_+} \left(|Q(\omega)| - \frac{\sqrt{G_n(\omega)}}{\lambda(E)} \right) \frac{|Q(\omega)|}{G_c(\omega)} d\omega. \end{aligned} \tag{3.29}$$

[1]© 2006 IEEE. Reprinted with permission, from "Energy Threshold Constraints in Transmit Waveform Design" by S. U. Pillai, K. Y. Li, B. Himed, *2006 International Waveform Diversity & Design Conference*, Kauai Hawaii, Jan. 22–27, 2006.

Finally making use of (3.27), the output $SINR_{\max}$ becomes

$$SINR_1 = \frac{1}{2\pi}\int\limits_{\Omega_+}\frac{|Q(\omega)|^2}{G_c(\omega)}d\omega - \frac{\left(\frac{1}{2\pi}\int\limits_{\Omega_+}\frac{|Q(\omega)|\sqrt{G_n(\omega)}}{G_c(\omega)}d\omega\right)^2}{E+\frac{1}{2\pi}\int\limits_{\Omega_+}\frac{G_n(\omega)}{G_c(\omega)}d\omega} \tag{3.30}$$

$$= a - \frac{c}{\lambda(E)} = a - \frac{c^2}{E+b} = \frac{aE + (ab - c^2)}{E+b}$$

where

$$a = \frac{1}{2\pi}\int\limits_{\Omega_+}\frac{|Q(\omega)|^2}{G_c(\omega)}d\omega, \tag{3.31}$$

$$b = \frac{1}{2\pi}\int\limits_{\Omega_+}\frac{G_n(\omega)}{G_c(\omega)}d\omega, \tag{3.32}$$

and

$$c = \frac{1}{2\pi}\int\limits_{\Omega_+}\frac{\sqrt{G_n(\omega)}|Q(\omega)|}{G_c(\omega)}d\omega. \tag{3.33}$$

Notice that $ab - c^2 \geq 0$.

Figure 3.5 shows the $SINR_{\max}$ dependency on transmit energy for the target and clutter spectra shown in Fig. 3.4. Notice that each point in Fig. 3.4(d)–(e) corresponds to a different operating bandwidth. Furthermore, to realize a specific performance level with a given bandwidth constraint, a certain amount of minimum transmit energy is required, thus indicating a threshold value on the transmit energy.

The optimization problem in (3.14) and (3.15) can be restated in terms of Ω_+ as follows: Given $Q(\omega)$, $G_c(\omega)$, $G_n(\omega)$, and the transmit energy E, how to partition the frequency axis into an "operating band" Ω_+ and a "no show" band Ω_o so that λ_+ obtained from (3.27) satisfies (3.25) and $SINR_{\max}$ in (3.29) and (3.30) is also maximized? In general maximization of $SINR_{\max}$ in (3.29) and (3.30) over Ω_+ is a highly nonlinear optimization problem for arbitrary $Q(\omega)$, $G_c(\omega)$, and $G_n(\omega)$.

In what follows two approaches to this problem are presented.

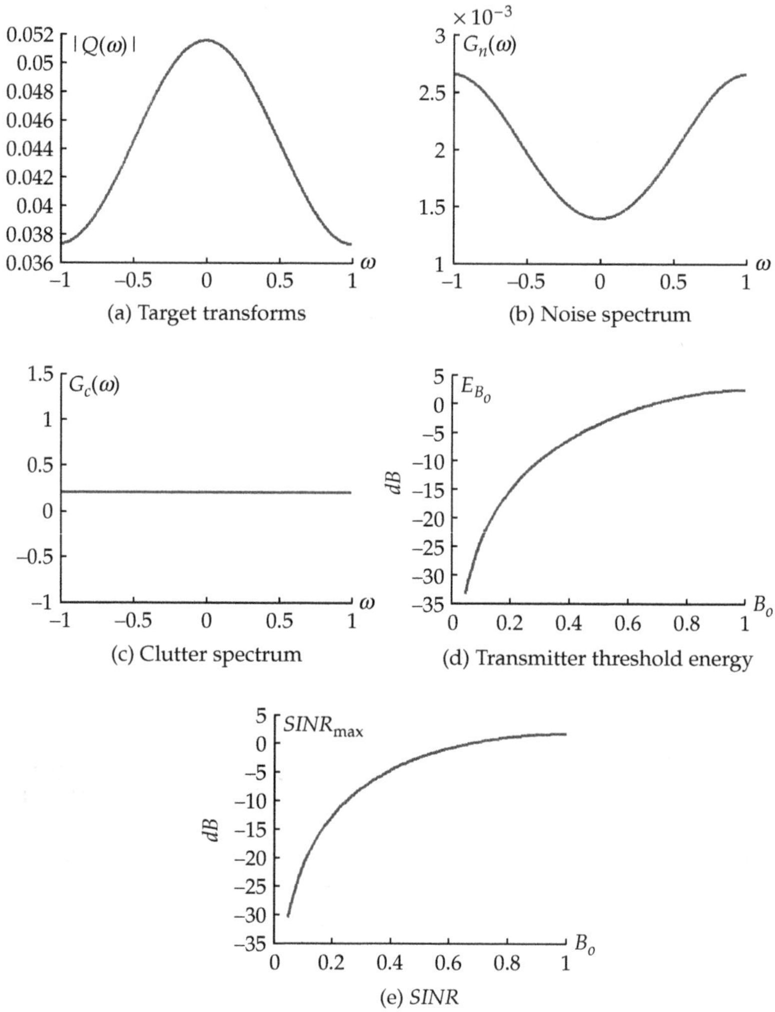

FIGURE 3.4 Transmitter threshold energy and the corresponding *SINR*. Lower pass target in high pass noise and flat clutter spectra are shown.

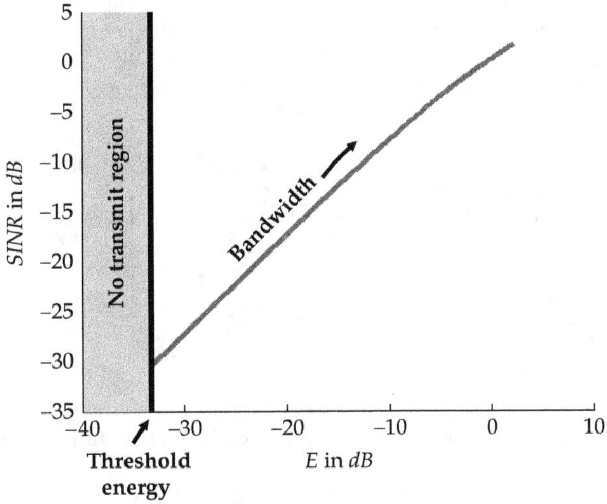

FIGURE 3.5 Transmitter threshold energy and corresponding $SINR_{\max}$ for the low pass target in Fig. 3.4.

3.3 Method I: Desired Band Approach[2]

One approach for maximizing the input *SINR* in (3.14) is to make use of the "desired frequency band" of interest B_o that is usually suggested by the target response $Q(\omega)$ (and the transmitter output filter) to determine the operating band Ω_+ [6]. The desired band B_o can represent a fraction of the total available bandwidth or the whole bandwidth itself. The procedure for determining Ω_+ is illustrated in Fig. 3.6 for two situations. In Fig. 3.6, the frequency band B_o represents the desired band, and because of the nature of the noise and clutter spectra, it may be necessary to operate on a larger region Ω_+ in the frequency domain. Thus the desired band B_o is contained always within the operating band Ω_+. To determine Ω_+, motivated by (3.25) we project the band B_o onto the spectrum $\sqrt{G_n(\omega)}/|Q(\omega)|$ and draw a horizontal line corresponding to

$$\lambda_{B_o} = \max_{\omega \in B_o} \frac{\sqrt{G_n(\omega)}}{|Q(\omega)|} \tag{3.34}$$

as shown there. Define $\Omega_+(B_o)$ to represent the frequency region where

$$\omega \in \Omega_+(B_o) : \frac{\sqrt{G_n(\omega)}}{|Q(\omega)|} \leq \lambda_{B_o} = \max_{\omega \in B_o} \frac{\sqrt{G_n(\omega)}}{|Q(\omega)|}. \tag{3.35}$$

[2] © 2008 IEEE. Reprinted with permission, from "Waveform Design Optimization using Bandwidth and Energy Considerations" by S. U. Pillai, K. Y. Li, and H. Beyer, *2008 IEEE Radar Conference*, Rome, Italy, May 26–30, 2008.

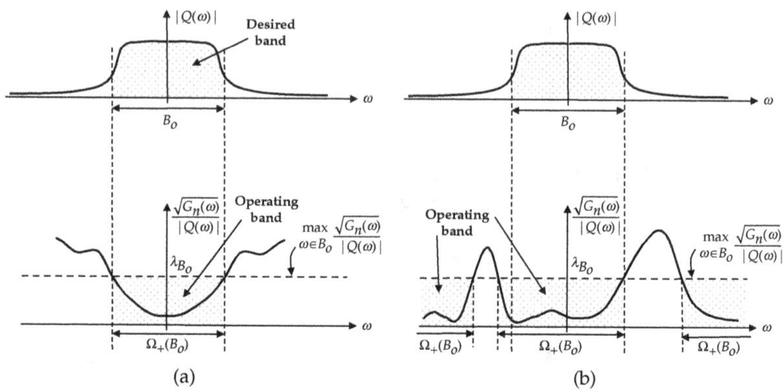

FIGURE 3.6 Partitioning of the frequency axis for the operating band.

This procedure can give rise to two situations as shown in Fig. 3.6(a) and (b). In Fig. 3.6(a), the operating band $\Omega_+(B_o)$ coincides with the desired band B_o, whereas in Fig. 3.6(b), the desired band B_o is a subset of $\Omega_+(B_o)$. Knowing $\Omega_+(B_o)$, one can compute $\lambda = \lambda(E)$ with the help of (3.27) over that region, and examine whether λ so obtained satisfies (3.25). If not, the transmitter energy E is insufficient to maintain the operating band $\Omega_+(B_o)$ given in (3.35), and either E must be increased, or $\Omega_+(B_o)$ must be decreased (by decreasing B_o) so that (3.25) is satisfied. Thus for a given desired band B_o [or an operating band $\Omega_+(B_o)$], as remarked earlier, there exists a *minimum* transmitter threshold energy E_{B_o}, below which it is *impossible* to maintain $|F(\omega)|^2 > 0$ over that entire operating band.

3.3.1 Threshold Energy

From (3.26) and (3.34), we obtain the minimum transmitter threshold energy in this case to be

$$E_{B_o} = \frac{\lambda_{B_o}}{2\pi} \int_{\Omega_+(B_o)} \frac{\sqrt{G_n(\omega)}\,|Q(\omega)|}{G_c(\omega)} d\omega - \frac{1}{2\pi} \int_{\Omega_+(B_o)} \frac{G_n(\omega)}{G_c(\omega)} d\omega$$
$$= \lambda_{B_o} c_o - b_o > 0. \tag{3.36}$$

With $E \geq E_{B_o}$, the $SINR_{\max}$ in (3.30) can be readily computed. In particular with $E = E_{B_o}$, we get

$$SINR_1 = SINR_1(B_o) = a_o - \frac{c_o^2}{E_{B_o} + b_o} = a_o - \frac{c_o^2}{E_{B_o} + \lambda_{B_o}}. \tag{3.37}$$

Here a_o, b_o, and c_o are as given in (3.31)–(3.33) with Ω_+ replaced by $\Omega_+(B_o)$. Equation (3.37) represents the performance level for bandwidth B_o using its minimum threshold energy. From (3.21), we also obtain the optimum transmit signal transform corresponding to energy E_{B_o} to be

$$|F(\omega)|^2 = \begin{cases} \dfrac{\sqrt{G_n(\omega)}\left(\lambda_{B_o}|Q(\omega)| - \sqrt{G_n(\omega)}\right)}{G_c(\omega)}, & \omega \in \Omega_+(B_o) \\ 0, & \omega \in \Omega_o \end{cases}$$

$$= \begin{cases} \sqrt{G_n(\omega)}\left(\left(\max\limits_{\omega \in B_o} \dfrac{\sqrt{G_n(\omega)}}{|Q(\omega)|}\right) - \dfrac{\sqrt{G_n(\omega)}}{|Q(\omega)|}\right)\dfrac{|Q(\omega)|}{G_c(\omega)}, & \omega \in \Omega_+(B_o) \\ 0, & \omega \in \Omega_o. \end{cases} \tag{3.38}$$

To summarize, to maintain a given desired band B_o, there exists an operating band $\Omega_+(B_o) \geq B_o$ over which $|F(\omega)|^2 > 0$ and to guarantee this, the transmit energy must exceed a minimum threshold value E_{B_o} given by (3.36).

Figures 3.7, 3.9 and 3.11 show the transmitter threshold energy E in (3.36) and the corresponding *SINR* in (3.37) as a function of the desired bandwidth B_o for various target, clutter, and noise spectra. Figures 3.8, 3.10 and 3.12 show the optimum transmit waveforms for selected target, clutter, and noise cases as indicated. In all of these figures, target to noise ratio (*TNR*) is set at 0 dB, and the clutter to noise power ratio (*CNR*) is set at 20 dB. The total noise power is normalized to unity. The desired bandwidth B_o is normalized with respect to the maximum available bandwidth (e.g., carrier frequency).

In Fig. 3.7, the noise and clutter have flat spectra and for the highly resonant target (solid line), the *SINR* and the required minimum energy thresholds reach a saturation value for small values of the bandwidth. In the case of the other two targets, additional bandwidth is required to reach the maximum attainable *SINR*. This is not surprising since for the resonant target, a significant portion of its energy is concentrated around the resonant frequency. Hence once the transmit signal bandwidth reaches the resonant frequency, it latches onto the target features, resulting in maximum *SINR* at a lower bandwidth. Similar conclusions can also be made for Figs. 3.9 and 3.11. Notice that in Fig. 3.11, the white noise-flat clutter case performance is uniformly inferior to the white noise-high-pass clutter situation as in the former case there is no frequency selectivity possible with respect to the total interference spectrum. In the extreme case, when target, clutter, and

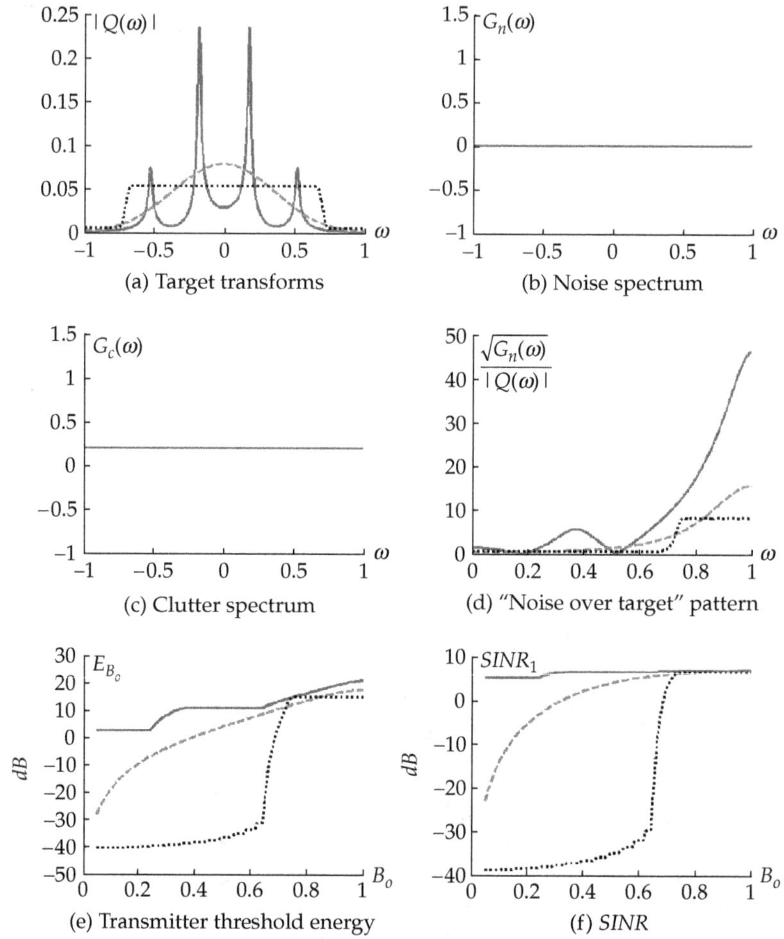

FIGURE 3.7 Transmitter threshold energy and the corresponding *SINR* for Method I. Three different target spectra (solid, dotted, and dashed) in white noise and flat clutter spectra are shown here.

noise all have flat spectra with identical bandwidth, transmit waveform optimization is difficult since there are no distinguishing spectral features to complete the design.

Figures 3.13–3.15 show the transmitter threshold energy E in (3.36) and the corresponding *SINR*, in (3.37) as a function of the desired bandwidth B_o for a different set of target, noise, and clutter spectra.

From Fig. 3.13(f), in the case of the resonant target (solid curve), the *SINR* reaches its peak value, resulting in saturation even when B_o is

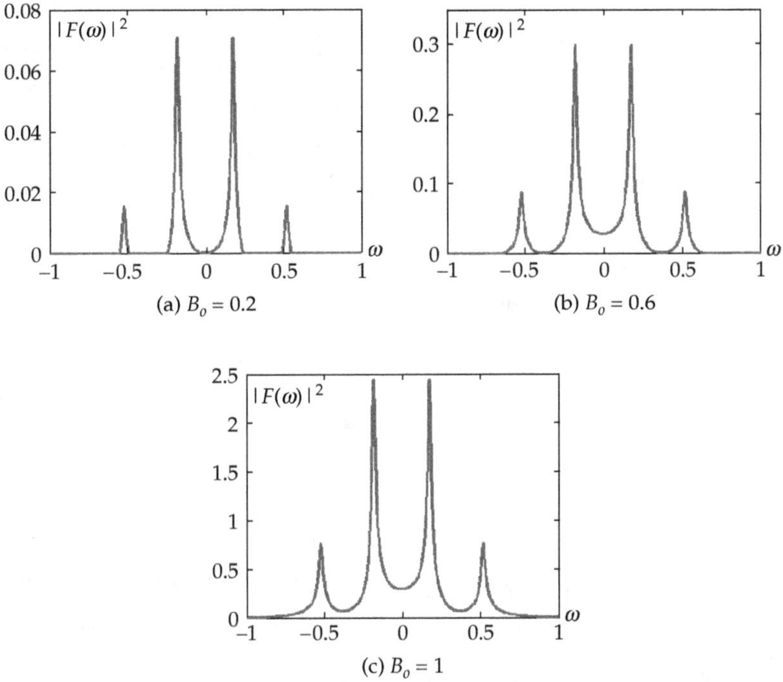

FIGURE 3.8 Transmit waveform transforms for three different bandwidths corresponding to the resonant target in Fig. 3.7.

a small fraction of the available bandwidth.[3] This is because in that case, the transmit waveform is able to latch onto the dominant resonant frequency of the target. On the other extreme, when the target has flat characteristics (dotted curve), there are no distinguishing frequencies to latch on, and the transmitter is unable to attain the above maximum *SINR* even when B_o coincides with the total available bandwidth. For a low pass target (dashed curve), the transmitter is indeed able to deliver the maximum *SINR* by making use of all the available bandwidth. Similar conclusions can be drawn from Figs. 3.14 and 3.15 as well [7–9].

As Fig. 3.6(b) shows, $\Omega_+(B_o)$ can consist of multiple disjoint frequency bands whose complement Ω_o represents the "no show" region.

[3]From Fig. 3.6, even when B_o is small, it is quite possible that the operating band $\Omega_+(B_o)$ can cover the entire frequency band. As a result, saturation for threshold energy E and *SINR* can occur even for small B_o. [See solid curves in Fig. 3.13(e) and (f).]

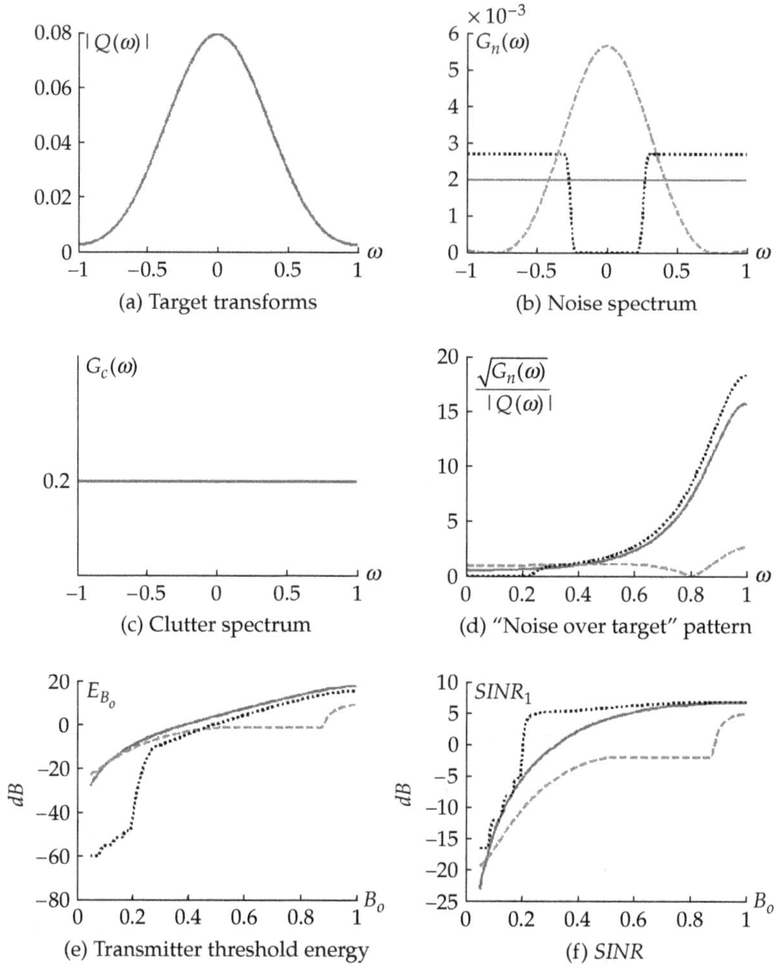

FIGURE 3.9 Transmitter threshold energy and the corresponding *SINR* for Method I. A target with low pass features in three different types of noise spectra and flat clutter are shown here.

Notice that the "no show" region Ω_o in the frequency domain in (3.38) for the optimum transmit signal can be controlled by the desired bandwidth B_o. By increasing B_o, these "no show" regions can be made narrower and this defines a minimum transmitter threshold energy E_∞ that allows $\Omega_+(B_o)$ to be the *entire* available frequency axis.

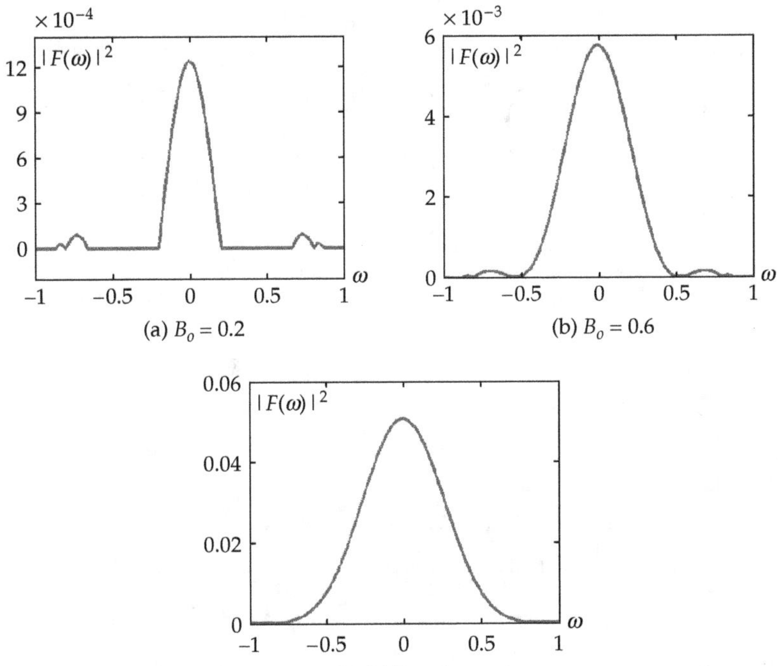

(a) $B_o = 0.2$ (b) $B_o = 0.6$ (c) $B_o = 1$

FIGURE 3.10 Transmit waveform transforms for three different bandwidth corresponding to the low pass noise in Fig. 3.9.

To determine E_∞, let λ_∞ represent the maximum in (3.25) over the *entire* frequency axis. Thus

$$\lambda_\infty = \max_{|\omega|<\infty} \frac{\sqrt{G_n(\omega)}}{|Q(\omega)|}, \tag{3.39}$$

and let $a_\infty, b_\infty, c_\infty$ refer to the constants a, b, c in (3.31)–(3.33) calculated with Ω_+ representing the entire frequency axis. Then from (3.26)

$$E_\infty = \lambda_\infty c_\infty - b_\infty = \frac{\lambda_\infty}{2\pi} \int_{-\infty}^{+\infty} \frac{\sqrt{G_n(\omega)}|Q(\omega)|}{G_c(\omega)} d\omega - \frac{1}{2\pi} \int_{-\infty}^{+\infty} \frac{G_n(\omega)}{G_c(\omega)} d\omega > 0 \tag{3.40}$$

represents the minimum transmit energy (threshold) required to avoid partitioning in the frequency domain. With E_∞ as given by (3.40), we

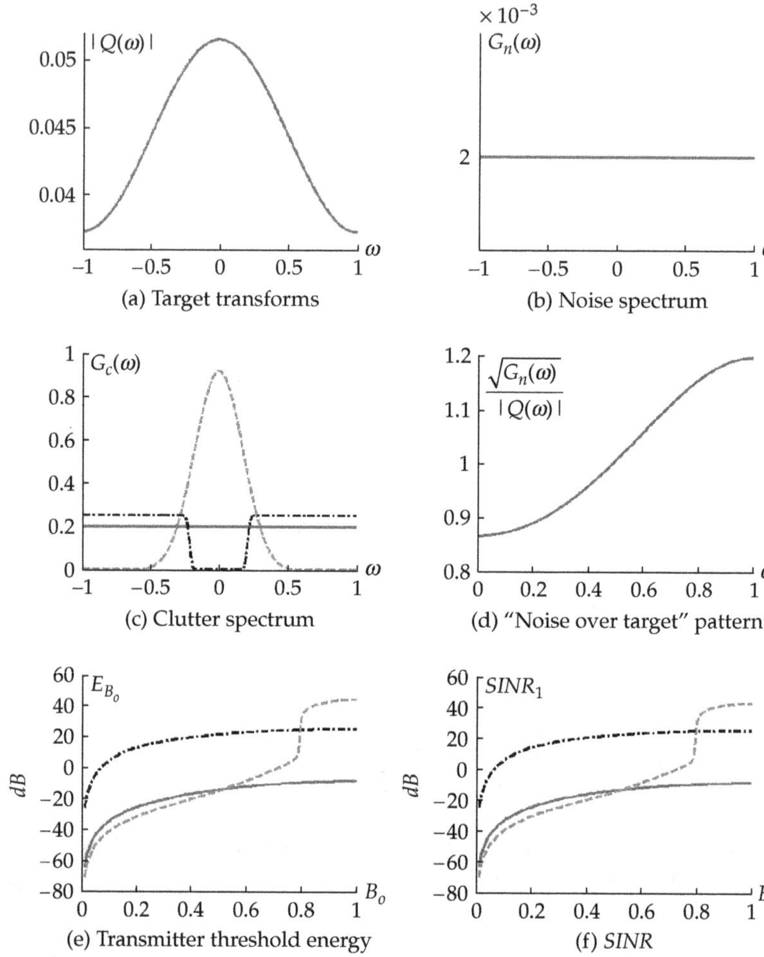

FIGURE 3.11 Transmitter threshold energy and the corresponding *SINR* for Method I. A low pass target in white noise and three different types of clutter spectra are shown here.

obtain $SINR_{\max}$ to be [use (3.30)]

$$SINR_1(\infty) = a_\infty - \frac{c_\infty}{\lambda_\infty} = a_\infty - \frac{c_\infty^2}{E_\infty + b_\infty} > 0 \tag{3.41}$$

and

$$|F(\omega)|^2 = \frac{\sqrt{G_n(\omega)}(\lambda_\infty |Q(\omega)| - \sqrt{G_n(\omega)})}{G_c(\omega)}, \quad |\omega| < \infty. \tag{3.42}$$

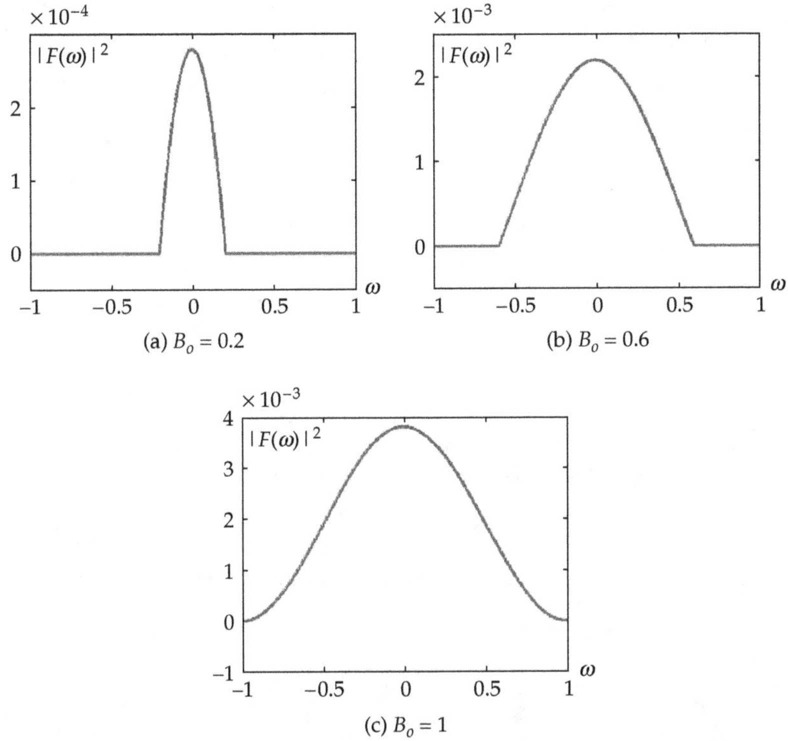

FIGURE 3.12 Transmit waveform transforms for three different bandwidth for the flat clutter shown in Fig. 3.11.

Clearly by further increasing the transmit energy in (3.41) beyond that in (3.40) we obtain

$$SINR_1 \rightarrow a_\infty = \frac{1}{2\pi}\int_{-\infty}^{-\infty} \frac{|Q(\omega)|^2}{G_c(\omega)}d\omega. \tag{3.43}$$

It follows that to avoid *any* restrictions in the frequency domain for the transmit signal, the transmitter energy E must exceed a minimum threshold value E_∞ given by (3.40), and (3.41) represents the maximum realizable SNR. By increasing E beyond E_∞, the performance can be improved up to that in (3.43).

In general from (3.36) for a given desired bandwidth B_o, the transmit energy E must exceed its threshold value E_{B_o}. With $E > E_{B_o}$ and $\lambda(E)$

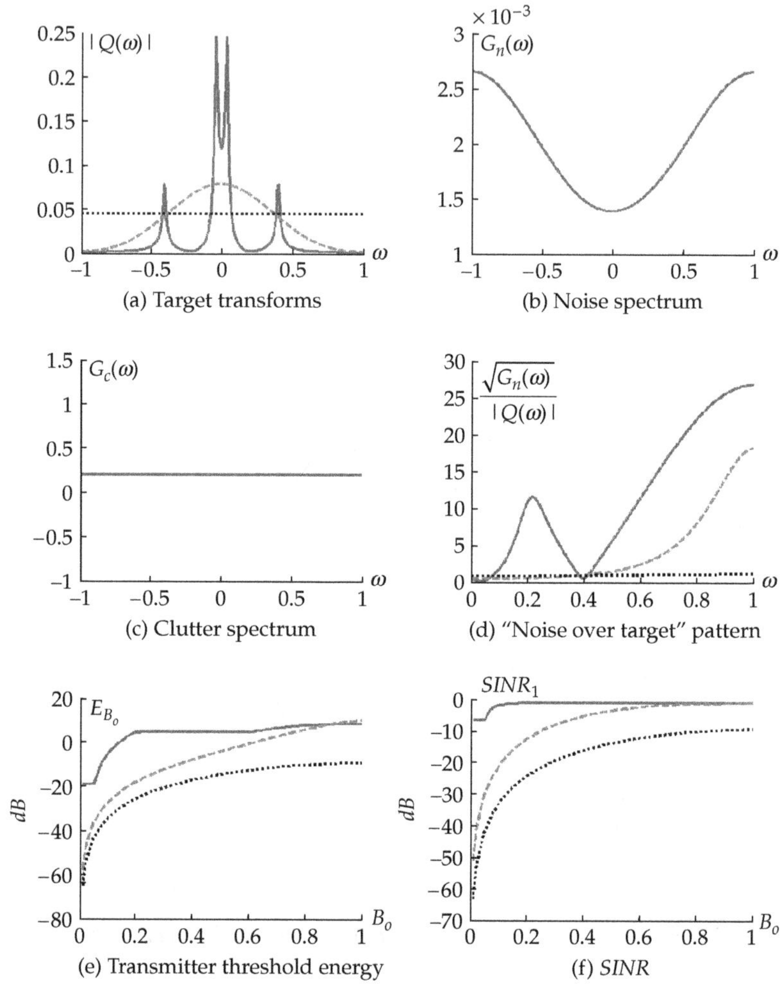

FIGURE 3.13 Transmitter threshold energy and the corresponding *SINR*. Three different target spectra (solid, dotted, and dashed) in high pass noise and flat clutter spectra are shown here.

as in (3.27), the corresponding optimum transmit signal transform is given by [see (3.21)–(3.24)]

$$|F(\omega)|^2 = \begin{cases} \dfrac{\sqrt{G_n(\omega)}\left(\lambda(E)|Q(\omega)| - \sqrt{G_n(\omega)}\right)}{G_c(\omega)}, & \omega \in \Omega_+(B_o) \\ 0, & \omega \in \Omega_o \end{cases} \tag{3.44}$$

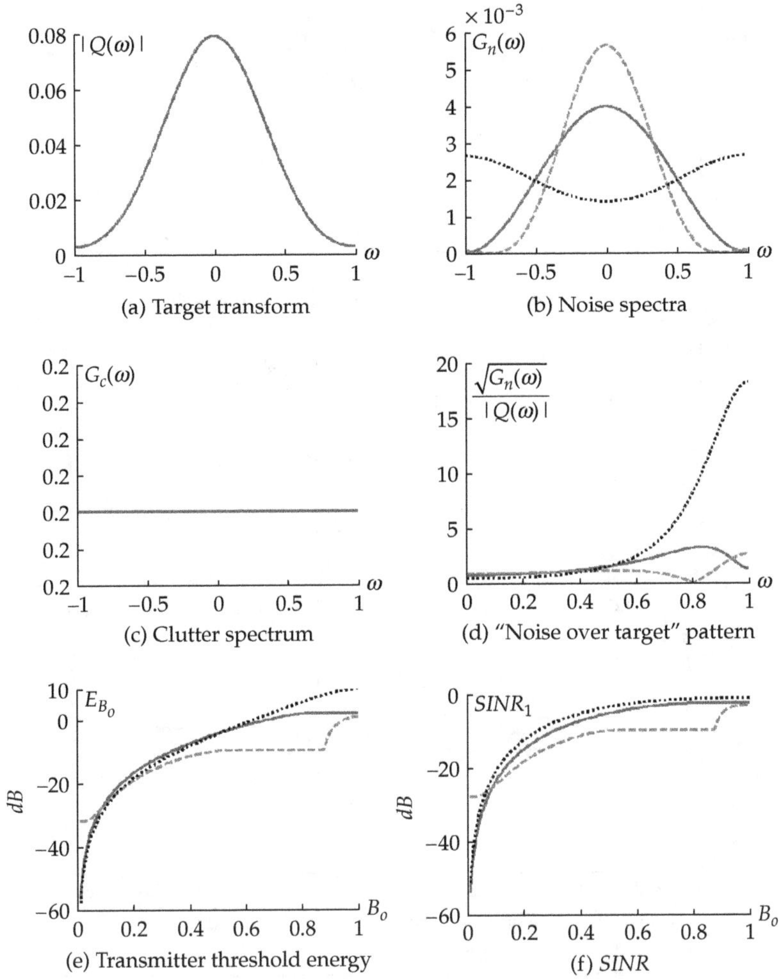

FIGURE 3.14 Transmitter threshold energy and the corresponding *SINR*. A target with low pass features in three different types of noise spectra and flat clutter are shown here.

and clearly this signal is different from the minimum threshold energy one in (3.38). From (3.30), the performance level $SINR_1(E, B_o)$ corresponding to (3.44) is given by (3.37) with E_{B_o} replaced by E. Thus

$$SINR_1(E, B_o) = a_o - \frac{c_o^2}{E + b_o} > SINR_1(B_o). \tag{3.45}$$

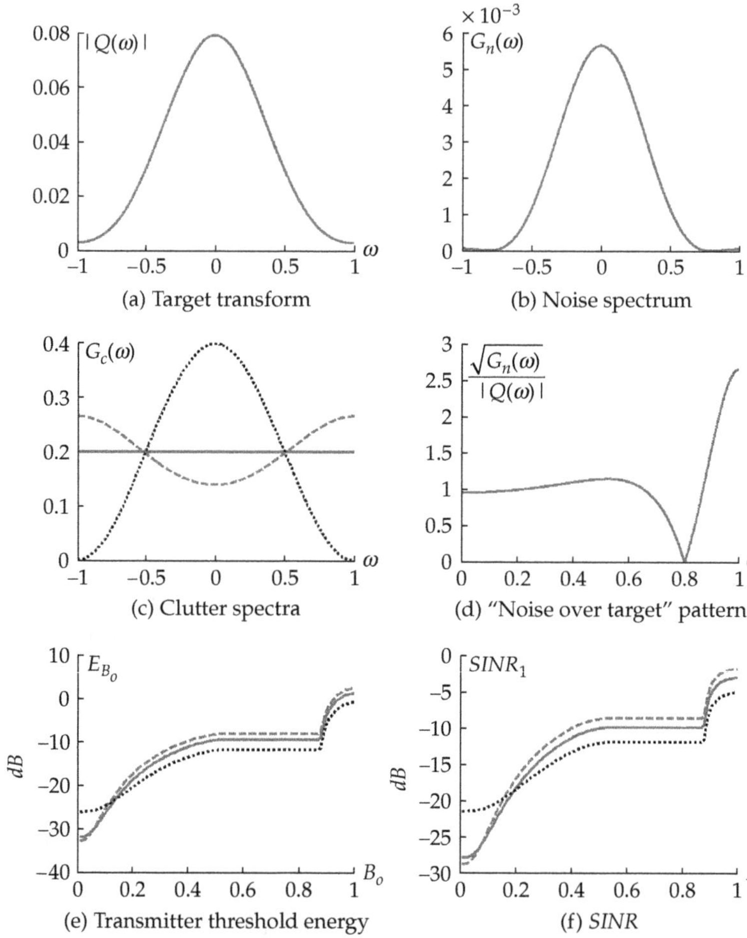

FIGURE 3.15 Transmitter threshold energy and the corresponding *SINR*. A low pass target in low pass noise and three different types of clutter spectra are shown here.

From (3.45), for a given bandwidth B_o, performance can be increased beyond that in (3.37) by increasing the transmit energy. Hence it follows that $SINR_1(B_o)$ represents the minimum performance level for bandwidth B_o that is obtained by using its minimum threshold energy. It is quite possible that this improved performance $SINR_1(E, B_o)$ can be equal to the minimum performance level corresponding to a *higher* bandwidth $B_1 > B_o$. This gives rise to the concept of *energy-bandwidth* tradeoff at a certain performance level [6, 10, 11]. Undoubtedly this is quite useful when bandwidth is at premium.

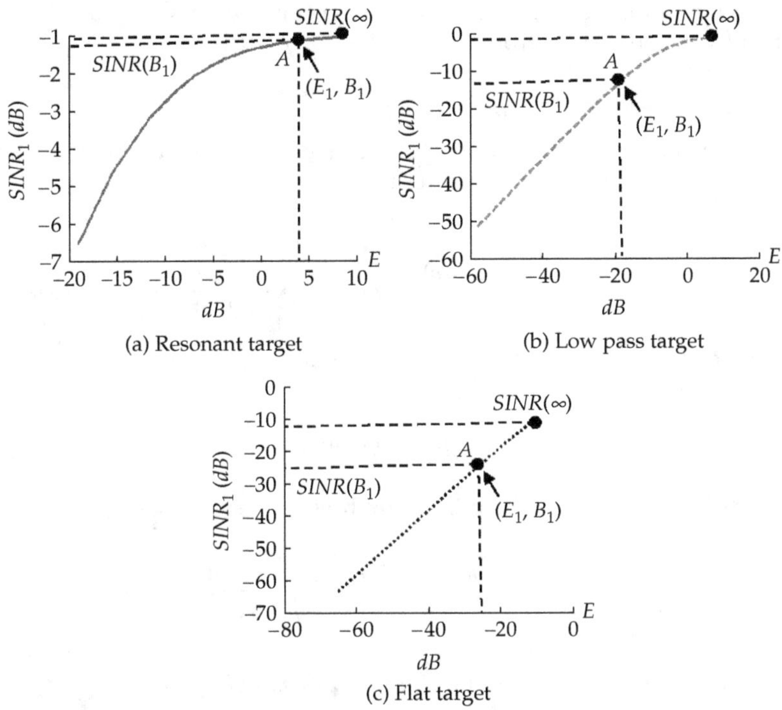

FIGURE 3.16 $SINR_1$ versus transmit threshold energy for the three target cases in Fig. 3.13. $B_1 = 0.2$.

3.3.2 Energy-Bandwidth Tradeoff [6]

Figure 3.13(e) and (f), and Fig. 3.15(e) and (f) exhibit the transmit threshold energy and maximum output $SINR_1(B_o)$ as a function of the desired bandwidth B_o. Combining these figures using (3.37), an *SINR* versus transmit threshold energy plot can be generated as in Fig. 3.13(f) for each target situation.

For example, Fig. 3.16 corresponds to the three different target situations considered in Fig. 3.13 with clutter and noise spectra as shown there. Notice that each point on the *SINR*-energy threshold curve for each target is associated with a specific desired bandwidth [12–14]. Thus for bandwidth B_1, the minimum threshold energy required is E_1, and the corresponding *SINR* equals $SINR_1(B_1)$ in (3.37). Let *A* represent the associated operating point in Fig. 3.16.[4] From (3.37),

[4]The operating point *A* corresponding to bandwidth B_1 has different threshold energies and different performance levels for different targets.

each operating point generates a distinct transmit waveform. As the bandwidth increases, from (3.41), $SINR \to SINR_1(\infty)$.

Monotonic Property of *SINR*

The threshold energy and *SINR* associated with a higher bandwidth are higher. To prove this, consider two desired bandwidths B_1 and B_2 with $B_2 > B_1$. Then from (3.34) we have

$$\lambda_2 = \max_{\omega \in B_2} \frac{\sqrt{G_n(\omega)}}{|Q(\omega)|} > \lambda_1 = \max_{\omega \in B_1} \frac{\sqrt{G_n(\omega)}}{|Q(\omega)|}, \tag{3.46}$$

and from Fig. 3.6, the corresponding operating bandwidths $\Omega_+(B_1)$ and $\Omega_+(B_2)$ satisfy

$$\Omega_+(B_2) \geq \Omega_+(B_1). \tag{3.47}$$

From (3.36) [or (3.26)], the minimum threshold energies are given by

$$E_i = \frac{1}{2\pi} \int_{\Omega_+(B_i)} \sqrt{G_n(\omega)} \left(\lambda_i - \frac{\sqrt{G_n(\omega)}}{|Q(\omega)|} \right) \frac{|Q(\omega)|}{G_c(\omega)} d\omega, \quad i = 1, 2 \tag{3.48}$$

and substituting (3.46) and (3.47) into (3.48) we get

$$E_2 > E_1. \tag{3.49}$$

Also from (3.29), the performance levels at threshold energy $SINR_1(B_i)$ equals

$$SINR_1(B_i) = \frac{1}{2\pi} \int_{\Omega_+(B_i)} \left(|Q(\omega)| - \frac{\sqrt{G_n(\omega)}}{\lambda_i} \right) \frac{|Q(\omega)|}{G_c(\omega)} d\omega \tag{3.50}$$

and an argument similar to (3.46) and (3.47) gives

$$SINR_1(B_2) \geq SINR_1(B_1) \tag{3.51}$$

for $B_2 > B_1$. Thus as Figs. 3.13–3.16 show, $SINR_1(B_i)$ is a monotonically nondecreasing function of both bandwidth and energy. Figure 3.17 illustrates this *SINR*-energy relation for the target with flat spectrum shown in Fig. 3.13(a). In Fig. 3.17, the two operating points A and B are associated with bandwidths B_1 and B_2, threshold energies E_1 and E_2, and performance levels $SINR_1(B_1)$ and $SINR_1(B_2)$, respectively. Since

$$B_2 > B_1 \Rightarrow E_2 \geq E_1 \quad \text{and} \quad SINR_1(B_2) \geq SINR_1(B_1). \tag{3.52}$$

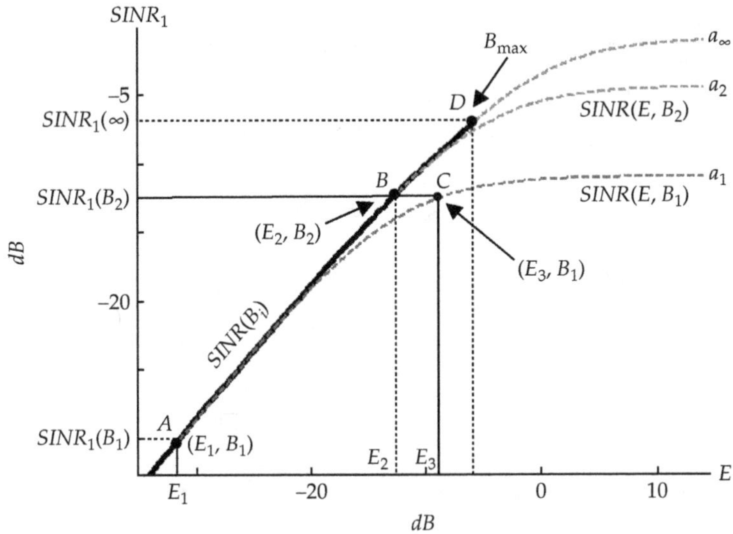

FIGURE 3.17 Energy-bandwidth tradeoff for the target with flat spectrum in Fig. 3.13. For $B_1 = 0.1$ the threshold energy $E_1 = -34.5\ dB$ and tradeoff gives $B_2 = 0.45$ with $E_3 = -14.2\ dB$.

The distinct transmit waveforms $|F_1(\omega)|^2$ and $|F_2(\omega)|^2$ associated with these operating points A and B are given by (3.38) and they are shown in Fig. 3.18(a) and (b).

Consider the operating point A associated with the desired bandwidth B_1. If the transmit energy E is increased beyond the corresponding threshold value E_1 with bandwidth *held* constant at B_1, the performance $SINR_1(E,\ B_1)$ increases beyond that at A since from (3.45)

$$SINR_1(E,\, B_1) = a_1 - \frac{c_1^2}{E + b_1} \geq a_1 - \frac{c_1^2}{E_1 + b_1} = SINR_1(B_1) \qquad (3.53)$$

and it is upper bounded by a_1. Here a_1 corresponds to the $SINR$ performance for bandwidth B_1 as the transmit energy $E \to \infty$. Note that a_1, b_1, and c_1 are the constants in (3.31)–(3.33) with Ω_+ replaced by $\Omega_+(B_1)$. The dashed curve Aa_1 in Fig. 3.17 represents $SINR_1(E,\ B_1)$ for various values of E. From (3.44), each point on the curve Aa_1 generates a new transmit waveform as well.

Interestingly, the dashed curves in Fig. 3.17 *cannot* cross over the optimum performance (solid) curve $SINR(B_i)$. If not, assume the performance $SINR_1(E,\ B_1)$ associated with the operating point A crosses over $SINR(B_i)$ at some $E_1^{'} > E_1$. Then from (3.49), there exists

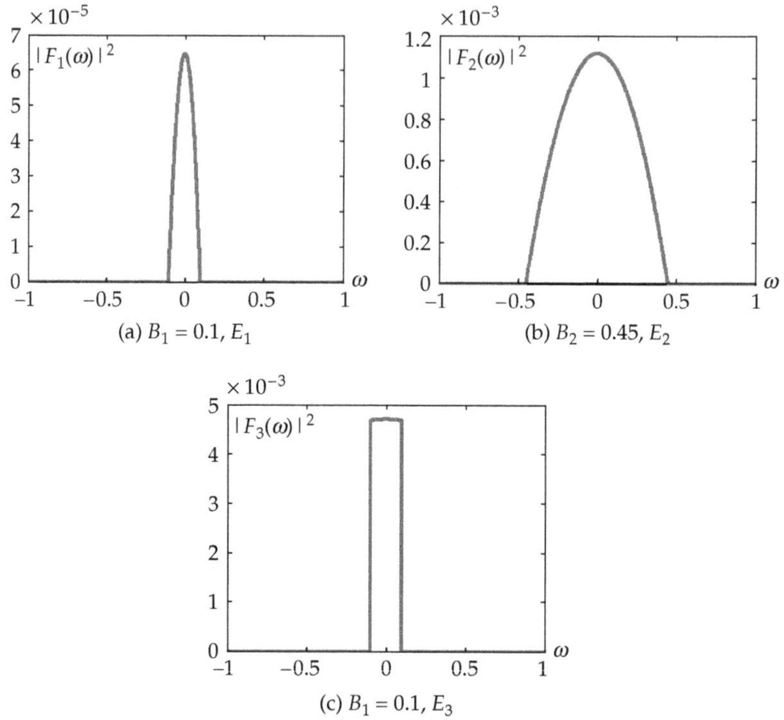

FIGURE 3.18 Transmit waveforms for operating points A, B, and C in Fig. 3.17.

a frequency point $B_1^{'} > B_1$ with threshold energy $E_1^{'}$ and optimum performance $SINR_1(B_1^{'})$. By assumption,

$$SINR_1(E_1^{'}, B_1) > SINR_1(B_1^{'}). \tag{3.54}$$

But this is impossible since $SINR_1(B_1^{'})$ corresponds to the maximum $SINR$ realizable at bandwidth $B_1^{'}$ with energy $E_1^{'}$, and hence performance at a lower bandwidth B_1 with the same energy *cannot* exceed it. Hence (3.54) *cannot* be true and we must have

$$SINR_1(E_1^{'}, B_1) \leq SINR_1(B_1^{'}), \tag{3.55}$$

that is, the curves Aa_1, $B\,a_2$, etc. do not cross over the optimum performance curve ABD.[5]

[5]The $SINR(E, B_1)$ can "hug" the $SINR(B_i)$ curve prior to saturating at a_i.

In Fig. 3.17, assume that the saturation performance value

$$a_1 \geq SINR_1(B_2), \tag{3.56}$$

i.e., the maximum performance level for bandwidth B_1 is greater than or equal to the performance level associated with the operating point B, that has a higher bandwidth B_2, and a higher threshold energy E_2. Draw a horizontal line through B to intersect the curve Aa_1 at C, and drop a perpendicular at C to intersect the x-axis at E_3. From (3.53) with $E = E_3$ we get

$$SINR_1(E_3, B_1) = SINR_1(B_2). \tag{3.57}$$

Thus the operating point C on the curve Aa_1 is associated with energy E_3, and bandwidth B_1, and corresponds to a performance level of $SINR_1(B_2)$ associated with a higher bandwidth. Notice that

$$E_3 > E_2 > E_1, \quad \text{and} \quad B_2 > B_1. \tag{3.58}$$

In other words, by increasing the transmit energy from E_1 to E_3 while holding the bandwidth constant at B_1, the performance equivalent to a higher bandwidth B_2 can be realized provided (3.56) is satisfied. As a result, energy-bandwidth tradeoff is possible within *reasonable* limits. The transmit waveform $|F_3(\omega)|^2$ associated with the operating point C is given by (3.44) and it is illustrated in Fig. 3.18(c). A comparison with Fig. 3.18(a) and (b) shows that the waveform at C is different from those associated with operating points A and B.

It is important to note that although the transmit waveform designs $|F_3(\omega)|^2$ and $|F_1(\omega)|^2$ correspond to the same bandwidth (with different energies E_3 and E_1), one is *not* a scaled version of the other. Changing transmit energy from E_1 to E_3 unleashes the whole design procedure and produces a new waveform $|F_3(\omega)|^2$ that maintains a performance level associated with a larger bandwidth B_2.

The question of how much bandwidth tradeoff can be achieved at an operating point is an interesting one. From the above argument, equality condition in (3.56) gives the upper bound on how much effective bandwidth increment can be achieved by increasing the transmit energy. Notice that for an operating point A, the desired bandwidth B_1 gives the operating bandwidth $\Omega_+(B_1)$ and from (3.31) the performance limit

$$a_1 = \frac{1}{2\pi} \int_{\Omega_+(B_1)} \frac{|Q(\omega)|^2}{G_c(\omega)} d\omega \tag{3.59}$$

for bandwidth B_1 can be computed. Assume $B_2 > B_1$, and from (3.50) $SINR_1(B_2)$ the minimum performance at B_2 also can be computed, and for maximum bandwidth swapping the nonlinear equation

$$a_1 = SINR_1(B_2) \tag{3.60}$$

must be solved for B_2. Then

$$\Delta B(B_1) = B_2 - B_1 \tag{3.61}$$

represents the maximum bandwidth enhancement that can be realized at B_1. This is illustrated in Fig. 3.19 for the target situation in Fig. 3.17. Notice that the maximum operating bandwidth, if finite in any system due to sampling considerations and after normalization, is represented by unity. Hence the upper limit in (3.61) must be min(1, B_2). This gives

$$\Delta B = \min(1, B_2) - B_1 \tag{3.62}$$

and this explains the linear nature of ΔB for larger value of B_1. In that case, bandwidth can be enhanced by $1 - B_1$ only.

Finally, Fig. 3.21 shows *SINR* versus energy performance curve for a low pass target with low pass clutter and high pass noise scenes as shown in Fig. 3.20. The four points *A*, *B*, *C*, and *D* marked there correspond to bandwidths 0.05, 0.2, 0.8, and 1, respectively. From there, the $SINR_1(E, B_i)$ curves "hug" the optimum performance curve $SINR(B_i)$ by different amounts depending on its location. Thus Aa_1 hugs $SINR(B_i)$ till the point B, whereas $C a_3$ departs from $SINR(B_i)$ at C itself, and the behavior of $B a_2$ is somewhere in between. Further, it is also clear from Fig. 3.21 that the percentage bandwidth gains depend on the location of the operating point on the frequency axis.

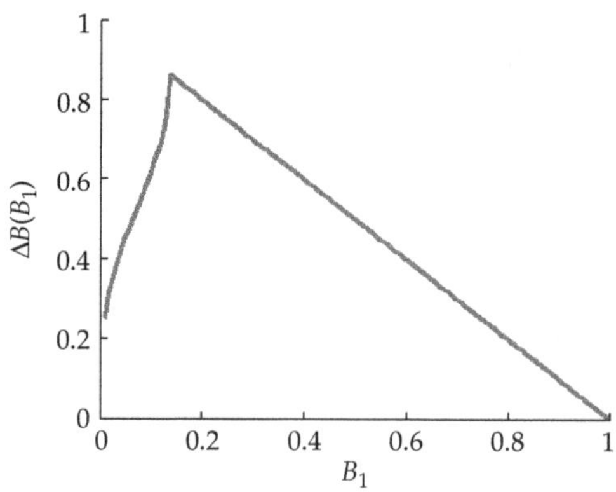

FIGURE 3.19 Maximum bandwidth tradeoff limit for the flat target in Fig. 3.13.

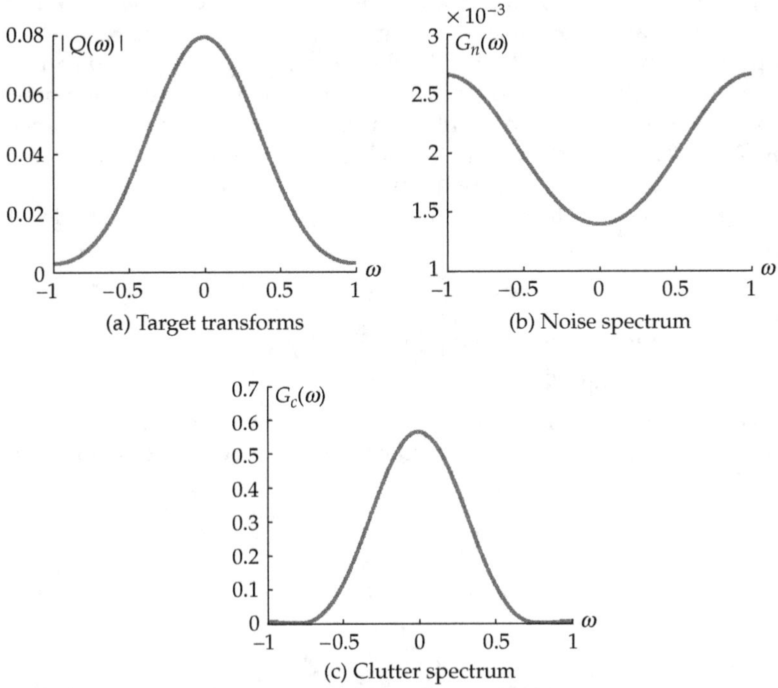

FIGURE 3.20 Target, noise, and clutter scenes for Fig. 3.21.

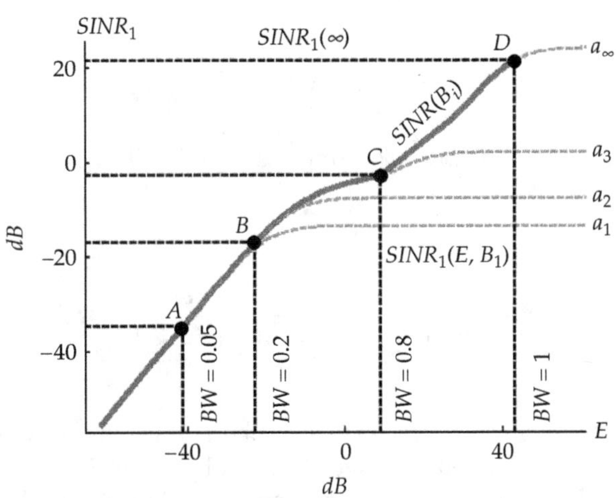

FIGURE 3.21 *SINR* versus energy performance for a low pass target and clutter and high pass noise scenes.

The waveform design procedure described here deals with saving bandwidth at the expense of energy or vice versa in the case of an optimal design. Interestingly the design procedure described can be used to simultaneously save both bandwidth and energy compared to a traditional waveform without sacrificing the *SINR* performance [6,7,15]. This is illustrated below where the optimum waveform design is used to simultaneously save on both bandwidth and energy compared to the chirp waveform for the same performance level [6,10–11,13–18].

3.3.3 Simultaneous Savings on Energy and Bandwidth

To realize this goal, let $F_o(\omega)$ represent the transform of any conventional waveform such as a chirp signal with energy E_o and bandwidth B_o. Then

$$E_o = \frac{1}{2\pi} \int_{-B_o}^{B_o} |F_o(\omega)|^2 d\omega \tag{3.63}$$

and for a given target with transform $Q(\omega)$, and clutter and noise with spectra $G_c(\omega)$ and $G_n(\omega)$, from equation (3.12), the optimum receiver is given by

$$H_o(\omega) = \frac{Q^*(\omega) F_o^*(\omega)}{G_c(\omega)|F_o(\omega)|^2 + G_n(\omega)} e^{-j\omega t_o} \tag{3.64}$$

and from (3.11), the corresponding maximum *SINR* output is given by

$$SINR_o = \frac{1}{2\pi} \int_{-B_o}^{B_o} \frac{|Q(\omega)|^2 |F_o(\omega)|^2}{G_c(\omega)|F_o(\omega)|^2 + G_n(\omega)} d\omega. \tag{3.65}$$

Notice that the output $SINR_o$ in (3.65) is a function of two independent parameters, the desired bandwidth B_o and energy E_o, and for comparison purposes these parameters can be made equal to their corresponding values obtained, using the optimum waveform design [19].

In the case of the optimum waveform design, for a given desired bandwidth B_o, the minimum required energy is given by E_{B_o} (3.36), and the corresponding maximum output $SINR$ equals $SINR_1$ which is given by

$$SINR_1 = SINR_1(B_o) = a_o - \frac{c_o^2}{E_o + b_o} \tag{3.66}$$

as in (3.37) where

$$E_o = E_{B_o} \tag{3.67}$$

and a_o, b_o, c_o are as described in (3.31)–(3.33) with Ω_+ there replaced by the operating bandwidth $\Omega_+(B_o)$ as defined in (3.35).

From Sec. 3.3.2, to compute the $SINR_1$ in (3.66) corresponding to the optimum method, the desired frequency bandwidth B_o is used first to define an operating bandwidth $\Omega_+(B_o)$, and then integration is carried out over $\Omega_+(B_o)$ in (3.29)–(3.36). As Fig. 3.4(b) shows the operating bandwidth $\Omega_+(B_o)$ in some cases can exceed the desired band $(-B_o, B_o)$. In the context of bandwidth comparison with respect to a given waveform $F_o(\omega)$ with bandwidth B_o, it is necessary to make sure that the operating bandwidth $\Omega_+(B_o)$ does not exceed the desired bandwidth $(-B_o, B_o)$. To realize this goal, a transmitter output filter $P_1(\omega)$ as in Fig. 3.2 whose passband bandwidth B_o coincides with the given desired bandwidth can be employed. From Fig. 3.3, the transmitter output filter $P_1(\omega)$ modifies the given target transform $Q(\omega)$ to $Q(\omega)P_1(\omega)$, and also modifies the given clutter spectrum $G_c(\omega)$ to $G_c(\omega)P_1(\omega)$. As a result, when Fig. 3.4(b) is redrawn using this modified target transform $Q(\omega)P_1(\omega)$ to determine the operating bandwidth $\Omega_+(B_o)$, it is easy to see that $\Omega_+(B_o)$ coincides with $(-B_o, B_o)$, and hence the operating bandwidth never exceeds the desired bandwidth B_o in this approach. This technique of inserting a transmitter output filter $P_1(\omega)$ with passband with B_o is employed whenever bandwidth savings comparison with respect to a given waveform is being investigated. Here onward, we will assume that the operating bandwidth $\Omega_+(B_o)$ and the desired band $(-B_o, B_o)$ are equal to each other.

The solid curve marked "$SINR_1(B_i)$" through B in Fig. 3.22 represents the optimum $SINR_1$ in (3.66) as a function of energy E_o for a hypothetical target, noise, and clutter spectra using the optimum waveform design in (3.38) with bandwidth set equal to B_o. By employing a transmitter output filter $P_1(\omega)$ with passband with B_o, the operating bandwidth $\Omega_+(B_o)$ is made to coincide with the desired bandwidth B_o. Thus each point on the solid curve $SINR_1(B_i)$ corresponds to an energy-bandwidth pair (E_i, B_i).

Using the same pair (E_o, B_o) for the given transmit signal such as the chirp waveform, the corresponding $SINR_o$ is computed using (3.65) and it is plotted as the solid curve marked "Chirp" in Fig. 3.22. The

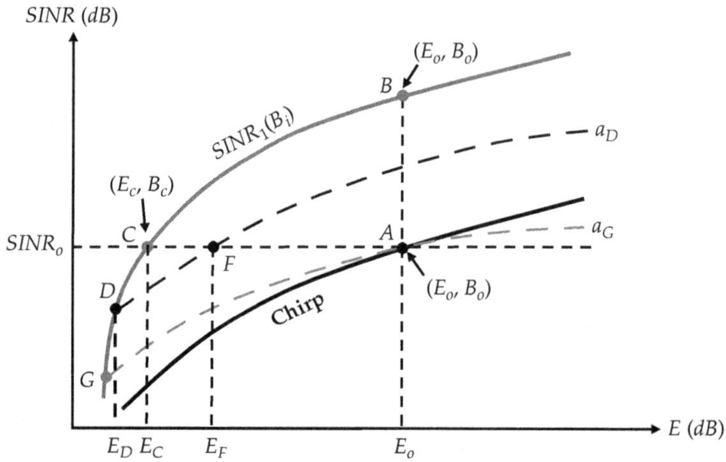

FIGURE 3.22 Simultaneous energy and bandwidth savings.

given waveform $F_o(\omega)$ being generally suboptimum, the $SINR_o$ curve marked "Chirp" will usually fall below the optimum $SINR_1$ curve through B as shown in Fig. 3.22. Thus points A and B in Fig. 3.22 correspond to the same bandwidth-energy pair (E_o, B_o) but with different performance levels $SINR_o$ and $SINR_1$ respectively with

$$SINR_o \leq SINR_1, \tag{3.68}$$

where $SINR_o$ represents the performance in (3.65) when the given transmit waveform is used, and $SINR_1$ represents the performance in (3.66) when the optimum transmit waveform given by (3.38) is used. Draw a horizontal line through A to meet the optimum solid curve at C as shown in Fig. 3.22. The operating point C has performance level equal to $SINR_o$ and corresponds to some energy-bandwidth pair (E_C, B_C) that can be solved by equating $SINR_1(B_C)$ with $SINR_o$. Thus B_C is obtained by solving

$$SINR_1(B_C) = a_C - \frac{c_C^2}{E_C + b_C} = SINR_o. \tag{3.69}$$

Here a_C, b_C, c_C are as described in equations in (3.31)–(3.33) with Ω_+ replaced by the operating bandwidth $\Omega_+(B_C) = B_C$, and E_C is given by (3.36) with B_o replaced by B_C. Since $SINR_o \leq SINR_1(B_o)$, and the operating point C falls below the point B in Fig. 3.22, from (3.52) we have

$$B_C < B_o, \quad E_C < E_o \tag{3.70}$$

As a result, the optimum waveform design $F_1(\omega)$ corresponding to the operating point C is given by (3.38) with B_o replaced by B_C with the passband for $P_1(\omega)$ in Fig. 3.2 extending only up to B_C. Hence $F_1(\omega)$ uses the energy-bandwidth pair (E_C, B_C) and from (3.70) this pair is uniformly superior (smaller bandwidth, smaller energy) to the given pair (E_o, B_o), and it performs at the same level $SINR_o$ as the given transmit waveform. Thus for a given waveform $F_o(\omega)$ associated with operating point A, simultaneous savings in both bandwidth and energy are possible by making use of the optimum waveform design $F_1(\omega)$ that corresponds to the operating point C in the optimum design curve. The operating point C corresponds to the minimum energy design for the same performance level and the maximum saving in energy equals $E_o - E_C > 0$, and the corresponding bandwidth savings equal $B_o - B_C$.

Interestingly, additional savings in bandwidth can be made at the expense of more transmit energy. To see this, consider an operating point D that is below C on the optimum design curve. From (3.52), the design point D operates with lower bandwidth $B_D < B_C$ and uses threshold energy E_D given by (3.36) with B_o replaced by B_D, and provides lower performance level $SINR_1(B_D)$. To increase the performance level above $SINR_1(B_D)$, the transmit energy E can be increased beyond its threshold value E_D, and from (3.45) and (3.53) increase in transmit energy E gives the performance

$$SINR_1(E, B_D) = a_D - \frac{c_D^2}{E + b_D} \tag{3.71}$$

where a_D, b_D, and c_D are as in (3.31)–(3.33) with Ω_+ replaced by $\Omega_+(B_D) = B_D$. The dashed curve Da_D in Fig. 3.22 represents $SINR_1(E, B_D)$ for various values of energy E. Each point on curve Da_D generates a new transmit waveform as given by (3.44). The dashed curve Da_D meets the horizontal line AC at F that corresponds to energy E_F. Hence from (3.71)

$$SINR_1(E_F, B_D) = SINR_o. \tag{3.72}$$

In other words, the optimum design $|F_2(\omega)|^2$ in (3.44) with energy-bandwidth pair (E_F, B_D) also maintains the same output performance as $SINR_o$. In this case, the useable bandwidth has further decreased from B_C to B_D, at the expense of an increase in transmit energy from E_C to E_F.

Proceeding in this manner, we observe that the dashed curve Ga_G in Fig. 3.22 that passes through the initial design point A uses minimum bandwidth B_G. Thus the operating point G on the optimum $SINR_1$ curve uses bandwidth B_G and energy E_G, and gives performance $SINR_1(B_G)$. By increasing energy from E_G to E_o, while bandwidth is held at B_G, the performance $SINR_1(E, B_G)$ defined as in (3.71) with B_D replaced by B_G, can be increased to $SINR_o$, and proceeding as in (3.72) by equating $SINR_1(E, B_G)$ to $SINR_o$ we obtain

$$SINR_1(E_o, B_G) = SINR_o, \tag{3.73}$$

that is, the energy required for the optimum design at point A on curve Ga_G is the same as the original transmit signal energy E_o. Thus B_G represents the minimum bandwidth that can be used without the transmit signal energy exceeding E_o for the optimum design, and $B_o - B_G$ represents the corresponding maximum savings in bandwidth. Notice that the design point A on the dashed curve generates a new transmit waveform $|F_3(\omega)|^2$ according to (3.44) with B_o replaced by B_G there.

If we are prepared to use additional transmit energy beyond E_o, then by starting with operating points that are below G on the optimum curve $SINR_1(B_i)$ and using additional transmit energy $E > E_o$, performance level equal to $SINR_o$ can be realized. Since operating points below G have lower bandwidth compared to B_G, these designs use even lower bandwidths at the expense of additional transmit energy.

In summary, optimum designs that start from operating points such as D located in between G and C result in simultaneous savings in energy and bandwidth for the same $SINR$ performance compared to the given transmit waveform such as at point A.

To validate the simultaneous savings in bandwidth and energy using waveform diversity, Figs. 3.23(a) and (b) show the actual $SINR$-energy plots employing optimum waveform design as well as a given chirp transmit signal. Both methods use a target transform $Q(\omega)$ as in Fig. 3.24(a), and noise and clutter spectra $G_n(\omega)$ and $G_c(\omega)$ as in Fig. 3.24(b) and (c), respectively. As described earlier, the target and clutter have been modified using the transmitter output filter $P_1(\omega)$ in Fig. 3.25 so that at every stage the operating band $\Omega_+(B_i)$ coincides with the desired bandwidth B_i. This is illustrated in Fig. 3.24(d) for a typical bandwidth B_i. Figure 3.23(b) shows an enlarged version for Fig. 3.23(a) with the features around the points of interest A, B, C, D, F, and G magnified.

The given chirp waveform $F_o(\omega)$ operating at point A in Fig. 3.23 uses the energy-bandwidth pair $(E_o, B_o) = (8.88\,dB, 0.15)$ with output performance given by $SINR_o = 6.15\,dB$. All bandwidths here are expressed as a fraction of the maximum available bandwidth dictated by the sampling period. The operating point C using waveform diversity on the optimum $SINR_1$ curve with performance equal to $SINR_o$, uses the energy-bandwidth pair $(E_C, B_C) = (5.875\,dB, 0.1193)$, resulting in simultaneous savings in both transmit energy and bandwidth compared to the chirp transmit waveform. Notice that compared to

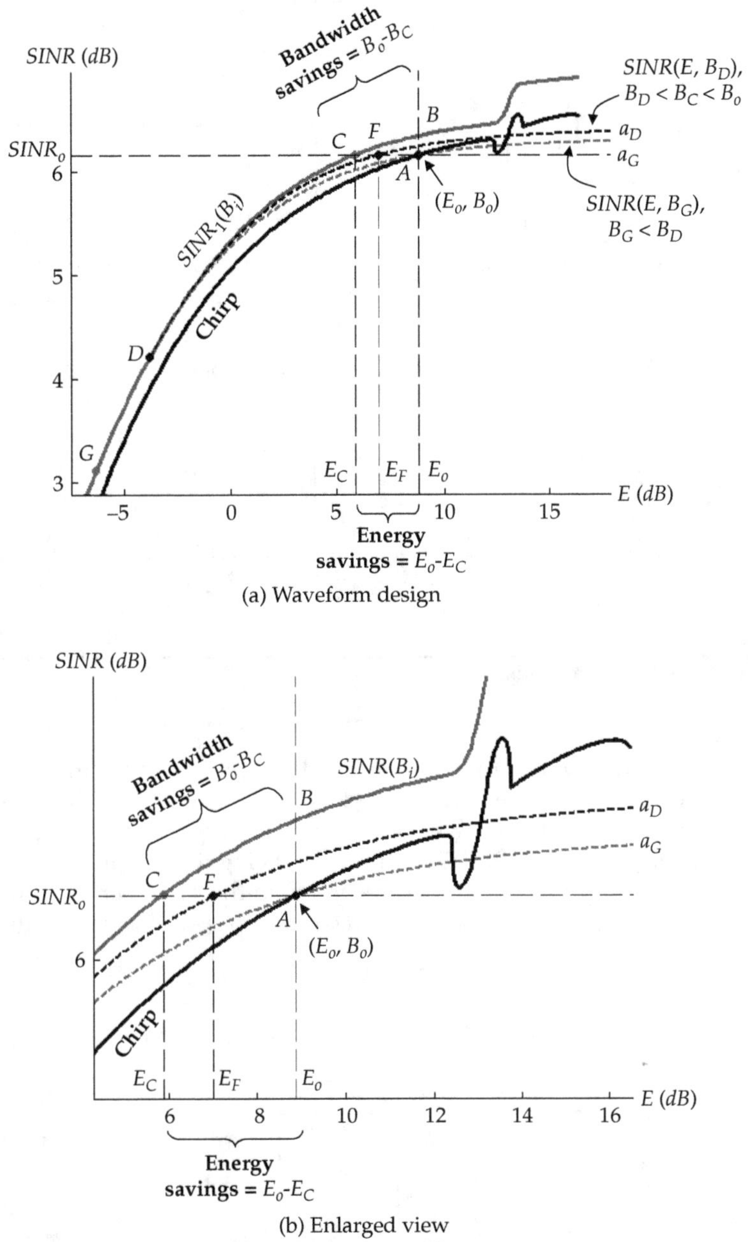

(a) Waveform design

(b) Enlarged view

FIGURE 3.23 Simultaneous savings in bandwidth and energy for the target interference scene shown in Fig. 3.24.

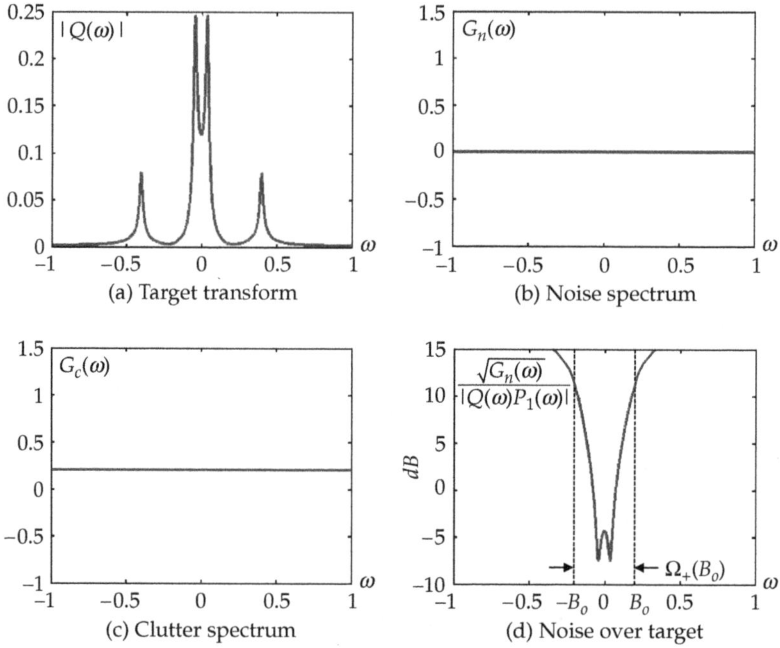

FIGURE 3.24 Target, noise, and clutter scene.

the given chirp waveform, the bandwidth requirement for the optimum waveform $|F_1(\omega)|^2$ given by (3.38) with $B_o = B_C$ has been reduced from 0.15 to 0.1193, whereas the energy requirement has been reduced from 8.88 to 5.875 *dB* without sacrificing the performance.

The operating point F also corresponds to the same performance level as $SINR_o$, and it lies on curve Da_D that passes through F and

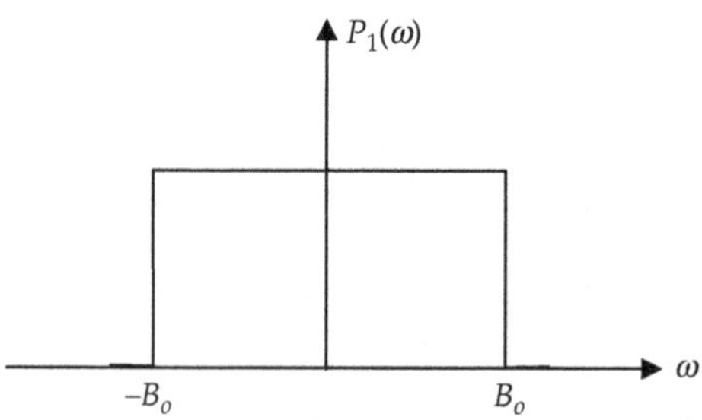

FIGURE 3.25 Transmitter output filter.

corresponds to $SINR_1(E, B_D)$ with $B_D = 0.065$. The performance level $SINR_o$ is met here with energy $E_F = 7.01\,dB$. In other words, the optimum waveform $|F_2(\omega)|^2$ given by (3.44) with $E = E_F$ and $B_o = B_D$ gives the same performance $SINR_o$ as the given chirp transmit waveform, but uses less bandwidth and less energy. In this case, the bandwidth requirement is reduced from 0.15 to 0.065, and the energy requirement is reduced from 8.88 to 7.01 dB. Notice that compared to the optimum waveform $|F_1(\omega)|^2$ at C, the waveform $|F_2(\omega)|^2$ at point F uses less bandwidth, but more energy. These waveform transform magnitude functions are plotted in Fig. 3.26(b) and (c) respectively along with the given transmit waveform transform magnitude $|F_o(\omega)|^2$ plotted in Fig. 3.26(a).

Finally, the curve Ga_G that starts at G passes through A; and the optimum waveform $|F_3(\omega)|^2$, given by (3.44) with $E = E_o$ and $B_o = B_G = 0.0589$, also maintains the same performance $SINR_o$. Notice that $|F_4(\omega)|^2$ design plotted in Fig. 3.26(d) uses the least amount of bandwidth (about 1/3rd bandwidth compared to the given chirp waveform) without the transmit energy exceeding the initial value E_o.

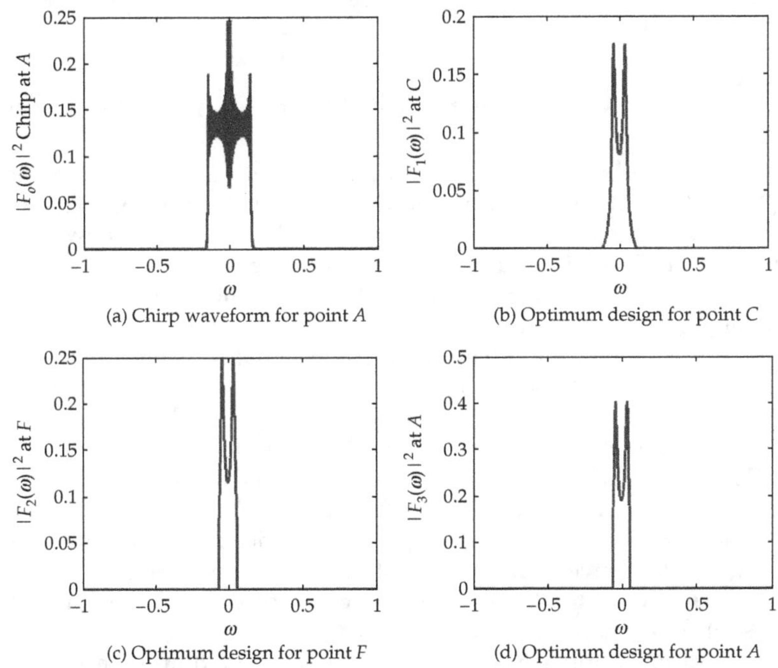

(a) Chirp waveform for point A

(b) Optimum design for point C

(c) Optimum design for point F

(d) Optimum design for point A

FIGURE 3.26 Optimum transmit signal transforms for various energy-bandwidth configurations.

Thus, the design procedure illustrated in Fig. 3.22 and the results of actual computation in Figs. 3.23 and 3.26 using data from Figs. 3.24 and 3.25 show that simultaneous savings in bandwidth and energy are indeed possible using optimum waveform design compared to a given transmit waveform.

Finally to convert the optimum Fourier transform magnitude information in Fig. 3.26(a)–(d) into a practically useful time domain signal, a suitable phase function $\psi_i(\omega)$ needs to be determined in each case. Let $A(\omega)$ represent the given Fourier transform magnitude $|F_i(\omega)|$ in Fig. 3.26(b)–(d), and together with the standard inverse Fourier transform operation,

$$f_i(t) \leftrightarrow |F_i(\omega)|e^{-j\psi_i(\omega)} \tag{3.74}$$

gives the desired time domain signal $f_i(t)$, $i = 1, 2, 3, 4$. The freedom present in the selection of the phase functions $\psi_i(\omega)$ in (3.74) can be used to "shape" the final time-domain signal $f_i(t)$ in (3.74) to be practically useful [16, 17].

One feature for the transmit waveform can be minimum phase. For minimum phase signals, the phase function is given by the Hilbert transmit of $\ln |F(\omega)|$

$$\psi(\omega) = \mathcal{H}\{\ln |F(\omega)|\} \tag{3.75}$$

where $\mathcal{H}$ represents the Hilbert transform. With the phase given in (3.75), the minimum phase time domain signals corresponding to the waveforms in Fig. 3.26 can be obtained. Figure 3.27 shows the minimum phase time domain signals corresponding to the waveforms in Fig. 3.26. Notice that the optimum time domain signals in Fig. 3.27 have all the transmit energy concentrated at the beginning. This undesired feature puts additional requirement on the transmitter amplifiers and may even lead to unrealizable transmit signals [22, 23].

In this context, one useful feature for time domain transmit signals is the constant envelope property that maximizes the transmit power efficiency. In that case, the final time domain waveform $f_i(t)$ will ideally have constant envelope or in practice possess an approximate constant envelope, thereby distributing the peak transmit power more evenly. Methods for obtaining constant envelope signals are described in Chap. 4.

The design approach described in this section requires the knowledge of the target characteristics (in addition to clutter and noise spectra), but often the target features may be unknown [24, 25]. In that case, the following approach is promising.

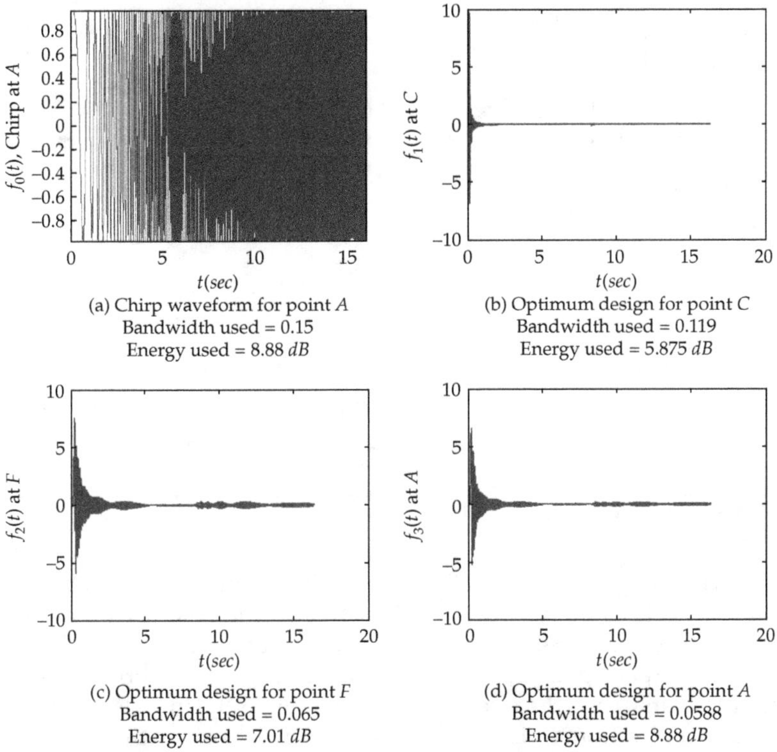

(a) Chirp waveform for point A
Bandwidth used = 0.15
Energy used = 8.88 dB

(b) Optimum design for point C
Bandwidth used = 0.119
Energy used = 5.875 dB

(c) Optimum design for point F
Bandwidth used = 0.065
Energy used = 7.01 dB

(d) Optimum design for point A
Bandwidth used = 0.0588
Energy used = 8.88 dB

FIGURE 3.27 Minimum phase time domain transmit waveforms corresponding to (a) chirp waveform for point A, (b) optimum design for point C, (c) optimum design for point F, (d) optimum design for point A.

3.4 Method II: Whitening Approach [11]

The total interference plus noise spectrum is given by $G_I(\omega)$ in (3.4). Suppose $G_I(\omega)$ can be made equal to a constant σ_o^2, then the problem reduces to the classic situation of a signal in white noise, in which case an ordinary matched filter receiver is optimum. To analyze whether it is indeed possible to realize this goal, let

$$G_c(\omega)|F(\omega)|^2 + G_n(\omega) = \sigma_o^2 \tag{3.76}$$

so that

$$|F(\omega)|^2 = \frac{\sigma_o^2 - G_n(\omega)}{G_c(\omega)}. \tag{3.77}$$

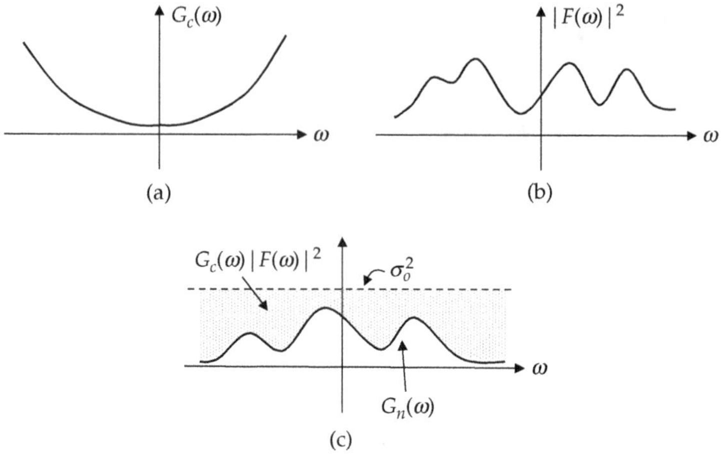

FIGURE 3.28 "Water filling effect."

The whitening procedure in (3.76) and (3.77) is equivalent to a "water filling effect" as shown in Fig. 3.28. In other words, the role of $|F(\omega)|^2$ is mainly to compress and reshape $G_c(\omega)$ so that, together with $G_n(\omega)$, the interference plus noise exhibits flat spectral characteristics as in (3.76).[6] In that case, the receiver observes a signal $s(t)$ in "white noise."

As a result, from (3.12) the optimum receiver is given by

$$H_{opt}(\omega) = \frac{S^*(\omega)e^{-j\omega t_o}}{\sigma_o^2} \tag{3.78}$$

or

$$h_{opt}(t) = \frac{1}{\sigma_o^2} s(t_o - t), \tag{3.79}$$

that is, the optimum receiver is simply matched to the incoming signal, which is clearly the case in white noise. Interestingly, $|F(\omega)|^2$ given by (3.77) ignores the target. Notice that the receiver design *does* involve target features since $S(\omega) = Q(\omega)F(\omega)$. Alternatively, one may use the suboptimum receiver (classical matched filter)

$$h_{sub}(t) = f(t_o - t); \tag{3.80}$$

[6]If $G_c(\omega)$ is zero at $\omega = \omega_o$, then maximum of $G_n(\omega)$ must occur at $\omega = \omega_o$. In that case, $\sigma_o^2 = G_n(\omega_o)$ in (3.77) so that common zeros cancel out in (3.77) and $|F(\omega)|^2$is well defined. It follows that if $G_n(\omega)$ and $G_c(\omega)$ have a common zero at $\omega = \omega_o$, then it will not be possible to realize the above whitening effect. However in that case, the transmitter should concentrate all its energy at that frequency for maximum *SINR*.

However, the corresponding *SINR* loss must be studied on a case by case basis.

The concept of "desired band" B_1 and operating band $\Omega_+(B_1)$ is useful here also (see Sec. 3.3). Clearly from (3.77), we must have

$$\sigma_o^2 \geq \max_{\omega \in B_1} G_n(\omega) \tag{3.81}$$

and the energy constraint in (3.26) gives

$$E = \frac{\sigma_o^2}{2\pi} \int_{\Omega_+(B_1)} \frac{1}{G_c(\omega)} d\omega - \frac{1}{2\pi} \int_{\Omega_+(B_1)} \frac{G_n(\omega)}{G_c(\omega)} d\omega, \tag{3.82}$$

where $\Omega_+(B_1)$ in (3.82) represents the operating frequency band. Given the desired band B_1, the operating band $\Omega_+(B_1)$ is the frequency region over which

$$G_n(\omega) \leq \max_{\omega \in B_1} G_n(\omega). \tag{3.83}$$

This gives rise to two situations as shown in Fig. 3.29. In Fig. 3.29(a), the operating band coincides with the desired band, whereas in

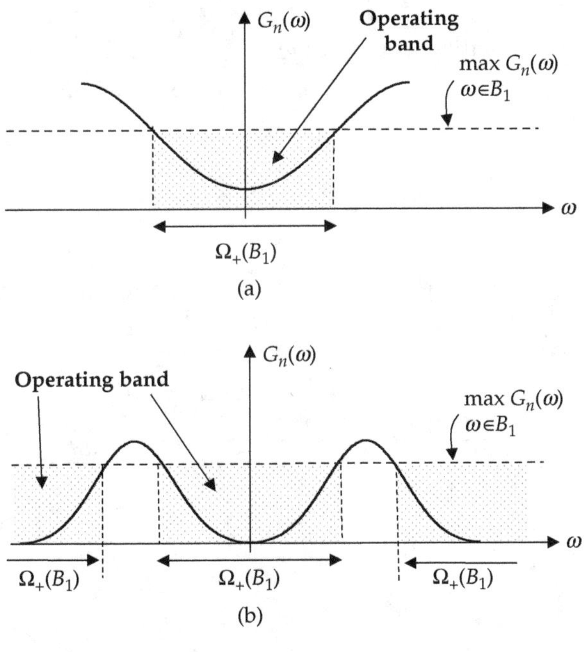

FIGURE 3.29 Partitioning of the frequency axis for a given desired band.

Fig. 3.29(b), the operating band $\Omega_+(B_1)$ is larger than the desired band B_1.

From (3.82) we get

$$\sigma_o^2(E) = \frac{E + \frac{1}{2\pi}\int_{\Omega_+(B_1)} \frac{G_n(\omega)}{G_c(\omega)}d\omega}{\frac{1}{2\pi}\int_{\Omega_+(B_1)} \frac{1}{G_c(\omega)}d\omega} \geq \max_{\omega\in\Omega_+(B_1)} G_n(\omega), \tag{3.84}$$

where the last inequality follows from (3.77). If this condition is not satisfied, then the energy E must be increased so that the inequality in (3.84) is satisfied.[7] Once again, this leads to a threshold energy condition for the transmitter energy.

Threshold Energy

Using (3.84) in (3.82), we obtain

$$E \geq \max_{\omega\in\Omega_+} G_n(\omega)\frac{1}{2\pi}\int_{\Omega_+(B_1)} \frac{1}{G_c(\omega)}d\omega - \frac{1}{2\pi}\int_{\Omega_+(B_1)} \frac{G_n(\omega)}{G_c(\omega)}d\omega \triangleq E_2. \tag{3.85}$$

Here E_2 represents the transmitter threshold energy. Using (3.84), the corresponding transmit waveform is given by

$$|F(\omega)|^2 = \begin{cases} \dfrac{\sigma_o^2(E) - G_n(\omega)}{G_c(\omega)}, & \omega \in \Omega_+(B_1) \\ 0, & \text{otherwise.} \end{cases} \tag{3.86}$$

Finally,

$$\begin{aligned} SINR_2 &= \frac{1}{2\pi}\int_{\Omega_+(B_1)} \frac{|Q(\omega)|^2|F(\omega)|^2}{\sigma_o^2(E)}d\omega \\ &= \frac{1}{2\pi}\int_{\Omega_+(B_1)} \frac{|Q(\omega)|^2}{G_c(\omega)}d\omega - \frac{1}{2\pi\sigma_o^2}\int_{\Omega_+(B_1)} \frac{|Q(\omega)|^2 G_n(\omega)}{G_c(\omega)}d\omega \\ &= \frac{1}{2\pi}\int_{\Omega_+(B_1)} \frac{|Q(\omega)|^2}{G_c(\omega)}d\omega - \frac{\frac{1}{2\pi}\int_{\Omega_+(B_1)} \frac{|Q(\omega)|^2 G_n(\omega)}{G_c(\omega)}d\omega \frac{1}{2\pi}\int_{\Omega_+(B_1)} \frac{1}{G_c(\omega)}d\omega}{E + \frac{1}{2\pi}\int_{\Omega_+(B_1)} \frac{G_n(\omega)}{G_c(\omega)}d\omega} \\ &= a_1 - \frac{d_1}{E + b_1} \leq SINR_1. \end{aligned} \tag{3.87}$$

[7]In the white noise case, $G_n(\omega) = \sigma_n^2$, and the inequality in (3.84) is always satisfied for any energy E.

Here

$$d_1 = \frac{1}{2\pi}\int_{\Omega_+(B_1)} \frac{|Q(\omega)|^2 G_n(\omega)}{G_c(\omega)} d\omega \cdot \frac{1}{2\pi}\int_{\Omega_+(B_1)} \frac{1}{G_c(\omega)} d\omega \tag{3.88}$$

and a_1, b_1 are as in (3.31) and (3.32) with Ω_+ replaced by $\Omega_+(B_1)$.

From (3.87) and (3.88), with bandwidth B_i and energy $E \geq E_2$ in (3.85), we get

$$SINR(E,\ B_i) = SINR_{\max} = a_i - \frac{d_i}{E + b_i} \tag{3.89}$$

which represents the maximum realizable *SINR* using this approach for a given bandwidth B_i, and transmitter energy E that exceeds $E_{\min} = E_2$ in (3.85). In particular, the maximum *SINR* realizable, using minimum energy is given by (3.89) with E replaced by $E_{\min} = E_2$ is given in (3.85). This gives

$$SINR(B_i) = a_i - \frac{d_i}{E_i + b_i}. \tag{3.90}$$

Equation (3.90) represents the performance level for bandwidth B_i using this minimum threshold energy E_i, and (3.89) represents the performance level for the same bandwidth using energy E that exceeds the minimum threshold energy. From (3.86), we also obtain the transmit signal transform corresponding to energy E_i to be

$$|F(\omega)|^2 = \begin{cases} \dfrac{\left(\max\limits_{\omega\in B_i} G_n(\omega)\right) - G_n(\omega)}{G_c(\omega)}, & \omega \in \Omega_+(B_i)\,, \\ 0, & \omega \in \Omega_o \end{cases} \tag{3.91}$$

where Ω_o represents the complement of the frequency region $\Omega_+(B_i)$.

To summarize, to maintain a given desired band B_i, there exists an operating band $\Omega_+(B_i) \geq B_i$ over which the transmit waveform transform $|F(\omega)|^2 > 0$ and to guarantee this, the transmit energy must be at least equal to a minimum threshold value E_2 given by (3.85) with B_1 replaced by B_i there.

On comparing (3.30) with (3.87), under identical values of Ω_+ and energy E, we have $SINR_2 < SINR_1$, indicating that the whitening approach is somewhat inferior to the first method. This is not surprising since the transmitter design in (3.77) ignores the target shape and the difference $SINR_1 - SINR_2$ may be taken as a penalty measure for ignoring the target. In the present case, the transmit spectrum can be determined *without* the knowledge of the target spectrum which is

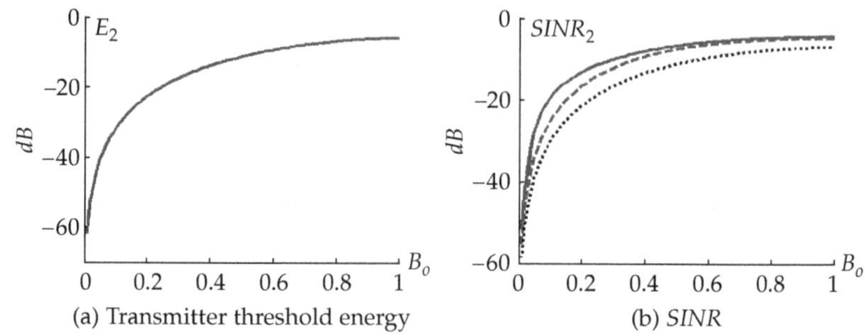

FIGURE 3.30 E_2 and $SINR_2$ as function of desired band corresponding to Fig. 3.13.

obviously an advantage for design purposes when the target characteristics are unknown.

Figure 3.30 shows the threshold energy E_2 in (3.85) and the corresponding $SINR_2$ in (3.87) as a function of the desired bandwidth B_o for $Q(\omega)$, $G_n(\omega)$, and $G_c(\omega)$ corresponding to Fig. 3.13(a)–(c), respectively. Similar diagrams corresponding to Figs. 3.14 and 3.15 are shown in Figs. 3.31 and 3.32, respectively.

To compare the two transmitter design methods discussed above, the threshold energy $E_{B_o} = E_1$ and E_2 as well as the two *SINR*s in (3.30) and (3.87) as a function of the desired bandwidth B_o are plotted in Figs. 3.33–3.36. As remarked earlier, under identical operating conditions, $SINR_1 - SINR_2$ may be taken as a penalty measure for ignoring the target shape in transmitter design. These figures [see (e) and (f) below] also exhibit the optimum transmitter designs given by (3.38) and (3.86) corresponding to the maximum available bandwidth

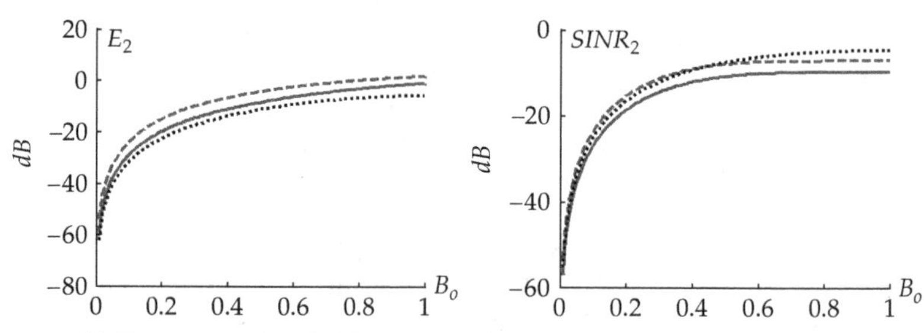

FIGURE 3.31 E_2 and $SINR_2$ as function of desired band corresponding to Fig. 3.14.

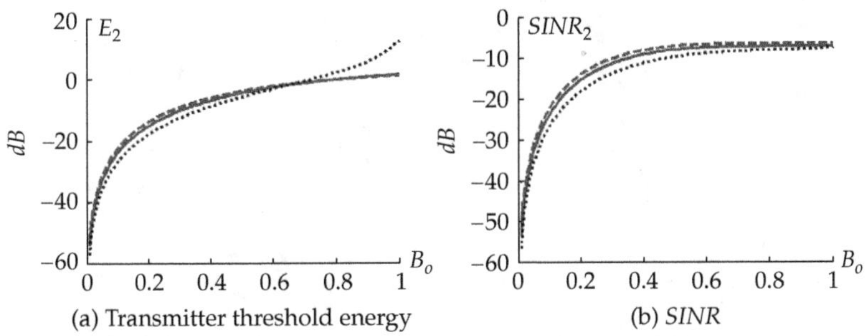

FIGURE 3.32 E_2 and $SINR_2$ as function of desired band corresponding to Fig. 3.15.

(with $B_o \equiv 1$). Interestingly, the two designs can give rise to entirely different transmit signals! This is not surprising since the first method makes use of the target features, whereas the second method ignores the target details. Thus, in Fig. 3.33(e), the transmitter $|F_1(\omega)|^2$ latches onto the dominant target mode in $|Q(\omega)|$ whereas in Fig. 3.33(f), the transmitter $|F_2(\omega)|^2$ is used to compress the clutter so that the total interference turns out to be flat as dictated by the design in (3.76). This leads to distinct transmitter features. The residual difference between $SINR_1$ and $SINR_2$ in this case represents the advantage in knowing the resonant features of the target and operating accordingly. From Fig. 3.33(h), under all available bandwidth, this corresponds to about 10 dB gain.

It is interesting also to note that entirely different transmitter shapes can lead to *similar* performance [26–28]. This is illustrated in Fig. 3.34(e) and (f), where the two designs lead to two distinct signals—a "low pass" and a "high pass" transmitter signal. However, the *SINRs* have comparable performance when all the available bandwidth is utilized. Similar conclusions can be seen in Fig. 3.35(e)–(h) as well. From Figs. 3.34 and 3.35(e) and (f), the transmitter shapes corresponding to both methods can be entirely different. Nevertheless, each method when operated optimally (using all available bandwidth and required energy) leads to comparable *SINR* performance levels. Hence dissimilar waveforms can achieve the same goal in terms of *SINR* realization [29].

Finally, Fig. 3.36 shows that the two different methods can lead to similar transmit waveforms as well.

To maintain a certain desired band B_o, the transmitter energy must exceed a certain threshold value given by (3.36) or (3.85), and the maximum achievable *SINR* in that case is given by either (3.30) or (3.87). Clearly, we must have $E \geq E_{B_o}$, and Fig. 3.37 shows the *SINR* as

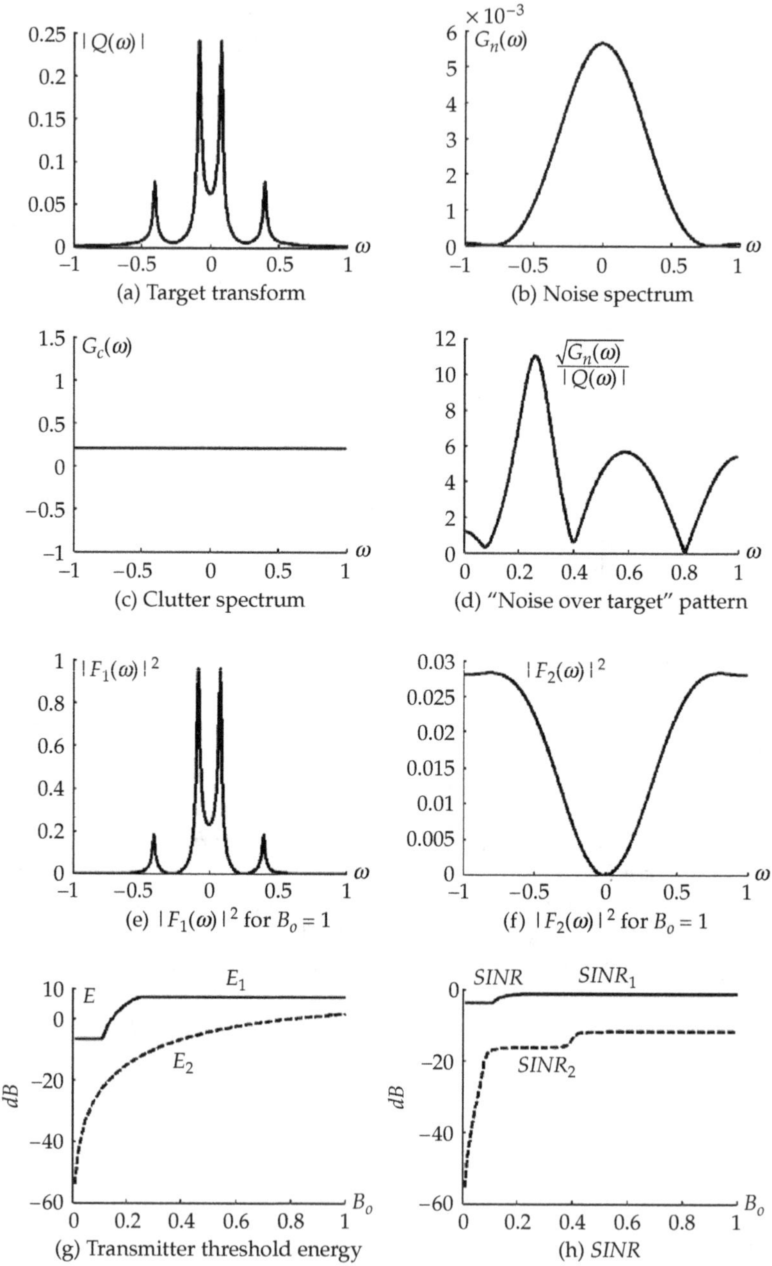

FIGURE 3.33 Transmitter design using Methods I and II. For the target, noise, and clutter spectra shown here, threshold energy E_1, E_2 and $SINR_1$, $SINR_2$ are plotted as functions of the operating band.

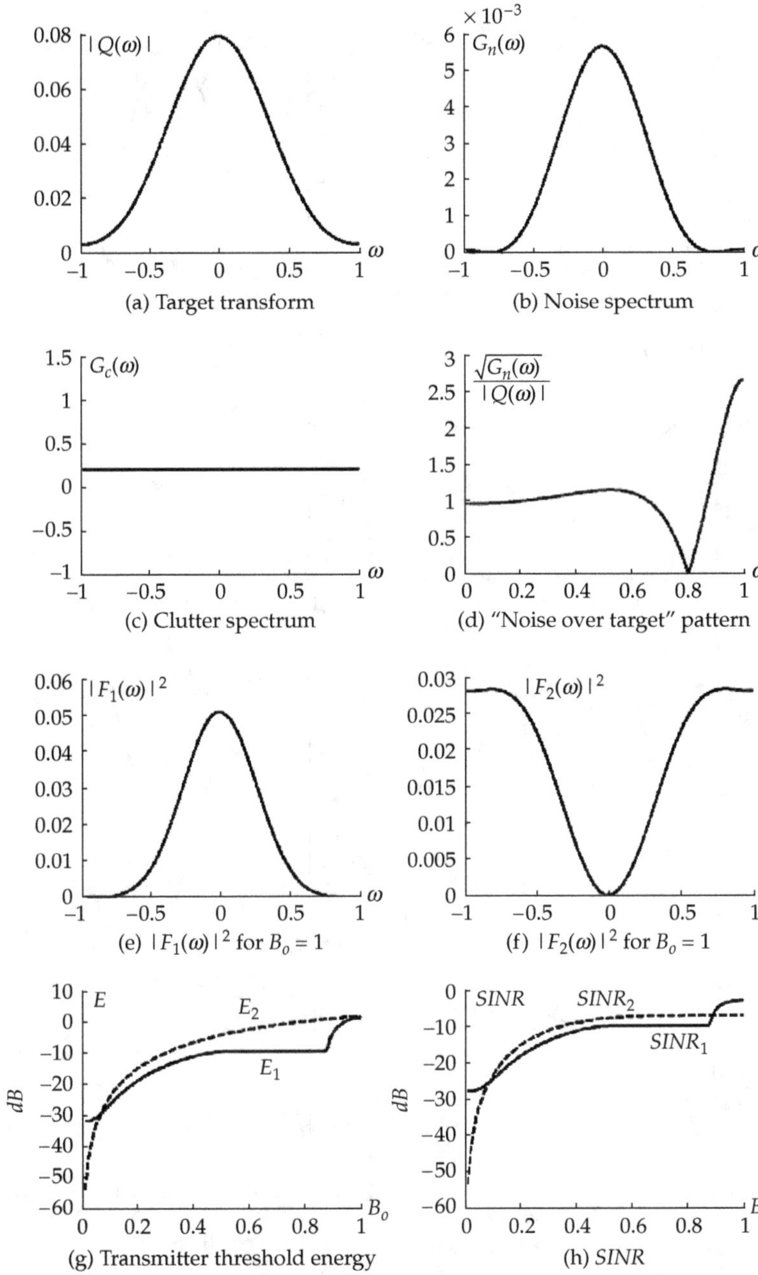

FIGURE 3.34 Transmitter design using Methods I and II. For the target, noise, and clutter spectra shown here, threshold energy E_1, E_2 and $SINR_1$, $SINR_2$ are plotted as functions of the operating band.

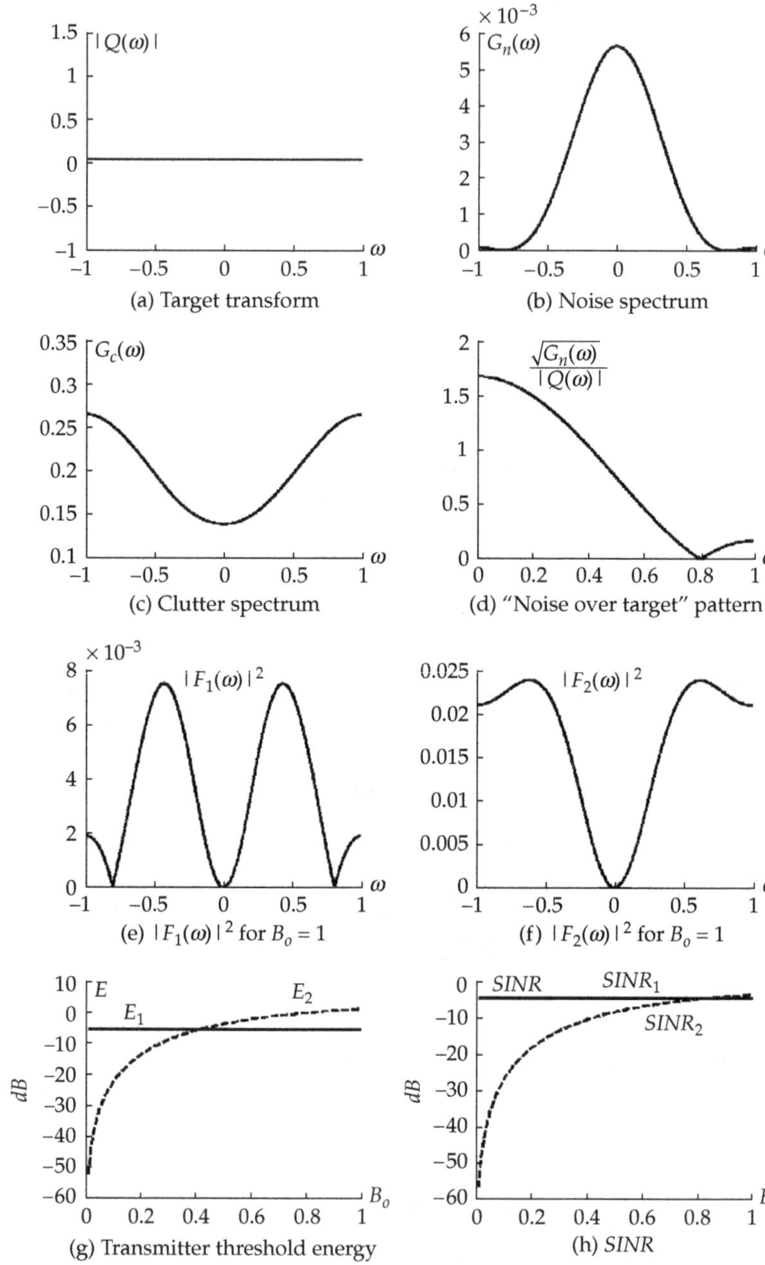

FIGURE 3.35 Transmitter design using Methods I and II. For the target, noise, and clutter spectra shown here, threshold energy E_1, E_2 and $SINR_1$, $SINR_2$ are plotted as functions of the operating band.

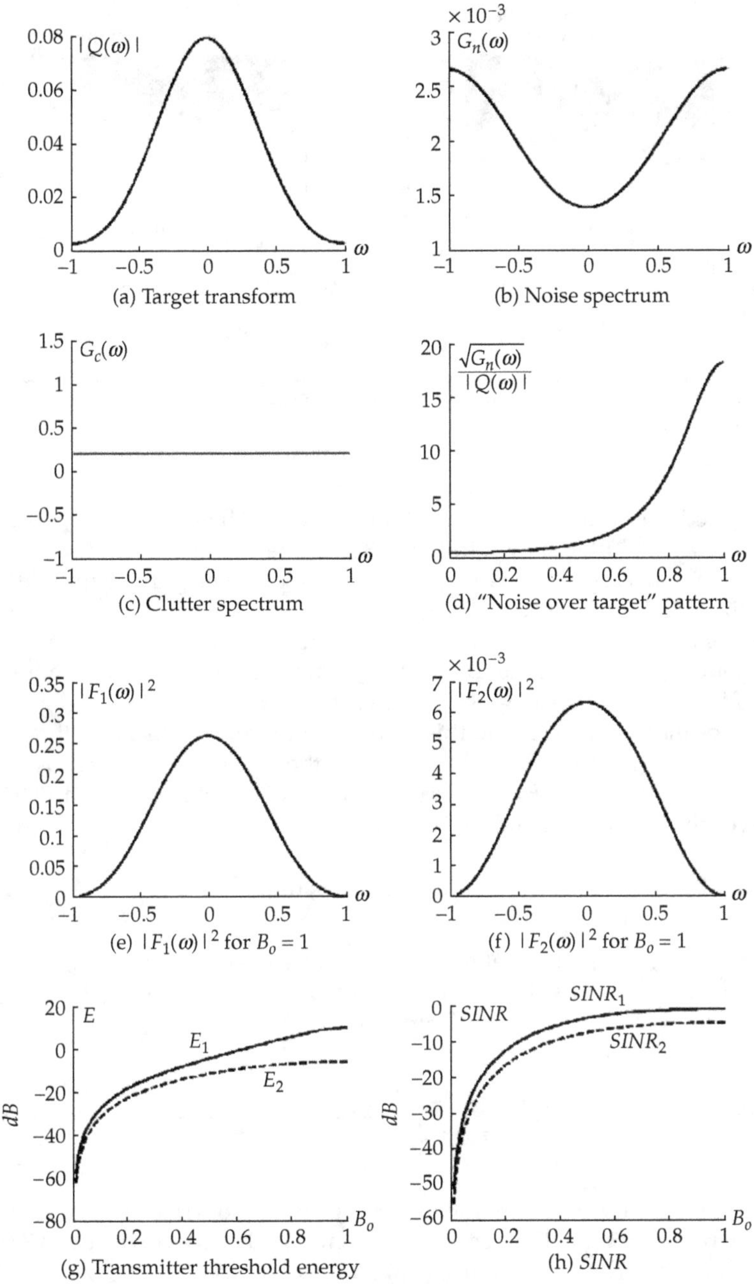

FIGURE 3.36 Transmitter design using Methods I and II. For the target, noise, and clutter spectra shown here, threshold energy E_1, E_2 and $SINR_1$, $SINR_2$ are plotted as functions of the operating band.

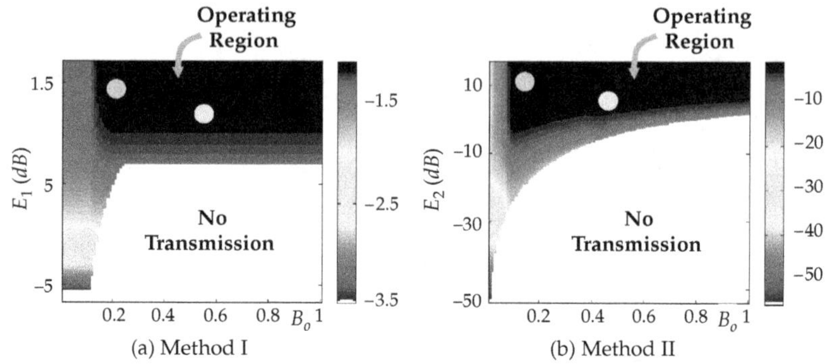

FIGURE 3.37 *SINRs* as functions of transmit energy E and the desired band B_o.

a function of transmit energy E and the desired band B_o for the target, noise, and clutter shown in Fig. 3.33 for the two methods discussed above. Once again, it is possible to tradeoff energy for bandwidth using Method-II also.

As Fig. 3.29(b) shows, $\Omega_+(B_1)$ can consist of multiple disjoint frequency bands whose complement Ω_o represents the "no show" region. As before, by increasing the bandwidth B_1, these "no show" regions can be made narrower and this defines a minimum transmitter threshold energy E_∞ that allows $\Omega_+(B_1)$ to be the entire available frequency axis. To determine E_∞, let $\sigma_o^2(\infty)$ represent the maximum in (3.81) over the entire frequency axis. Thus

$$\sigma_o^2(\infty) = \max_{|\omega|<\infty} G_n(\omega), \tag{3.92}$$

and let a_∞, b_∞, and d_∞ refer to the constants a, b, and d in (3.31), (3.32), and (3.88) respectively, calculated with $\Omega_+(B_1)$ representing the entire frequency axis. Then from (3.85)

$$E_\infty = \left(\max_{|\omega|<\infty} G_n(\omega)\right) \frac{1}{2\pi} \int_{-\infty}^{+\infty} \frac{1}{G_c(\omega)} d\omega - \frac{1}{2\pi} \int_{-\infty}^{+\infty} \frac{G_n(\omega)}{G_c(\omega)} d\omega > 0 \tag{3.93}$$

represents the minimum transmit energy (threshold) required to avoid any partitioning in the frequency domain. With E_∞ as given by (3.93), we obtain $SINR_{\max}$ to be [use (3.87)]

$$SINR_2(\infty) = a_\infty - \frac{c_\infty}{E_\infty + b_\infty} > 0 \tag{3.94}$$

and from (3.86) with $\sigma_o^2(E) = \sigma_o^2(\infty)$, we get

$$|F(\omega)|^2 = \frac{\left(\max\limits_{|\omega|<\infty} G_n(\omega)\right) - G_n(\omega)}{G_c(\omega)}, \qquad |\omega| < \infty \tag{3.95}$$

which represents the corresponding transmit waveform. Clearly by further increasing the transmit energy in (3.94) beyond that in (3.93), we obtain

$$SINR_2 \to a_\infty = \frac{1}{2\pi} \int_{-\infty}^{-\infty} \frac{|Q(\omega)|^2}{G_c(\omega)} d\omega. \tag{3.96}$$

It follows that to avoid any restrictions in the frequency domain for the transmit signal, the transmitter energy E must be at least equal to a minimum threshold value E_∞ given by (3.93) and (3.94) that represents the maximum realizable *SINR* in that case. By increasing E beyond E_∞, the performance can be improved up to that in (3.96).

In general from (3.85) for a given desired bandwidth B_i, the transmit energy E must exceed its threshold value $E_i = E_2(B_i)$. With $E > E_i$ and $\sigma_o^2(E)$ as in (3.84), the corresponding transmit signal transform is given by (3.86) and clearly this signal is different from the minimum threshold energy one given by (3.91). From (3.87), the performance level $SINR(E, B_1)$ corresponding to (3.86) with transmit energy exceeding the threshold energy is given by (3.89) (with $B_i = B_1$), and the performance level that corresponds to (3.91) with transmit energy equal to the threshold energy is given by (3.90) (with $B_i = B_1$). Thus with $E > E_1$, we have

$$SINR(E, B_1) = a_1 - \frac{d_1}{E + b_1} > SINR(B_1). \tag{3.97}$$

From (3.97), for a given bandwidth B_1, performance can be increased beyond that in (3.90) by increasing the transmit energy and it is upper bounded by a_1. Thus

$$SINR(\infty, B_1) = a_1. \tag{3.98}$$

Hence it follows that $SINR(B_1)$ represents the minimum performance level for bandwidth B_1 that is obtained by using its minimum threshold energy. It is quite possible that this improved performance $SINR(E, B_1)$ for bandwidth B_1 can be equal to the minimum performance level corresponding to a higher bandwidth $B_2 > B_1$. This gives rise to the concept of energy-bandwidth tradeoff at a certain performance level [6,30].

3.4.1 Energy-Bandwidth Tradeoff

Figure 3.38 (d) and (e) exhibits the transmit threshold energy in (3.85) and the corresponding output $SINR(B_1)$ in (3.90) as a function of the desired bandwidth B_1. Combining these figure using (3.85) and (3.90), an *SINR* versus transmit threshold energy plot can be generated for each of the target situations shown in Fig. 3.38 (a).

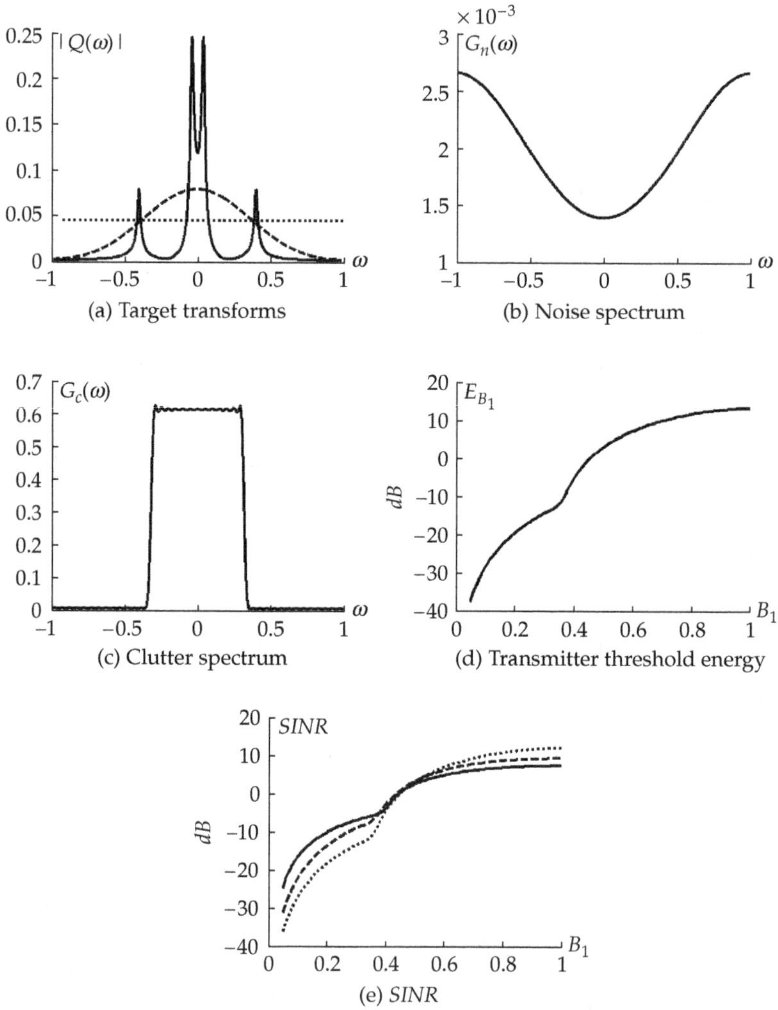

FIGURE 3.38 Transmitter threshold energy and *SINR* as function of bandwidth using whitening approach.

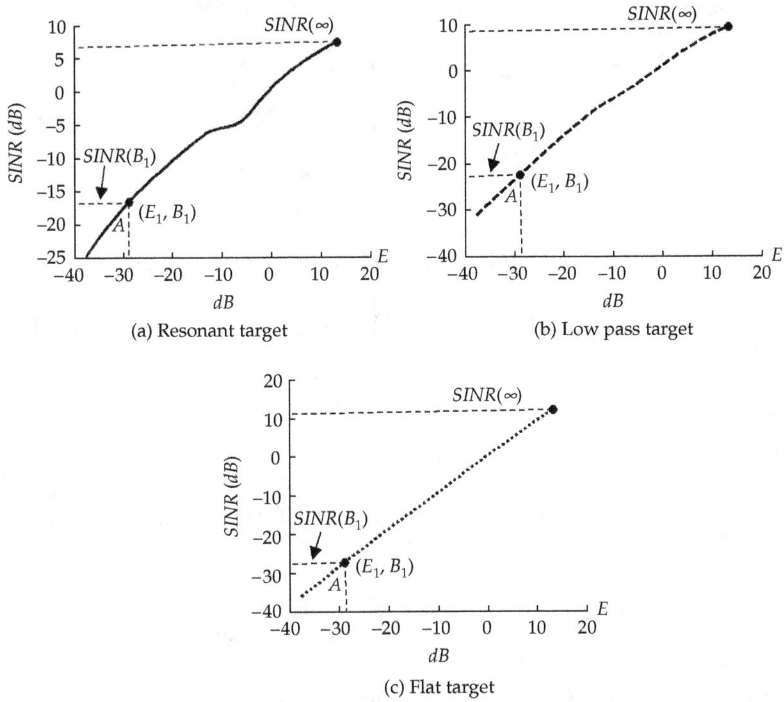

FIGURE 3.39 *SINR* versus threshold energy for three different targets shown in Fig. 3.38.

For example, Fig. 3.39 (a)–(c) corresponds to the three different target situations considered in Fig. 3.38 with clutter and noise spectra as shows there. Notice that each point on the *SINR*-energy threshold curve for each target is associated with a specific desired bandwidth. Thus for bandwidth B_1, the minimum threshold energy required is E_1 in (3.85), and corresponding *SINR* equals $SINR(B_1)$ in (3.90). Let *A* represent the associated operating point in Fig. 3.39. Notice that the operating point *A* corresponding to a bandwidth B_1 has different threshold energies and different performance levels for different targets. From (3.95), each operating point generates a distinct transmit waveform. As the bandwidth increases, from (3.94), $SINR \to SINR(\infty)$.

Monotonic Property of *SINR*

The threshold energy and *SINR* associated with a higher bandwidth can be shown to be higher. To prove this, consider two desired bandwidths B_1 and B_2 with $B_2 > B_1$. Then from (3.84), we have

$$\sigma_o^2(E_2) = \max_{\omega \in B_2} G_n(\omega) \geq \sigma_o^2(E_1) = \max_{\omega \in B_1} G_n(\omega), \qquad (3.99)$$

and from Fig. 3.29, the corresponding operating bandwidths $\Omega_+(B_1)$ and $\Omega_+(B_2)$ satisfy

$$\Omega_+(B_2) \geq \Omega_+(B_1). \qquad (3.100)$$

From (3.85), the minimum threshold energies are given by

$$E_i = \frac{1}{2\pi} \int_{\Omega_+(B_i)} \left(\sigma_o^2(E_i) - G_n(\omega)\right) \frac{1}{G_c(\omega)} d\omega, \quad i = 1, 2 \qquad (3.101)$$

and substituting (3.99) and (3.100) into (3.101), we get

$$E_2 \geq E_1. \qquad (3.102)$$

Also from (3.87), the performance levels at threshold energy $SINR(B_i)$ equal

$$SINR(B_i) = \frac{1}{2\pi} \int_{\Omega_+(B_i)} \left(1 - \frac{G_n(\omega)}{\sigma_o^2(E_i)}\right) \frac{|Q(\omega)|^2}{G_c(\omega)} d\omega \qquad (3.103)$$

and an argument similar to (3.99) and (3.100) gives

$$SINR(B_2) \geq SINR(B_1) \qquad (3.104)$$

for $B_2 > B_2$. Thus as Figs. 3.39(a)–(c) show, $SINR(B_i)$ is a monotonically nondecreasing function of both bandwidth and energy. Figure 3.40(a) illustrates this *SINR*-energy relation for the target with flat spectrum (dotted line) shown in Fig. 3.38(a) and Fig. 3.40(b) illustrates this *SINR*-energy relation for the resonant target (solid line with sharp spikes) shown in Fig. 3.38(a). In Fig. 3.40(a) and (b), the solid curve represents $SINR(B_i)$ in (3.90) and the two operating points A and B are associated with bandwidths B_1 and B_2, threshold energies E_1 and E_2, and performance levels $SINR(B_1)$ and $SINR(B_2)$, respectively.

Since

$$B_2 > B_1 \Rightarrow E_2 \geq E_1 \quad \text{and} \quad SINR(B_2) \geq SINR(B_1). \qquad (3.105)$$

The distinct transmit waveforms $|F_1(\omega)|^2$ and $|F_2(\omega)|^2$ associated with these operating points A and B for Fig. 3.40(a) are given by (3.91) with $i = 1, 2$ there, and they are shown in Fig. 3.41(a) and (b), respectively. Notice that these waveforms are the same for all three target cases for each operating point A or B in Fig. 3.40(b), since from (3.91), these waveforms depend only on the noise and clutter spectra which are the same in these two figures.

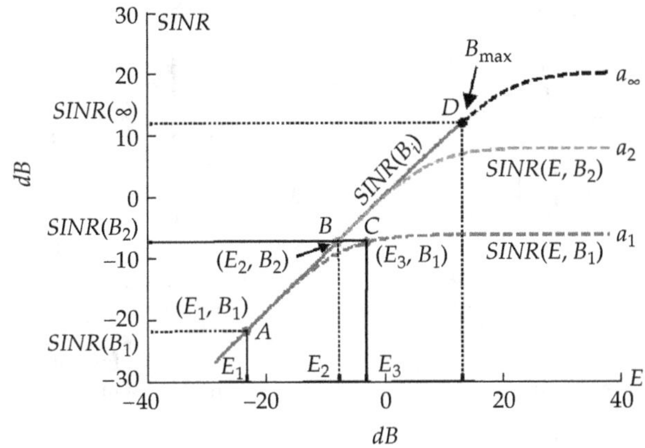

(a) Flat target

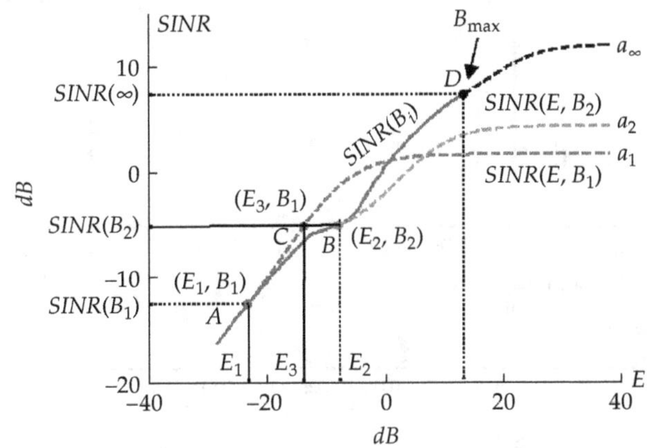

(b) Resonant target

FIGURE 3.40 Energy-bandwidth tradeoff for (a) flat target and (b) resonant target shown in Fig. 3.38.

Consider the operating point A associated with the desired bandwidth B_1. If the transmit energy E is increased beyond the corresponding threshold value E_1 with bandwidth held constant at B_1, the performance $SINR_1(E, B_1)$ increases beyond that at A since from (3.97)

$$SINR(E, B_1) = a_1 - \frac{d_1}{E + b_1} \geq a_1 - \frac{d_1}{E_1 + b_1} = SINR(B_1) \qquad (3.106)$$

and it is upper bounded by a_1. Here a_1 corresponds to the $SINR_1(\infty, B_1)$ performance in (3.98) for bandwidth B_1 as the transmit energy $E \to \infty$. Note that a_1, b_1, and c_1 are the constants in (3.31),

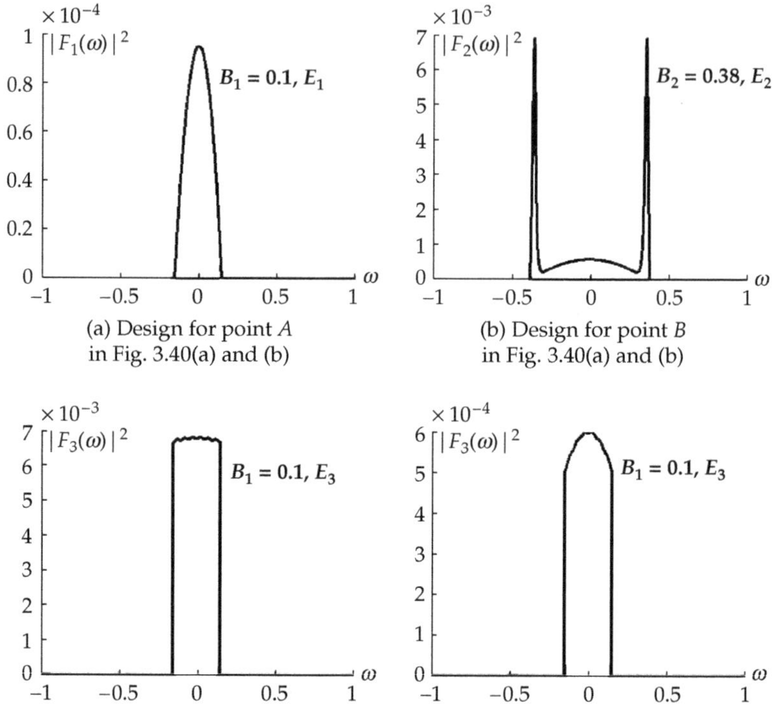

(a) Design for point A in Fig. 3.40(a) and (b)

(b) Design for point B in Fig. 3.40(a) and (b)

(c) Design for point C in Fig. 3.40(a)

(d) Design for point C in Fig. 3.40(b)

FIGURE 3.41 Transmit waveforms correspond to points A, B, and C in Fig. 3.40 (a) and (b).

(3.32), and (3.88) with $\Omega_+(B_i)$ replaced by $\Omega_+(B_1)$. The dashed curves Aa_1 in Fig. 3.40(a) and (b) represent $SINR_1(E, B_1)$ for various values of E. From (3.86), each point on the curve Aa_1 generates a new transmit waveform as well.

Depending on the target characteristics, the dashed curves $SINR_1(E, B_1)$ may or may not cross over the performance (solid) curve $SINR(B_i)$. This is illustrated in Fig. 3.40(a) and (b) for these two different situations. In Fig. 3.40(a), the dashed curve *does not* cross over the solid curve $SINR(B_i)$, whereas in Fig. 3.40(b), the dashed curve crosses over the solid curve $SINR(B_i)$.

In Fig. 3.40(a) and (b), assume that the saturation performance value a_1 for bandwidth B_1 satisfies

$$a_1 \geq SINR(B_2), \tag{3.107}$$

that is, the maximum performance level for bandwidth B_1 (obtained with $E \to \infty$) is greater than or equal to the performance level associated with the operating point B with a higher bandwidth B_2 and a higher threshold energy E_2. In Fig. 3.40(a) and (b), draw a horizontal line through B to intersect the curve Aa_1 at C, and drop a perpendicular at C to intersect the x-axis at E_3. From (3.106) with $E = E_3$ by construction we get

$$SINR(E_3, B_1) = SINR(B_2). \tag{3.108}$$

Thus the operating point C on the curve Aa_1 is associated with energy E_3 and lower bandwidth B_1, and corresponds to a performance level of $SINR_1(B_2)$ associated with a higher bandwidth. Notice that in Fig. 3.40(a), we have

$$E_3 > E_2 > E_1, \quad \text{and} \quad B_1 < B_2 \tag{3.109}$$

and in Fig. 3.40(b)

$$E_2 > E_3 > E_1, \quad \text{and} \quad B_1 < B_2. \tag{3.110}$$

In other words, by increasing the transmit energy from E_1 to E_3 while holding the bandwidth constant at B_1, the performance equivalent to a higher bandwidth B_2 can be realized in both situations provided B_2 satisfies (3.107). As a result, an energy-bandwidth tradeoff is possible within reasonable limits. Observe that E_3 is always higher than E_1; however, it may be higher or lower than E_2, the minimum threshold energy corresponding to bandwidth B_2 depending on the situation at hand. The transmit waveform $|F_3(\omega)|^2$ associated with the operating point C is obtained using (3.86) by replacing E with E_3 and it is illustrated in Fig. 3.41(c) for the situation in Fig. 3.40(a) and in Fig. 3.41(d) for the situation in Fig. 3.40(b). Notice that since the clutter and noise spectra are the same in both Fig. 3.40(a) and (b), the waveforms corresponding to either point A or B for all three targets are identical in Fig. 3.40(b) and they are given by those in Fig. 3.41(a) and (b). A comparison with Fig. 3.41(a) and (b) shows that the waveforms at C in Fig. 3.41(c) and (d) are different from those associated with operating points A and B. Since E_3 is different in Fig. 3.40(a) and (b), the waveforms in Fig. 3.41(c) and (d) are also different.

It is important to note that although the transmit waveform designs $|F_3(\omega)|^2$ and $|F_1(\omega)|^2$ correspond to the same bandwidth (with different energies E_3 and E_1), one is not a scaled version of the other. Changing transmit energy from E_1 to E_3 produces a new waveform $|F_3(\omega)|^2$ that maintains a performance level associated with a larger bandwidth B_2. Finally, to obtain the transmit waveforms in time domain an appropriate phase function can be used with the magnitude transforms so obtained, prior to their inverse Fourier transform operation. These phase functions can be used to make the transmit waveforms causal in which case a minimum phase transmit waveform can be generated. The phase functions can also be selected so as to retain chirp signal like properties for the transmit signal [27,28,31].

The question of how much bandwidth tradeoff can be achieved at an operating point is an interesting one. From the above argument the equality condition in (3.107) gives the upper bound on how much effective bandwidth increment can be achieved by increasing the transmit energy. Notice that for an operating point A, the desired bandwidth B_1 gives the operating bandwidth $\Omega_+(B_1)$ and from (3.89) with $i = 1$ the performance limit (as $E \to \infty$)

$$a_1 = \frac{1}{2\pi} \int_{\Omega_+(B_1)} \frac{|Q(\omega)|^2}{G_c(\omega)} d\omega \tag{3.111}$$

for bandwidth B_1 can be computed. Assume $B_2 > B_1$, and from (3.90) using $B_i = B_2$, $SINR_1(B_2)$ the minimum performance at B_2 also can be computed, and for maximum bandwidth swapping the nonlinear equation

$$a_1 = SINR_1(B_2) \tag{3.112}$$

must be solved for B_2. Then

$$\Delta B(B_1) = B_2 - B_1 \tag{3.113}$$

represents the maximum bandwidth enhancement that can be realized at B_1.

The transmit waveform specified by both methods discussed here puts constraints only in its magnitude $|F(\omega)|$ in the frequency domain [see (3.38), (3.42), or (3.77)] and hence the phase information $\psi(\omega)$ can be used as a free parameter to complete the design of $f(t)$. Recall that from Sec. 2.4, to maximize the $SINR$ up to the decision instant, the transmit waveform must be a minimum phase signal. Clearly, to maintain such a minimum phase transmit signal, a certain threshold energy E_o [see (3.40) for example] must be available at the transmitter. In that case $f(t)$ will not be time-limited. However, because of its minimum phase nature, most of its energy will be concentrated up front, making it an acceptable alternative to a pulse waveform [32–35].

As noted earlier, minimum phase signals tend to concentrate their energy up front, and in this context, constant envelope signals can be generated following the procedure described in Chap. 4.

3.5 Optimization of a Functional

Consider maximizing the functional

$$\Lambda(y) = \int_{-\infty}^{+\infty} F\left(y(\omega), \omega\right) d\omega \tag{3.114}$$

by varying over all $y(\omega)$. Suppose $y = y_o(\omega)$ maximizes $\Lambda(y)$, then

$$\Lambda(y_o) = \int_{-\infty}^{+\infty} F\left(y_o(\omega), \omega\right) d\omega \tag{3.115}$$

is the maximum value of (3.114). For any arbitrary function $\delta(\omega)$, let

$$y(\omega) = y_o(\omega) + \varepsilon\, \delta(\omega) \tag{3.116}$$

in (3.114) and define

$$\Lambda_\varepsilon(y) = \int_{-\infty}^{+\infty} F\left(y_o(\omega) + \varepsilon\delta(\omega), \omega\right) d\omega. \tag{3.117}$$

Then from (3.115), $\Lambda_\varepsilon(y)$ is maximum at $\varepsilon = 0$. Hence we must have

$$\left.\frac{\partial \Lambda_\varepsilon(y)}{\partial \varepsilon}\right|_{\varepsilon=0} = 0 \tag{3.118}$$

and

$$\left.\frac{\partial^2 \Lambda_\varepsilon(y)}{\partial \varepsilon^2}\right|_{\varepsilon=0} < 0. \tag{3.119}$$

But $\Lambda_\varepsilon(y)$ in (3.117) can be expanded using the Taylor series as

$$\Lambda_\varepsilon(y) = \int_{-\infty}^{+\infty} \left\{ \begin{array}{l} F\left(y_o(\omega), \omega\right) + \varepsilon\delta(\omega) \left.\dfrac{\partial F(y,\omega)}{\partial y}\right|_{y=y_o} \\ + \dfrac{\varepsilon^2\delta^2(\omega)}{2} \left.\dfrac{\partial^2 F(y,\omega)}{\partial y^2}\right|_{y=y_o} + \cdots \end{array} \right\} d\omega = 0. \tag{3.120}$$

Using (3.120) in (3.118), we get

$$\int_{-\infty}^{+\infty} \delta(\omega) \left.\frac{\partial F(y,\omega)}{\partial y}\right|_{y=y_o(\omega)} d\omega = 0, \tag{3.121}$$

for all $\delta(\omega)$. Since (3.121) must be true for all arbitrary $\delta(\omega)$, we have

$$\left.\frac{\partial F(y,\omega)}{\partial y}\right|_{y=y_o(\omega)} = F'(y_o(\omega),\omega) = 0 \tag{3.122}$$

where the maximum is attained, provided (3.119) is also satisfied. Proceeding as before using (3.120), (3.119) translates into

$$\int_{-\infty}^{+\infty} \delta^2(\omega) \left.\frac{\partial^2 F(y,\omega)}{\partial y^2}\right|_{y=y_o(\omega)} d\omega < 0, \tag{3.123}$$

for all $\delta(\omega)$. Once again since (3.123) must be true for all $\delta(\omega)$, we must have

$$\left.\frac{\partial^2 F(y,\omega)}{\partial y^2}\right|_{y=y_o(\omega)} = F''(y_o(\omega),\omega) < 0. \tag{3.124}$$

References

[1] P. Stoica, J. Li, X. Zhu, "Waveform Synthesis for Diversity-Based Transmit Beampattern Design," *IEEE Transactions on Signal Processing*, vol. 56, no. 6, June 2008.
[2] K.W. Forsythe, D.W. Bliss, "Waveform Correlation and Optimization Issues for MIMO Radar," *Proc. 39th Asilomar Conf. Signals, Systems, Computer*, Pacific Grove, CA, pp. 1306–1310, Nov. 2005.
[3] B. Friedlander, "Waveform Design for MIMO Radars," *IEEE Transactions on Aerospace and Electronic Systems*, vol. 43, no. 3, pp. 1227–1238, Oct. 2007.
[4] Y. Chi, A. Pezeshki, R. Calderbank, S. Howard, "Range Sidelobe Suppression in a Desired Doppler Interval," *IEEE International Waveform Diversity & Design Conference*, Orlando, FL, Feb. 8–13, 2009.
[5] A. Papoulis, S.U. Pillai, *Probability, Random Variables and Stochastic Processes*, New York: McGraw-Hill Higher Education, 2002.
[6] S.U. Pillai, "Simultaneous Savings in Bandwidth and Energy Using Waveform Design in Presence of Interference and Noise," U.S. Patent 7,538,720 B2, May 26, 2009.
[7] HI. Demirkiran, D. Weiner, A. Drozd, I. Kasperovich, "Equivalent Based Analysis for Narrow Weakly Nonlinear Bandpass Systems for Waveform Diversity Applications," *IEEE International Waveform Diversity & Design Conference*, Orlando, FL, Feb. 8–13, 2009.
[8] M. Stuff, B. Thelen, N. Subotic, J. Parker, J. Browning, "Optimization and Waveforms for Compressive Sensing Applications in the Presence of Interference," *IEEE International Waveform Diversity & Design Conference*, Orlando, FL, Feb. 8–13, 2009.
[9] H. Mendelson, P. Zulch, "Mitigation of Radar Range Ambiguities via Waveform Diversity," *IEEE International Waveform Diversity & Design Conference*, Orlando, FL, Feb. 8–13, 2009.
[10] S.U. Pillai, "Apparatus and Method for Providing Energy-Bandwidth Tradeoff and Waveform Design in Interference and Noise," U.S. Patent 7,711,057 B2, May 4, 2010.
[11] S.U. Pillai, K.Y. Li, "Energy-Bandwidth Tradeoff and Transmit Waveform Design Using Interference and Noise Whitening Method," U.S. Patent 7,720,132 B2, May 18, 2010.
[12] J. Goodman, B. Miller, M. Herman, M. Wai, P. Monticciolo, "Extending the Dynamic Range of RF Receiver with Nonlinear Equalization," *IEEE International Waveform Diversity & Design Conference*, Orlando, FL, Feb. 8–13, 2009.
[13] J. Zhang, A. Papandreou-Suppappola, "MIMO Radar with Frequency Diversity," *IEEE International Waveform Diversity & Design Conference*, Orlando, FL, Feb. 8–13, 2009.
[14] P. Gonzalez-Blano, E.D. Diego, E. Millan, B. Errasti, I. Montiel, "Stepped-Frequency Waveform in Radar Demonstrator and Its Jamming," *IEEE International Waveform Diversity & Design Conference*, Orlando, FL, Feb. 8–13, 2009.
[15] P.F. Sammartino, C. Baker, "The Frequency Diverse Bistatic System," *IEEE International Waveform Diversity & Design Conference*, Orlando, FL, Feb. 8–13, 2009.
[16] T. Sturman, "Enabling Chaotic Waveform Diversity," *IEEE International Waveform Diversity & Design Conference*, Orlando, FL, Feb. 8–13, 2009.
[17] J. Li, J.R. Guerci, L. Xu, "Signal Waveform's Optimal-Under-Restriction Design for Active Sensing," *IEEE Signal Process. Lett.*, vol. 13, no. 9, pp. 565–568, Sep. 2006.

[18] D.A. Garren, M.K. Osborn, A.C. Odom, J.S. Goldstein, S.U. Pillai, J.R. Guerci, "Enhanced Target Detection and Identification via Optimized Radar Transmission Pulse Shape," *Proc. IEEE*, vol. 148, no. 3, pp. 130–138, June 2001.
[19] M. Inggs, M. Brooker, "Extensible simulator for Waveform Diversity Testing," *IEEE International Waveform Diversity & Design Conference*, Orlando, FL, Feb. 8–13, 2009.
[20] N. Bharadwaj, V. Chandrasekar, "Distributed Waveforms for Networked Meteorological Radar," *IEEE International Waveform Diversity & Design Conference*, Orlando, FL, Feb. 8–13, 2009.
[21] T. Higgins, S. Blunt, "Analysis of Range-Angle Coupled Beamforming with Frequency-Diverse Chirps," *IEEE International Waveform Diversity & Design Conference*, Orlando, FL, Feb. 8–13, 2009.
[22] Y. Paichard, M. Brooker, M. Inggs, "A Signal Level Simulator for Netted Radar Waveforms Evaluation," *IEEE International Waveform Diversity & Design Conference*, Orlando, FL, Feb. 8–13, 2009.
[23] H.B. Manjunath, J. Zhang, A. Papandreou-Suppappola, D. Morrell, "Waveform-Agile Sensing for Range and DoA Estimation in MIMO Radars," *IEEE International Waveform Diversity & Design Conference*, Orlando, FL, Feb. 8–13, 2009.
[24] M.C. Wicks, "History of Waveform Diversity and Future Benefits to Military Systems," *NATO Lecture Series on Waveform Diversity (SET119)* held in UK, Italy, Germany,and France, Sep. 8–19, 2008.
[25] G. Genello, "Information Processing for Waveform Diversity Control," *IEEE International Waveform Diversity & Design Conference*, Orlando, FL, Feb. 8–13, 2009.
[26] I. Bradaric, G. Capraro, M. Wicks "Multistatic Ambiguity Function—A Tool for Waveform Selection in Distributed Radar Systems," *IEEE International Waveform Diversity & Design Conference*, Orlando, FL, Feb. 8–13, 2009.
[27] D.D. Weiner, M.C. Wicks, G.T. Capraro, "Waveform Diversity and Sensors as Robots in Advanced Military Systems," *1st International Waveform Diversity and Design Conference*, Edinburgh, UK, November 2004.
[28] I. Bradaric, G.T. Capraro, P. Zulch, "Signal Processing and Waveform Selection Strategies in Multistatic Radar Systems," *International Waveform Diversity and Design Conference*, Pisa, Italy, June 2007, invited paper.
[29] I. Akduman, A. Yapar, F. Akleman, H. Sahinturk, "Electromagnetic Imaging of Objects Buried under a Dielectric Rough Surface," *IEEE International Waveform Diversity & Design Conference*, Orlando, FL, Feb. 8–13, 2009.
[30] I. Shirokov, I. Serdyuk, "The Investigation of Phase Synchronization of Reference Oscillators Through Atmospheric Channel," *IEEE International Waveform Diversity & Design Conference*, Orlando, FL, Feb. 8–13, 2009.
[31] R. Bonneau, "A Coherence Gradient Method in Waveform Design for Apertures," *IEEE International Waveform Diversity & Design Conference*, Orlando, FL, Feb. 8–13, 2009.
[32] Y. Paichard, "Orghogonal Multicarrier Phased Coded Signal for Netted Radar Systems," *IEEE International Waveform Diversity & Design Conference*, Orlando, FL, Feb. 8–13, 2009.
[33] J. Stiles, J. Jenshak, "Sparse Array Construction Using Marginal Fisher's Information," *IEEE International Waveform Diversity & Design Conference*, Orlando, FL, Feb. 8–13, 2009.
[34] T. Ch. Tang, C.C. Ho, "EMI Suppression for High-Speed PCB," *IEEE International Waveform Diversity & Design Conference*, Orlando, FL, Feb. 8–13, 2009.
[35] L.L. Monte, R. Ansari, D. Erricolo, M. wicks, "The Use of Geometric Diversity for Spectral Dominance in Underground Imaging," *IEEE International Waveform Diversity & Design Conference*, Orlando, FL, Feb. 8–13, 2009.

CHAPTER 4

Constant Envelope Transmit Signals

One useful feature for transmit signals in time domain is the constant envelope property that maximizes the transmit power efficiency. In that case, the time domain waveform will ideally have a constant envelope property or in practice possess an approximately constant envelope, thereby distributing the peak transmit power more evenly in time.

In addition, for transmission purposes, the signal is of finite duration and there is always some restriction in terms of the signal bandwidth as well as its spectral content. The transmit signal is also expected to have good pulse compression properties since the return signal is usually a delayed version of the transmit signal and upon matched filtering, the resulting output should separate out any closely spaced target returns that may be present in the signal. These requirements are illustrated Fig. 4.1. These requirements are conflicting since finite duration and finite bandwidth are certainly not compatible. Moreover, pulse compression property for the transmit signal demands that the matched filter output $y(t)$ be ideally a time-delayed version of the impulse function. But from Fig. 4.1

$$y(t) = f(t) * f^*(t_o - t) \leftrightarrow |F(\omega)|^2 e^{-j\omega t_o} \tag{4.1}$$

so that the requirement

$$y(t) = \delta(t_o - t) \tag{4.2}$$

forces the transmit signal to have a wide bandwidth of equal intensity contradicting the requirement for adapting the transmit waveform to

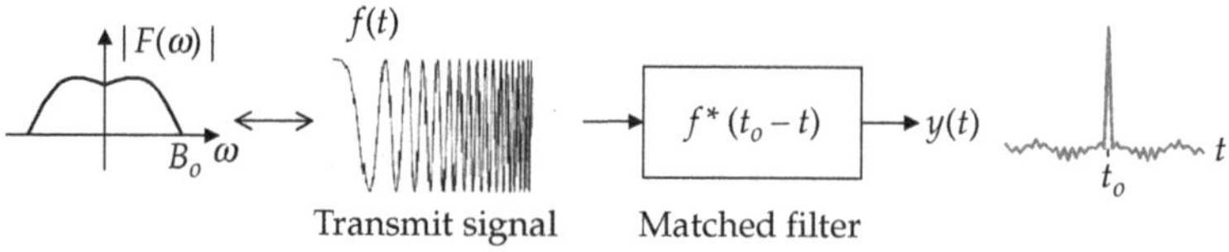

FIGURE 4.1 Finite duration constant envelope signal with finite bandwidth and good pulse compression properties.

have some known spectral characteristics. Hence if adaptivity in the spectral domain is desired, the matched filter may not be the optimum receiver.

Hence, the challenge is to design waveforms that accommodate these conflicting requirements in some compromised manner. Toward this, we will pursue several approaches, of which the method of alternating convex projections is reviewed next.

4.1 Convex Sets[1]

Many useful notions in signal processing can be expressed in terms of convex sets. Members of a convex set satisfy the convexity constraint which states that if C is a convex set with $f_1 \in C$ and $f_2 \in C$, then their convex linear combination also belongs to C. Thus (Fig. 4.2)

$$f = \alpha f_1 + (1-\alpha) f_2 \in C, \quad 0 \le \alpha \le 1. \tag{4.3}$$

Convex sets have the remarkable property that for any point x outside C, there exists a "unique" nearest neighbor x_o that belongs to C; that is, for any arbitrary x, there exists a unique vector $x_o \in C$ such that [1]

$$\|x - x_o\| \le \|x - x_1\|, \quad \text{for all } x_1 \in C. \tag{4.4}$$

In other words, such an x_o satisfies the infimum condition

$$\inf_{x_1 \in C} \|x - x_1\| = \|x - x_o\| \tag{4.5}$$

where x is any arbitrary vector. This unique member x_o is known as the projection of x onto C. Thus in Fig. 4.3

$$x_o = Px \tag{4.6}$$

[1] © 2009 IEEE. Reprinted with permission, from "Reconstruction of Constant Envelope Signals with Given Fourier Transform Magnitude" by S. U. Pillai, K. Y. Li, and H. Beyer, *2009 IEEE Radar Conference*, Pasadena, CA, May 4–8, 2009.

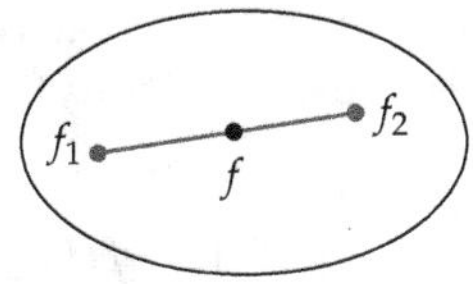

FIGURE 4.2 Convex set.

where P is the projection operator associated with the closed convex set C that assigns a unique nearest neighbor Px in C for any x. Projection operators onto closed convex sets are nonexpansive operators, that is,

$$\|Px - Py\| \leq \|x - y\|. \tag{4.7}$$

Interestingly, many useful notions in signal processing form convex sets. For example, the set of all signals that are band limited to B_o forms a convex set C_{B_o}, since if $f_1(t)$ and $f_2(t)$ are both band limited to B_o, then $\alpha f_1(t) + (1 - \alpha) f_2(t)$ is also band limited to B_o. As a result, given any signal $f(t)$, it is easy to show that its projection onto C_{B_o} is given by band limiting its Fourier transform $F(\omega)$ to B_o (setting $F(\omega) = 0$, $|\omega| > B_o$). Thus [1,2]

$$P_{B_o} f(t) \leftrightarrow \begin{cases} F(\omega), & |\omega| \leq B_o \\ 0, & |\omega| > B_o. \end{cases} \tag{4.8}$$

Similarly, the set of all signals with bounded envelope property also forms a convex set C_c and for any signal $f(t)$ its projection onto C_c is given by

$$P_c f(t) = \begin{cases} f(t), & |f(t)| \leq c \\ +c, & f(t) > c \\ -c, & f(t) < c, \end{cases} \tag{4.9}$$

where c represents the desired upper bound on the envelope.

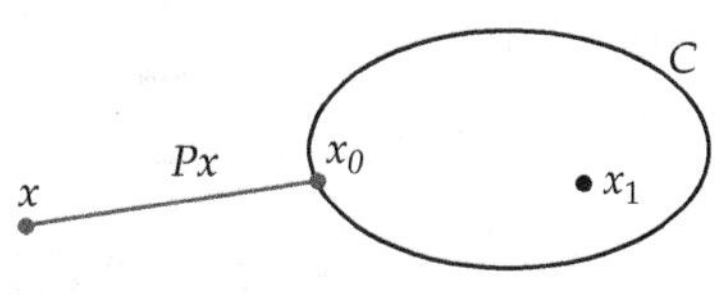

FIGURE 4.3 "Unique" neighbor property.

In a similar manner, the set of all nonnegative signals forms a convex set C_+. Given any $f(t)$, its projection onto C_+ is given by setting its negative part to zero, that is,

$$P_+ f(t) = \begin{cases} f(t), & f(t) \geq 0 \\ 0, & f(t) < 0. \end{cases} \tag{4.10}$$

The set of all signals with finite nonzero support $(0, T)$ also forms a convex set C_T. Given any $f(t)$, its projection onto C_T is given by setting its nonzero part outside the desired interval to zero. Thus [3]

$$P_T f(t) = \begin{cases} f(t), & 0 < t < T \\ 0, & \text{otherwise.} \end{cases} \tag{4.11}$$

Another example is the set of all signals with a common phase function $\varphi(\omega)$ in the frequency domain. All such signals form a convex set since if $f_1(t) \leftrightarrow F_1(\omega) = A_1(\omega)e^{j\varphi(\omega)}$ and $f_2(t) \leftrightarrow F_2(\omega) = A_2(\omega)e^{j\varphi(\omega)}$, then their convex combination

$$\begin{aligned} f(t) &= \alpha f_1(t) + (1-\alpha) f_2(t) \leftrightarrow \alpha F_1(\omega) + (1-\alpha) F_2(\omega) \\ &= \{\alpha A_1(\omega) + (1-\alpha) A_2(\omega)\} e^{j\varphi(\omega)} = A(\omega) e^{j\varphi(\omega)} \end{aligned} \tag{4.12}$$

preserves the same phase function $\varphi(\omega)$ [4].

If the signal $f(t)$ satisfies convex properties associated with several closed convex sets $C_1, C_2, \ldots C_n$, then $f(t)$ belongs to their intersection C, a closed convex set, that is,

$$f(t) \in C_o = \bigcap_{i=1}^{n} C_i. \tag{4.13}$$

In this case, the problem is to find a signal $f(t)$ that belongs to C_o given the various convex properties and their projection operators, satisfied by the unknown signal. In this context, the method of alternating projections has been shown to be very useful [5–7].

4.1.1 Methods of Alternating Projections

Consider the closed convex sets $C_1, C_2, \ldots C_n$ and their associated projection operators $P_1, P_2 \ldots P_n$. If the desired function belongs to each C_i, then it belongs to their intersection

$$C_o = C_1 \cap C_2 \cap \ldots C_n. \tag{4.14}$$

Starting from an outside point x (usually an approximant), the problem is to determine its "nearest neighbor" x_o within the desired convex set C_o; if not *at least* a point within C_o. If the projection operator P_o

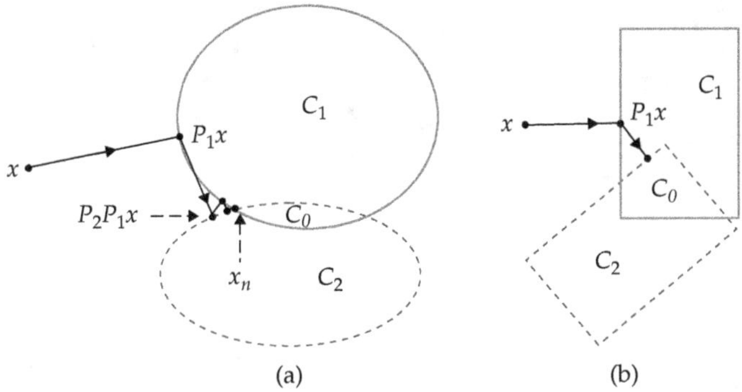

FIGURE 4.4 Method of iterative projections.

corresponding to the new convex set C_o is known, then that nearest neighbor to x equals $x_o = P_o x$ as in (4.6). However, P_o is usually unknown, since C_o is defined only through (4.14), and in general only $P_1, P_2 \ldots P_n$ are known. Thus the problem reduces to finding a neighbor (not necessarily the nearest) in C_o starting at x given the projection operators $P_1, P_2 \ldots P_n$ [5–7].

The method of alternating projections originally proposed by John Neumann [8] has been generalized to address this problem [1]. Define the alternating projection operator

$$P = P_n P_{n-1} \ldots P_2 P_1 \tag{4.15}$$

and consider the iteration

$$x_i = P\, x_{i-1}, \quad i = 1, 2, \ldots \tag{4.16}$$

Starting with $x_o = x$. Then it has been shown that the iteration in (4.16) converges *weakly* to a point in the intersection C_o in (4.14)[2] [1,6,7]. This is illustrated in Fig. 4.4 for two convex sets [1].

The iterations in (4.16) can be accelerated in their convergence rate by making use of the relaxed projection operators as shown.

[2]A sequence $\{f_n\}$ converges to f if $\lim_{k\to\infty} \|f_k - f\| = 0$. It is said to converge weakly to f if $\lim_{k\to\infty} (f_k, g) = (f, g)$ for every g in the vector space, where (x, y) denotes the inner product of x and y.

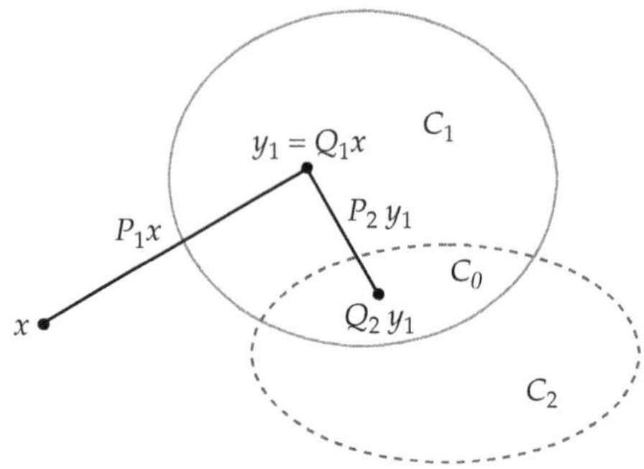

FIGURE 4.5 Relaxed projection method.

4.1.2 Relaxed Projection Operators

The projection operators P_k can be "relaxed" to generate

$$Q_k = I + \lambda_k (P_k - I), \quad k = 1, 2, \ldots n \tag{4.17}$$

that are nonexpansive[3] operators for $0 \leq \lambda_k \leq 2$, with I representing the identity operator. The quantity λ_k in (4.17) represents the relaxation parameter and it can be chosen to form a sequence of adaptive coefficients for faster convergence as in Fig. 4.5.

Thus, relaxed projection operators can be substituted in (4.16) and by controlling λ_k's, the convergence rate can be controlled as well [9]. Thus more generally with Q_k as in (4.17) and with

$$Q = Q_n Q_{n-1} \cdots Q_2 Q_1, \tag{4.18}$$

the iteration

$$x_i = Q x_{i-1}, \quad i = 1, 2, \ldots \tag{4.19}$$

converges weakly to a point in the intersection C in (4.13).

Fourier Transform Magnitude

Although signals with the same Fourier transform phase do form a closed convex set as seen in (4.12), signals with the same Fourier transform magnitude function *do not* form a convex set. This follows

[3]If T is a nonexpansive operator, then $\|Tx - Ty\| \leq \|x - y\|$. If on the other hand $\|Tx - Ty\| < \|x - y\|$ then T represents a contraction (strict inequality).

easily since if

$$f_1(t) \leftrightarrow M(\omega)e^{j\psi_1(\omega)}, \quad f_2(t) \leftrightarrow M(\omega)e^{j\psi_2(\omega)} \tag{4.20}$$

then for any $0 < \alpha < 1$ we get [use (4.3)]

$$\begin{aligned} f(t) = \alpha f_1(t) + (1-\alpha) f_2(t) &\leftrightarrow M(\omega)\left(e^{j\psi_1(\omega)} + (1-\alpha)e^{j\psi_2(\omega)}\right) \\ &= M(\omega)A(\omega)e^{j\psi(\omega)}. \end{aligned} \tag{4.21}$$

Since the convex combination above has its transform magnitude equal to $M(\omega)A(\omega) \neq M(\omega)$, it does not preserve the original magnitude function $M(\omega)$. As a result, signals with the same Fourier transform magnitude functions over a prescribed frequency region do not form a convex set and there exists no unique projection operator as in (4.6).

Let C_M denote the set of functions $\{g(t)\}$ that have the prescribed Fourier transform magnitude $M(\omega)$ over a prescribed frequency set Ω. Although C_M is not a closed convex set, it is nevertheless possible to define an operator P_M that will assign to every arbitrary function $f(t)$ a "nearest neighbor" $P_M f(t)$ that belongs to C_M, such that there exists *no other element* $g \in C_M$ for which

$$\|f - g\| < \|f - P_M f\| \tag{4.22}$$

is satisfied. In the case of closed convex sets, the above operator is the projection operator and the "nearest neighbor" is unique as in (4.5). In the case of the magnitude substitution operator, the uniqueness is not preserved and in general we only have

$$\|f - P_M f\| \leq \|f - g\|, \quad \text{for all } g \in C_M. \tag{4.23}$$

Given an arbitrary $f(t)$, to determine $P_M f(t)$, with the prescribed Fourier transform magnitude $M(\omega)$, we need $g(t) \in C_M$ such that $\|f(t) - g(t)\|^2$ is minimum over Ω. With

$$f(t) \leftrightarrow F(\omega) = |F(\omega)|e^{j\psi(\omega)}, \quad g(t) \leftrightarrow G(\omega) = M(\omega)e^{j\phi(\omega)} \tag{4.24}$$

we have

$$\begin{aligned} \|f(t) - g(t)\|^2 &= \frac{1}{2\pi}\int_{-\infty}^{+\infty} |F(\omega) - G(\omega)|^2\, d\omega \\ &= \frac{1}{2\pi}\int_{\Omega} |F(\omega) - G(\omega)|^2 d\omega + \frac{1}{2\pi}\int_{\Omega'} |F(\omega) - G(\omega)|^2\, d\omega \end{aligned} \tag{4.25}$$

where Ω' denotes the complementary set of the prescribed frequency set Ω. To minimize (4.25), we set the unknown $G(\omega) = F(\omega)$ for $\omega \in \Omega'$ so that (4.25) reduces to

$$\begin{aligned} 2\pi \| f(t) - g(t) \|^2 &= \int_{\Omega} |F(\omega) - G(\omega)|^2 \, d\omega \\ &= \int_{\Omega} |F(\omega)|^2 d\omega + \int_{\Omega} |M(\omega)|^2 \, d\omega \\ &\quad - 2 \int_{\Omega} |F(\omega)| M(\omega) \cos(\psi(\omega) - \phi(\omega)) \, d\omega \end{aligned} \tag{4.26}$$

and the right side is minimized by selecting the free phase function $\phi(\omega)$ equal to $\psi(\omega)$ so that $\cos(\psi(\omega) - \phi(\omega)) = 1$ for all $\omega \in \Omega$. Hence for any $f(t) \leftrightarrow F(\omega) = |F(\omega)| e^{j\psi(\omega)}$,

$$P_M f(t) \leftrightarrow \begin{cases} M(\omega) e^{j\psi(\omega)}, & \omega \in \Omega \\ F(\omega), & \omega \in \Omega' \end{cases} \tag{4.27}$$

gives the magnitude substitution operator. Using the operator P_M, we can define the "alternating" iteration

$$f_{k+1} = P P_M f_k \tag{4.28}$$

where P represents *any* one of the projection operators discussed earlier (P_B, P_c, P_+, P_T, etc.). Although iteration (4.28) is not shown to be convergent, it possesses the following "error reduction" property [2,10]

$$\sigma_{k+1} = \| f_{k+1} - P_M f_{k+1} \| \leq \| f_k - P_M f_k \| = \sigma_k. \tag{4.29}$$

This follows by observing that by the definition of P_M in (4.23) we have (with $g = P_M f_k \in C_M$)

$$\| f_{k+1} - P_M f_{k+1} \| \leq \| f_{k+1} - P_M f_k \| = \| P P_M f_k - P_M f_k \|, \tag{4.30}$$

where the last equality follows from (4.28). But $P P_M f_k = P x$ is the projection of $x = P_M f_k$ onto a closed convex set C and from (4.28) since $f_k = PP_M f_{k-1}$ is also in that convex set, using (4.4), we obtain (with $x_o = PP_M f_k$, $x = P_M f$, $x_1 = f_k \in C$)

$$\|PP_M f_k - P_M f_k\| \leq \|f_k - P_M f_k\|. \tag{4.31}$$

Hence from (4.30) and (4.31) we obtain

$$\|f_{k+1} - P_M f_{k+1}\| \leq \|f_k - P_M f_k\|, \tag{4.32}$$

indicating that the distance from the iterate f_k to its nearest neighbor $P_M f_k$ in C_M with the given Fourier transform magnitude is a decreasing function. Hence

$$\lim_{k\to\infty} \|f_k - P_M f_k\| = d_o \geq 0. \tag{4.33}$$

The value of d_o can be greater than or equal to zero. Even if $d_o = 0$, that may not guarantee that the iterate in (4.28) converges to a point in C_M, since they can "run around" the boundary getting closer to C_M but not converging necessarily to any point [2].

Interestingly, constant envelope signals also share properties similar to the Fourier transform magnitude situation.

Constant Envelope Signals [11–13]

A constant envelope signal $f(t)$ can be expressed as

$$f(t) = Ae^{j\theta(t)} \tag{4.34}$$

where A is a suitable positive constant that can be used to maintain a prescribed energy level for $f(t)$. Notice that [14, 15]

$$|f(t)| = A \quad \text{for all } t. \tag{4.35}$$

It is easy to see that, signals with the same constant envelope level *do not* form a convex set; however, their behavior is similar to the signal set with given Fourier transform magnitude [16, 17]. Hence if C_A denotes the set of functions $\{g(t)\}$ that have constant envelope level A, although C_A is not a convex set, it is possible to define an operator P_A that will assign to every arbitrary function $f(t)$ a nearest neighbor $P_A f(t)$ that belongs to C_A such that no other element $g \in C_A$ satisfies [see also (4.22)]

$$\|f(t) - g(t)\| < \|f(t) - P_A f(t)\|. \tag{4.36}$$

Given an arbitrary signal $f(t) = a(t)e^{j\theta(t)}$ to determine $P_A f(t)$, we need $g(t) = Ae^{j\psi(t)}$ such that $\|f(t) - g(t)\|$ is minimum. In this case, we can proceed as in (4.24)–(4.27). Thus with

$$f(t) = a(t)e^{j\theta(t)} \tag{4.37}$$

$$P_A f(t) = \begin{cases} Ae^{j\theta(t)}, & t \in T \\ f(t), & \text{otherwise} \end{cases} \tag{4.38}$$

where T represents the temporal duration over which the constant envelope property is to be maintained [18]. Notice that the constant envelope operator P_A defined above also enjoys the error reduction property when employed in an iteration similar to (4.28). Thus with

$$f_{k+1} = PP_A f_k \tag{4.39}$$

where P represents any projection operator, from (4.32) and (4.33), we also get

$$\|f_{k+1} - P_A f_{k+1}\| \leq \|f_k - P_A f_k\|. \tag{4.40}$$

Thus the error between f_k and its constant envelope version $P_A f_k$ also reduces as $k \to \infty$.

Gerchberg-Saxton Algorithm [9]

More interestingly, if the iteration combines both Fourier transform substitution operation as well as the constant envelope property, then we have

$$f_{k+1}(t) = P_A P_M f_k(t) = P_A g_k(t) \tag{4.41}$$

where we define

$$g_k(t) = P_M f_k(t) \tag{4.42}$$

and in that case also the error reduction property

$$\begin{aligned} d_{k+1} &= \|f_{k+1} - g_{k+1}\| = \|f_{k+1} - P_M f_{k+1}\| \leq \|f_k - P_M f_k\| \\ &= \|f_k - g_k\| = d_k \end{aligned} \tag{4.43}$$

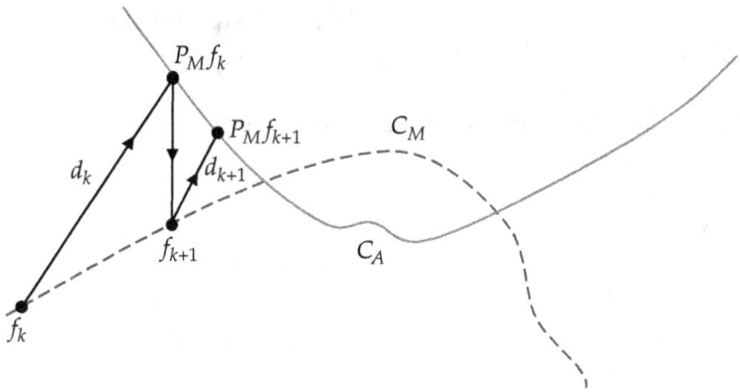

FIGURE 4.6 Error reduction property.

is maintained since (Fig. 4.6)

$$\begin{aligned} d_{k+1} &= \|f_{k+1} - P_M f_{k+1}\| \leq \|f_{k+1} - P_M f_k\| \\ &= \|P_A P_M f_k - P_M f_k\| \leq \|f_k - P_M f_k\| = d_k. \end{aligned} \tag{4.44}$$

Observe that the iteration in (4.41) is an excellent way to prescribe a given Fourier transform magnitude while maintaining a constant envelope in the time domain, and together they result in iterates that "get closer" in the sense of (4.43). From (4.41), f_{k+1} satisfies the constant modulus property while only approximating the given magnitude transform $M(\omega)$, whereas $g_k = P_M f_{k+1}$ possesses the given magnitude transform while only approximating the constant modulus property. Nevertheless, after a large number of iterates, as (4.43) shows, the difference between these signals decreases. Observe that since there is no unique solution, the final solution is sensitive to the starting point f_o, and a careful selection of the initial solution based on other factors is essential [19, 20].

Equation (4.41) represents the classic Gerchberg-Saxton algorithm [9], and, in particular, the discrete version of the Gerchberg-Saxton algorithm described above can be summarized as follows [21]:

Let $\{f_i\}_{i=1}^N$ and $\{F_k\}_{k=1}^N$ form an N-point DFT pair at $\omega_k = \frac{2\pi k}{N}$, $k = 1 \rightarrow N$,

whose magnitude values $|f_i| = a_i$ and $|F_k| = A_k$ are specified both in the time and frequency domains. Here $a_i = a$ for example will refer to the unimodular case. Perform the N-point DFT of the given sequence and substitute the frequency magnitude sequence by A_k, then perform the inverse DFT and substitute the temporal magnitude values by a_i

and repeat the procedure. For example, Wikipedia summarizes the above algorithm as follows [21]:

```
Gerchberg--Saxton Algorithm(Source, Target,
Retrieved_Phase)
 A = IFT(Target)
 while error criterion is not satisfied
   B = Amplitude(Source) * exp(i*Phase(A))
   C = FT(B)                                        (4.45)
   D = Amplitude(Target) * exp(i*Phase(C))
   A = IFT(D)
 end while
 Retrieved_Phase = Phase(A)
end Gerchberg—Saxton Algorithm
```

The error reduction property in (4.44) is guaranteed in this case also. Figure 4.7 shows a given magnitude transform and results of iteration using (4.41) to preserve the constant envelope. Observe that the error reduction property for d_k in (4.43) is satisfied in Fig. 4.7(b) [20,21].

As mentioned before, the iterate f_k in Fig. 4.7 preserves the constant envelope property, but its transform magnitude is only approximately close to the given transform as indicated by Fig. 4.7(c) and (d). On the other hand, $g_k = P_M f_k$ preserves the transform magnitude, but its envelope is only approximately constant as shown in Fig. 4.7(e) and (f). However, as the iterations proceed, the error between f_k and g_k decreases and hence after a sufficient number of steps, either the output f_k or $g_k = P_M f_k$ may be used as the desired constant envelope signal having the given transform magnitude. Notice that only one condition is exactly satisfied and the other one is approximately satisfied by both these signals. Depending on which condition is more important, the user can select either f_k or g_k as the desired signal.

Figure 4.8 shows another example of given magnitude transform and results of iterations using (4.41) to preserve the constant envelope. Once again, after a sufficiently large number of iterations, either f_k or $P_M f_k$ may be selected as the desired signal.

Figure 4.9 shows the time domain signals $f_k(t)$ and $g_k(t)$ and their transforms corresponding to the point C in Fig. 3.23(b) using the iteration in (4.41). Here Fig. 3.26(b) is used as the prescribed magnitude transform, and the iteration in (4.41) is initiated using $f_o(t) \leftrightarrow M(\omega)e^{j\psi_o\omega}$, with a random phase function $\psi_o(\omega)$. The waveform $f_n(t)$ is constant envelope, while its spectrum hugs the original one up to about -40 dB, whereas $g_n(t)$ exactly matches the spectrum,

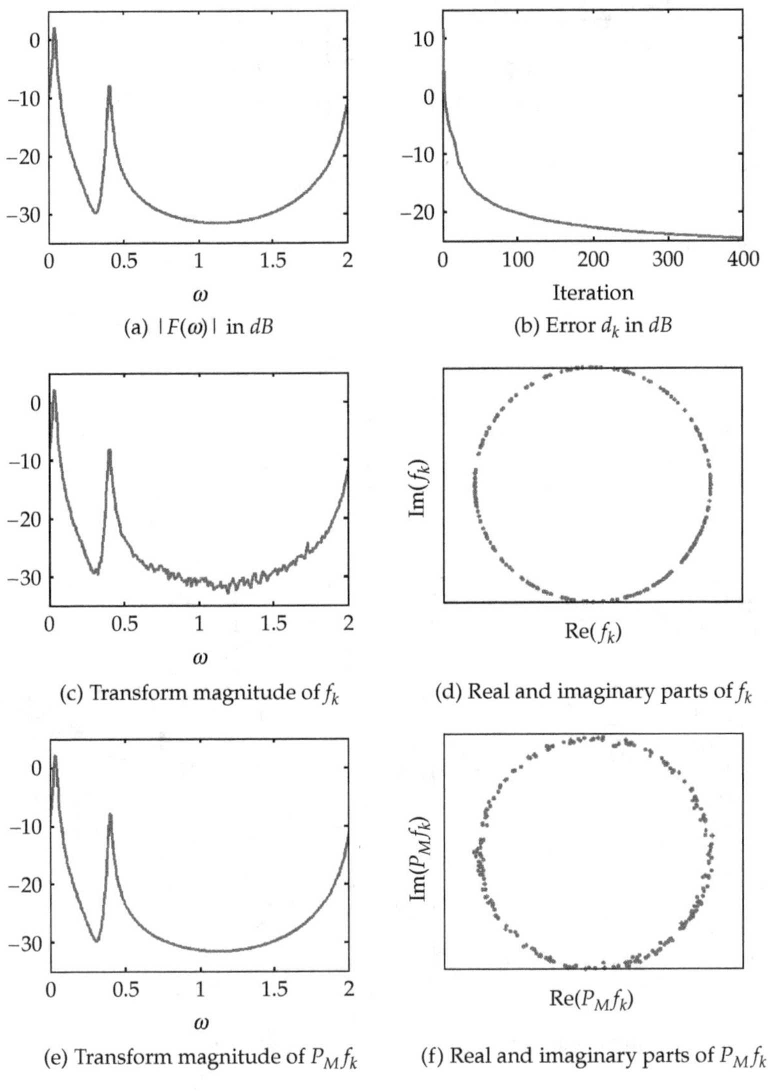

FIGURE 4.7 Constant envelope preservation along with given magnitude transform after 400 iterations.

but it has only an approximately constant envelope. Figure 4.10 shows the similar results for the point F marked in Fig. 3.23(b).

Interestingly this procedure can be coupled with the optimum transmit waveform design methods described in Chap. 3, that prescribe the

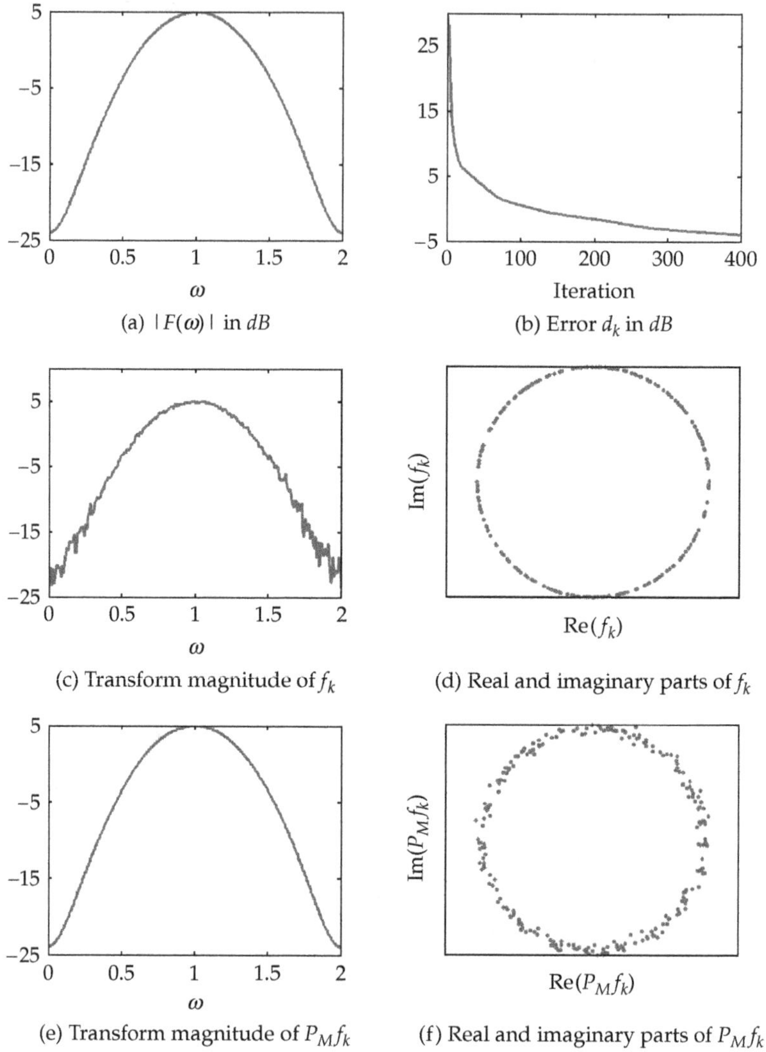

FIGURE 4.8 Constant envelope preservation along with given magnitude transform after 400 iterations.

transmit waveform magnitude information in the frequency domain. The magnitude information so obtained can be used to generate a constant envelope signal or an almost constant envelope signal, and can be further scaled to maintain the prescribed energy level in each case [20, 22].

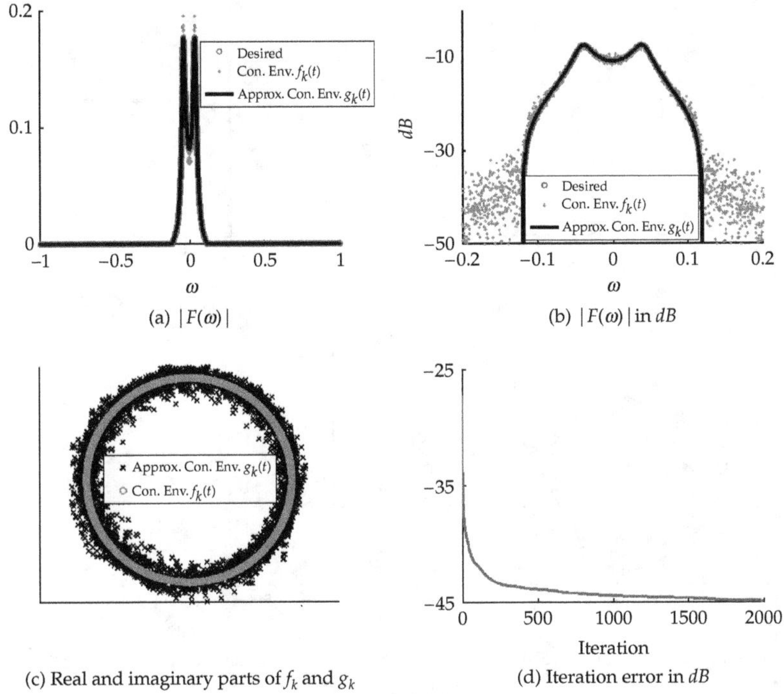

FIGURE 4.9 Constant envelope signal and the approximate constant envelope signal corresponding to point C in Fig. 3.23(b).

Another approach to constant envelope signal design is outlined in Sec. 4.2.

4.2 Unimodular Sequences Using Cyclic Algorithms[4]

In a series of excellent papers, Stoica et al. have recently addressed the problem of designing unimodular sequences with good auto/cross-correlation properties [23]. The problem is to design a single or multiple sequences with constant modulus in the time domain such that their respective matched filter outputs ideally resemble delta functions while the cross-matched filter outputs are zero. Since

[4]© 2010 IEEE. Reprinted with permission, from "Design of Unimodular Sequences using Generalized Receivers" by S. U. Pillai, K. Y. Li, R. Zheng, and B. Himed, *2010 IEEE International Radar Conference*, Washington, DC, USA, May 10–14, 2010.

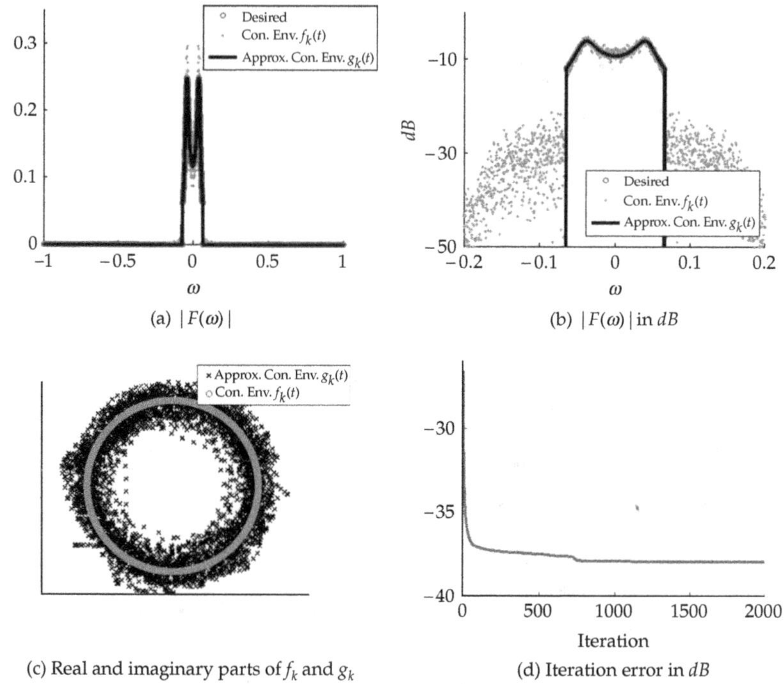

FIGURE 4.10 Constant envelope signal and the approximate constant envelope signal corresponding to point F in Fig. 3.23(b).

the matched filter outputs represent auto and cross-correlations of the original sequences, the problem is to design constant modulus time series whose autocorrelations approximate the delta function, and the cross correlations approximate the all-zero function. In this context CAN (cyclic-algorithm new) and WeCAN (Weighted CAN) have been proposed for designing such sequences with good auto/cross correlation properties [23]. After a brief review of the CAN/WeCAN algorithms, the equivalence of the CAN algorithm and the classic Gerchberg-Saxton algorithm involving the sequential magnitude substitution operations both in the time and frequency domains is demonstrated [9]. In the multichannel case, the CAN/WeCAN algorithms coincide with the Gerchberg-Saxton algorithm in the time domain, and generalize it to the frequency domain in an interesting manner [23–25].

The design of unimodular sequences can be further generalized by considering the receiver design to be something other than the respective matched filters. The receivers need not have the unimodular

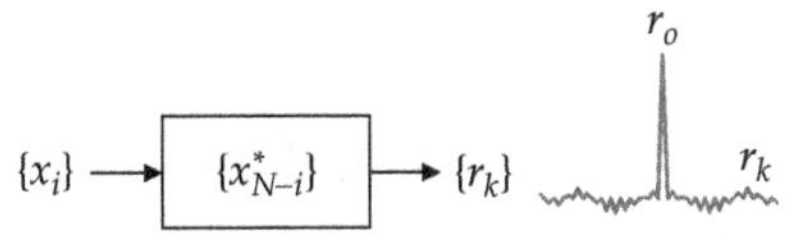

FIGURE 4.11 Matched filter.

property, and instead they can be designed so as to further minimize the sidelobe levels. This approach gives rise to more efficient receiver design that realizes the desired output sequences, and the freedom present at the input can be used to further suppress the sidelobe levels [26,27].

In the single channel case, as Fig. 4.11 shows, the problem is to design a sequence $\{x_i\}_{i=1}^{N}$ such that $|x_i| = 1$ for $i = 1 \to N$, and their matched filter outputs approximate an ideal delta function for pulse compression and sidelobe suppression. Since the matched filter outputs $\{r_k\}_{k=-(N-1)}^{N-1}$ are the autocorrelations of the input sequence, we have

$$r_k = \sum_{i=k+1}^{N} x_i x^*_{i-k} = r^*_{-k}, \quad k = 0, \pm 1, \pm (N-1). \tag{4.46}$$

Hence for pulse compression and sidelobe suppression, ideally we should have a delta function at the output, that is,

$$r_o = 1, \qquad r_k = 0, \quad k \geq 1 \tag{4.47}$$

or in this context, minimization of the total sidelobe error [23]

$$\varepsilon = (r_o - 1)^2 + 2\sum_{k=1}^{N-1} |r_k|^2 \tag{4.48}$$

has been suggested as a suitable criterion. Interestingly, the minimization of the above error is closely related to that of the integrated sidelobe level (*ISL*)

$$ISL = \sum_{k=1}^{N-1} |r_k|^2 \tag{4.49}$$

or the maximization of the merit factor (*MF*)

$$MF = \frac{|r_o|^2}{2ISL} \tag{4.50}$$

proposed in the literature [25,28]. It is easy to relate the error in (4.48) or the *ISL* in (4.49) to the classic discrete-time Fourier transform of the desired sequence $\{x_i\}$ given by

$$X(\omega) = \sum_{i=1}^{N} x_i \, e^{-ji\omega}. \tag{4.51}$$

In fact, using (4.46), the corresponding spectrum $S(\omega)$ is given by

$$S(\omega) = |X(\omega)|^2 = \sum_{k=-(N-1)}^{N-1} r_k \, e^{-jk\omega} \tag{4.52}$$

and hence when the ideal conditions in (4.47) are substituted into (4.52) we obtain the new requirement

$$S(\omega) \equiv 1 \tag{4.53}$$

for all ω in the frequency domain. Equation (4.53) represents the well known result that to achieve excellent pulse compression, the input sequence must exhibit flat spectrum. Since the requirement in (4.53) is impossible to maintain over all frequencies by nontrivial finite sequences, to generate unimodular finite length sequences that satisfy (4.53), the optimization problem

$$\Delta_o = \min_{|x_i|=1} \| S(\omega) - I \|^2 \tag{4.54}$$

is more meaningful. Equations (4.53) and (4.54) also involve the difficult task of minimizing the error over *all* frequencies. However, the more relaxed criterion

$$\Delta_1 = \frac{1}{2N} \sum_{n=1}^{2N} \| S(\omega_n) - I \|^2, \quad \omega_n = \frac{2\pi n}{2N} \tag{4.55}$$

obtained by averaging the spectral error only at $2N$ equally spaced discrete points is more appealing since it is also equal to the total sidelobe error ε described in (4.48). In fact using (4.51) and (4.55), by direct expansion [23]

$$\Delta_1 = \frac{1}{2N} \sum_{n=1}^{2N} \| S(\omega_n) - I \|^2 = \frac{1}{2N} \sum_{n=1}^{2N} \left| \sum_{k=-(N-1)}^{N-1} r_k e^{-jk\omega_n} - 1 \right|^2$$

$$= (r_o - 1)^2 + 2 \sum_{k=1}^{N-1} \| r_k \|^2 = \varepsilon, \tag{4.56}$$

so that minimizing the total sidelobe error in (4.48) is equivalent to minimizing the relaxed spectral error Δ_1 in (4.55). But using (4.52), we also have

$$\Delta_1 = \frac{1}{2N}\sum_{n=1}^{2N}\left(|X(\omega_n)|^2 - 1\right)^2 \tag{4.57}$$

and clearly (4.57) is minimized by letting [23]

$$X(\omega_n) = e^{j\psi_n}, \quad n = 1 \to 2N, \tag{4.58}$$

where ψ_n are arbitrary real constants. From (4.51), $X(\omega_n)$ represents the DFT coefficients at frequencies $\omega_n = 2\pi n/2N$ and constraint (4.58) implements the unity magnitude substitution requirement in the discrete frequency domain as in (4.45) . Using the $2N \times 2N$ DFT matrix **D** whose (i, j)th element is given by

$$D_{i,k} = \frac{1}{\sqrt{2N}} e^{j2\pi ik/2N} \tag{4.59}$$

we can rewrite the $2N$ equations in (4.58) compactly as

$$\mathbf{D}^*\underline{X} = \left[e^{j\psi_1}, e^{j\psi_2}, \ldots e^{j\psi_{2N}}\right]^T \triangleq \underline{V}, \text{ or } \underline{X} = \mathbf{D}\underline{V} \tag{4.60}$$

where the $2N \times 1$ vector $\underline{X}$

$$\underline{X} = [x_1, x_2, \ldots x_N, 0 \cdots 0]^T, \quad |x_i| = 1 \tag{4.61}$$

represents the extended data vector with $x_i, i = 1 \to N$ representing the unimodular entries. Equations (4.60) and (4.61) suggest that for the above minimization problem, the $2N$-point DFT of the constant modulus data vector $\underline{X}$ in (4.61) in the temporal domain should generate another constant modulus vector $\underline{V}$ in the frequency domain and the CAN/WeCAN algorithms perform these operations by employing the magnitude substitution operation, both in the frequency domain and the time domain sequentially until the desired accuracy is achieved [23]. But this is the same as a Gerchberg-Saxton algorithm described in (4.41)–(4.45). Notice that the CAN algorithm in (4.60) and (4.61) resets the second half of the temporal coefficients to zero in (4.61) at every stage. This corresponds to the projection operator P associated with the time-limiting operator [29].

Multichannel Case

The multichannel generalization of the CAN approach also uses the Gerchberg-Saxton algorithm in the time domain and generalizes it to the frequency domain. In fact, the $N \times M$ matrix [23]

$$\mathbf{X} \triangleq \left[\underline{x}_1, \underline{x}_2, \cdots \underline{x}_M\right] \triangleq \begin{bmatrix} \underline{y}_1^T \\ \underline{y}_2^T \\ \vdots \\ \underline{y}_N^T \end{bmatrix} \tag{4.62}$$

contains the M temporal sequences

$$\underline{x}_i = [x_{i,1}, x_{i,2}, \cdots x_{i,N}]^T, \quad i = 1 \to M, \ |x_{i,j}| = 1 \tag{4.63}$$

that represent M orthonormal sequences each with N unimodular samples, and the nth row

$$\underline{y}_n^T = [x_1(n), x_2(n), \cdots x_M(n)] \tag{4.64}$$

in (4.62) corresponds to the various channel outputs at the nth time instant. Under stationary assumptions, cross-correlations of different channel waveforms give rise to the auto/cross-correlation coefficients

$$r_{i,j}(k) = \sum_{n=1}^{N-k} x_i(n) x_j^*(n-k). \tag{4.65}$$

Using (4.64), the $M \times M$ auto/cross-correlation coefficient matrices generated using (4.65) can be compactly expressed also using the $M \times 1$ column vector $\underline{y}_n$ as

$$\mathbf{R}_k = \sum_{n=1}^{N-k} \underline{y}_n \underline{y}_{n-k}^* = (r_{i\,j}(k)), \quad k = 0 \to N-1 \tag{4.66}$$

where $\mathbf{R}_k$ represents the $M \times M$ array output data vector cross-covariance matrix corresponding to lag k. For orthonormal input sequences we must have

$$\mathbf{R}_o = \mathbf{I}, \quad \mathbf{R}_k = 0, \quad k \geq 1 \tag{4.67}$$

as in (4.47). Following (4.49), if we define the total sidelobe error criterion

$$\varepsilon = \|\mathbf{R}_o - \mathbf{I}\|^2 + 2 \sum_{k=1}^{N-1} \|\mathbf{R}_k\|^2, \tag{4.68}$$

then, as before with

$$\underline{Z}(\omega) = \sum_{i=1}^{N} \underline{y}_i \, e^{-ji\omega} \tag{4.69}$$

representing the discrete time Fourier transform of the data, we have the power spectral density matrix

$$\mathbf{S}(\omega) \triangleq \underline{Z}(\omega)\, \underline{Z}^*(\omega) = \sum_{k=-(N-1)}^{N-1} \mathbf{R}_k \, e^{-jk\omega} \tag{4.70}$$

so that once again the desired error criterion in (4.67) can be expressed as $\mathbf{S}(\omega) \equiv \mathbf{I}$. As in (4.54) and (4.55), the relaxed criterion

$$\Delta_1 = \frac{1}{2N} \sum_{k=1}^{2N} \|\mathbf{S}(\omega_k) - \mathbf{I}\|^2 = \varepsilon, \quad \omega_k = \frac{2\pi k}{2N} \tag{4.71}$$

is more useful in this context since it relates the point power spectral error with the total sidelobe error in (4.68). Expanding (4.71) by using (4.70), we also obtain [23]

$$\varepsilon = \frac{1}{2N} \sum_{k=1}^{2N} \left(\|\underline{Z}(\omega_k)\|^2 - 1\right)^2 + M - 1 \tag{4.72}$$

so that minimization of ε is achieved by setting the DFT vectors

$$\underline{Z}(\omega_k) = \underline{\alpha}_k, \quad \|\underline{\alpha}_k\|^2 = 1, \quad k = 1 \to 2N \tag{4.73}$$

in (4.72). Notice that $\underline{\alpha}_k$ in (4.73) are the $2N$-point DFT vectors, and their normalization requirement is a magnitude constraint in the frequency domain. Once again, Eq. (4.73) can be compactly expressed as

$$\begin{bmatrix} \underline{Z}^T(\omega_1) \\ \underline{Z}^T(\omega_2) \\ \vdots \\ \underline{Z}^T(\omega_{2N}) \end{bmatrix} = \mathbf{D} \begin{bmatrix} \mathbf{X} \\ \hline \mathbf{0} \end{bmatrix} = \begin{bmatrix} \underline{\alpha}_1^T \\ \underline{\alpha}_2^T \\ \vdots \\ \underline{\alpha}_{2N}^T \end{bmatrix} = \mathbf{V}, \quad |x_{i\,j}| = 1, \quad \|\underline{\alpha}_k\|^2 = 1 \tag{4.74}$$

where $\mathbf{D}$ is the $2N \times 2N$ DFT matrix defined in (4.59), and $\mathbf{X}$ is defined as in (4.62) with $|x_{i,j}| = 1$. Observe that the normalization of $\underline{\alpha}_k$ in the frequency domain in (4.73) and (4.74) is a generalization of the scalar version of the Gerchberg-Saxton algorithm since

$$\|\underline{\alpha}_k\|^2 = \sum_{m=1}^{M} |Z_m(\omega_k)|^2 = M(\omega_k) \tag{4.75}$$

represents the sum of the magnitudes of the various channel components of the DFT coefficients $\underline{Z}(\omega_k)$ as defined in (4.69) corresponding to the discrete frequency $\omega_k = \frac{2\pi k}{2N}$. Following (4.73) and (4.74), the multichannel version of the CAN algorithm performs the magnitude substitution operation sequentially both in the frequency domain ($\|\underline{\alpha}_k\|^2 = 1$) and in the time domain ($|x_{ij}| = 1$) as in the Gerchberg-Saxton algorithm in (4.45).

Figure 4.12 shows the matched filter receiver outputs of a 10-channel unimodular sequence of length 200 generated using the CAN algorithm in (4.73) and (4.74). Both Fig. 4.12(a) and (b) contain 10 autocorrelation sequences and 90 cross-correlation sequences. Notice that the auto/cross-correlation terms are at about 20 dB below the peak term. The sidelobe levels here can be further improved by considering various weighting sequences as well [23].

4.2.1 General Unimodular Sequences

The design of unimodular sequences can be further generalized by considering the receiver design to be independent from the actual matched filter sequences. This approach is useful in many applications including radar, where the receiver does not have to be unimodular and can be something other than the matched filter sequence. In this context, the problem is to design M input unimodular sequences $\underline{x}_i, i = 1 \rightarrow M, |x_{i,j}| = 1$ each of length N as in (4.62), and their corresponding receivers

$$\underline{h}_i = [h_{i,1}, h_{i,2}, \ldots h_{i,K}], \quad i = 1 \rightarrow M \tag{4.76}$$

such that the receiver outputs should satisfy the "desired" output sequence properties. Thus with

$$z_{i,m}(n) = \sum_{k=1}^{N} x_i(k)\, h_m(n-k+1), \quad n = 1 \rightarrow N+K-1 \tag{4.77}$$

representing the output at the mth receiver due to the ith sequence at time instant n, then ideally, we require (Fig. 4.13)

$$z_{i,m}(n) = \begin{cases} \delta_{i,m}(n-n_o), & i = m \\ 0, & i \neq m \end{cases} \tag{4.78}$$

where n_o corresponds to the desired peak term at the output. In general, the receiver length K is a free parameter left to other design considerations.

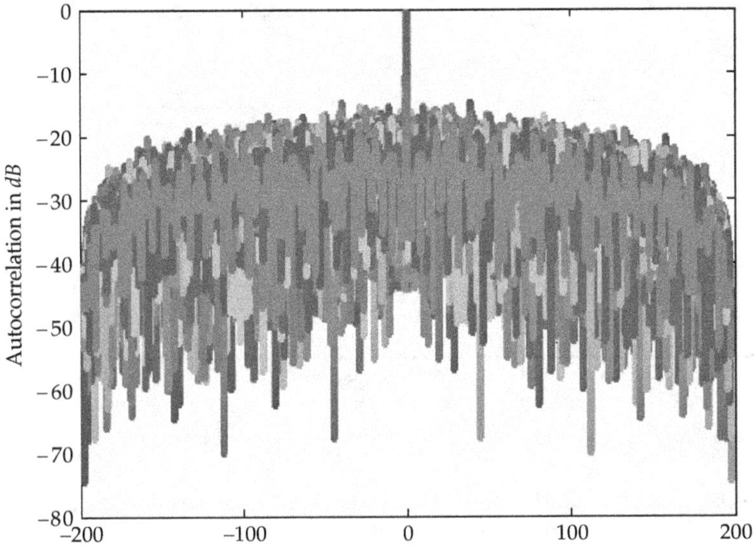

(a) Front view

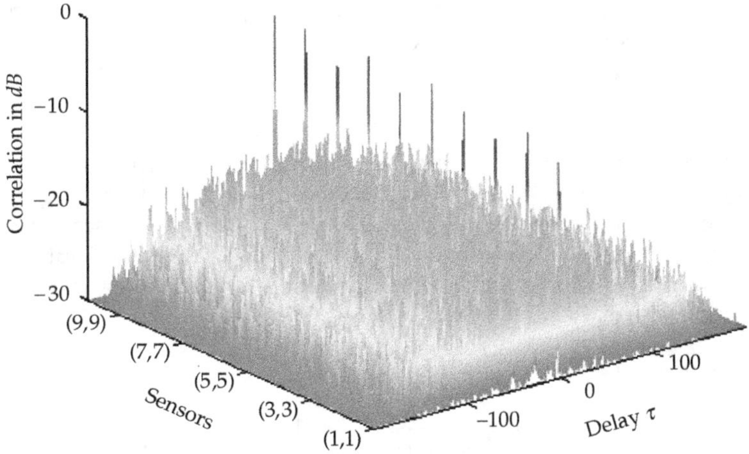

(b) Side view

FIGURE 4.12 A 10-channel unimodular sequence design using the multichannel CAN algorithm in (4.73) and (4.74). Each figure contains 10 autocorrelation sequences and 90 cross-correlation sequences.

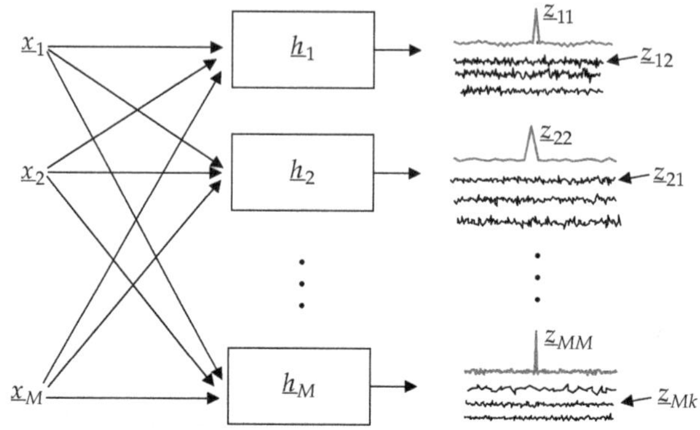

FIGURE 4.13 Generalized unimodular sequences.

One approach in this context is to use a criterion similar to the merit factor (*MF*) in (4.50). Thus if we define

$$\eta_i = \frac{|z_{ii}(n_o)|^2}{\sum\limits_{n \neq n_o} |z_{ii}(n)|^2 + \sum\limits_{j \neq i} \sum\limits_{n=1}^{N+K-1} |z_{ij}(n)|^2}, \quad i = 1 \rightarrow M \tag{4.79}$$

to represent the ratio of the receiver output peak value power to the total undesired sidelobe power at each receiver output, then the receiver design problem is to maximize η_i, $i = 1 \rightarrow M$. Observe that the criterion in (4.79) is well known and it is the same as the approach behind the classical matched filter. Equation (4.79) can be expressed in matrix form as

$$\eta_i = \frac{|\underline{b}_{n_o} \underline{h}_i|^2}{\underline{h}_i^* \mathbf{A}_i \underline{h}_i}, \quad i = 1 \rightarrow M \tag{4.80}$$

where

$$\mathbf{A}_i = \mathbf{F}_{i,0}^* \mathbf{F}_{i,0} + \sum \mathbf{F}_j^* \mathbf{F}_j > 0 \tag{4.81}$$

represents a $K \times K$ positive-definite matrix, and $\mathbf{F}_i$ represents the $(N + K - 1) \times K$ matrix

$$\mathbf{F}_i = \begin{pmatrix} x_i(1) & 0 & \cdots & 0 & 0 \\ x_i(2) & x_i(1) & \ddots & \vdots & 0 \\ \vdots & x_i(2) & \ddots & 0 & \vdots \\ x_i(N) & \vdots & \ddots & x_i(1) & 0 \\ 0 & x_i(N) & \ddots & \vdots & x_i(1) \\ 0 & 0 & \ddots & x_i(N-1) & \vdots \\ \vdots & \vdots & \vdots & x_i(N) & x_i(N-1) \\ 0 & 0 & 0 & 0 & x_i(N) \end{pmatrix} \tag{4.82}$$

generated from the ith input sequence $\underline{x}_i$. In (4.81), $\mathbf{F}_{i,0}$ represents the $(N + K - 2) \times K$ matrix generated from $\mathbf{F}_i$ with its n_o row deleted. Finally in (4.80), $\underline{b}_{n_o}$ represents the row vector corresponding to the n_o row of $\mathbf{F}_i$. Thus in general

$$\underline{b}_{n_o} = [0, \cdots 0, \; x_i(N), \; x_i(N-1), \; \cdots \; x_i(1), \; 0, \; \cdots \; 0] \tag{4.83}$$

for $n_o > N$. The number of zeros in (4.83) depends on the value of n_o with respect to N. For example, with $n_o = N$ and $K = N$, the vector $\underline{b}_{n_o}$ is the same as the flipped version of $\underline{x}_i$ in (4.64). The Schwarz inequality applied to (4.80) gives the optimum ith receiver filter to be

$$\underline{h}_i = \mathbf{A}_i^{-1} \underline{b}_{n_o}^*. \tag{4.84}$$

Observe that the receivers $\underline{h}_i$ in (4.84) are obtained noniteratively and the freedom present at the input makes the unimodular constraint trivial to implement at the input. Although each receiver design depends on *all* input sequences through the matrix $\mathbf{A}_i$ in (4.81), they can be determined in an uncoupled manner as in (4.84). Furthermore, the explicit design for the receiver allows freedom in selecting the output instant n_o at which the peak should be observed. This freedom allows the various peak outputs to be time aligned, a desired property in some radar applications.

Figure 4.14 shows the receiver output using the new design approach in (4.84) for a 10-channel set of randomly generated

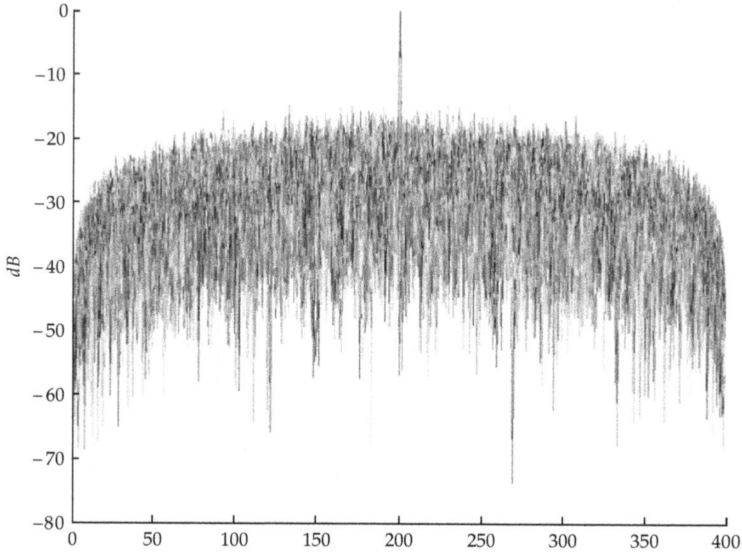

(a) Front view

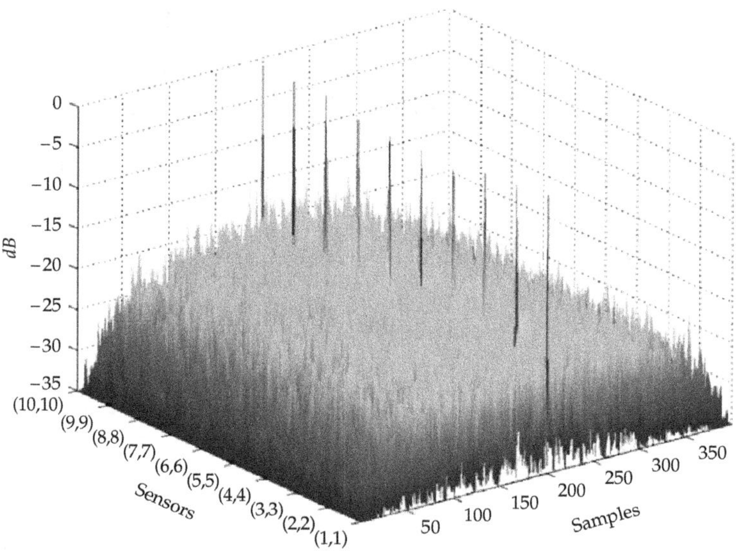

(b) Side view

FIGURE 4.14 A 10-channel unimodular sequence auto and cross terms at the output of the sidelobe minimizer receiver in (4.84). (a) Front view, (b) side view.

unimodular sequences of length 200. For comparison purposes, here the receiver length is set equal to 200. Notice that the auto/cross side-lobe terms are about 20 dB below the peak value, similar to that in Fig. 4.12. This represents performance similar to the CAN sequences obtained using (4.73) and (4.74). In this new approach, additional optimization can be carried out at the input to further reduce the sidelobe level and for compression in the Doppler domain.

4.3 Constant Envelope Signals with Prescribed Discrete Fourier Transform Magnitude[5]

Consider the problem of determining a constant envelope signal $f(t)$ whose Fourier transform magnitude function $|F(\omega)|$ matches a given magnitude function $P(\omega)$ (reasonably accurately). If the constant envelope property is restricted to be only over the interval of interest $0 < t < T$, then the problem reduces to determining $f(t)$ of the form

$$f(t) = Ae^{j\psi(t)}, \quad 0 < t < T \tag{4.85}$$

whose transform magnitude matches the given nonnegative function $P(\omega)$ in some accurate manner. Here A is a suitable constant to be determined, and $\psi(t)$ is the unknown phase function to be determined.

The finite duration property in (4.85) can be exploited by considering the Fourier series representation of $f(t)$ in the interval $(0, T)$. From (4.85), we get

$$f(t) = Ae^{j\psi(t)} = \sum_{n=-\infty}^{+\infty} A_n\, e^{jn\omega_o t}, \quad 0 < t < T \tag{4.86}$$

where

$$\omega_o = \frac{2\pi}{T} \tag{4.87}$$

and

$$A_n = \frac{1}{T}\int_0^T f(t)\, e^{-jn\omega_o t} dt = F(n\omega_o) = \frac{A}{T}\int_0^T e^{j(\psi(t)-n\omega_o t)} dt. \tag{4.88}$$

From (4.88) since $A_n = F(n\omega_o) = |F(n\omega_o)|e^{j\theta_n}$, a necessary condition for the Fourier transform magnitude $|F(\omega)|$ of $f(t)$ in (4.85) to match

[5]© 2011 IEEE. Reprinted with permission, from "Constant Envelope Signals with Prescribed Discrete Fourier Transform magnitude" by S. U. Pillai, K. Y. Li, B. Himed, *2011 IEEE Radar Conference*, Kansas City, USA, May 23–27, 2011.

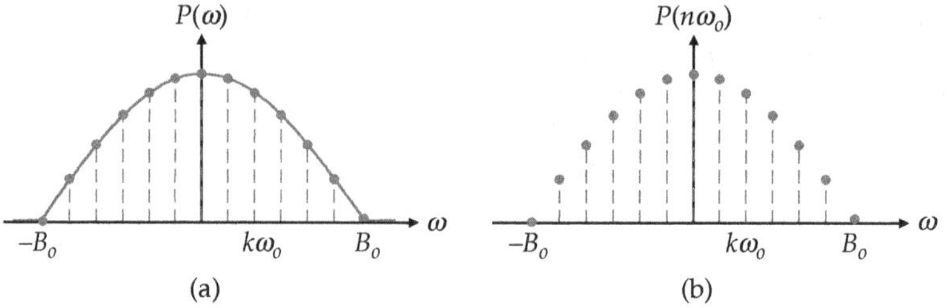

FIGURE 4.15 Matches in the frequency domain.

the given magnitude function $P(\omega)$ in Fig. 4.15(a) is that the Fourier coefficients A_n in (4.88) must have the form

$$A_n = P(n\omega_o)e^{j\theta_n} \tag{4.89}$$

for some suitable phase θ_n as shown in Fig. 4.15(b), where

$$p_k = P(k\omega_o), \tag{4.90}$$

the given magnitude function.

Interestingly, it is possible to realize this requirement by following a procedure similar to that of reference [28]. Toward this, define the unknown phase function $\psi(t)$ in (4.85) to be piece-wise linear as

$$\psi(t) = \theta_k + k\omega_o t, \quad 0 < t_{k-1} \le t < t_k \le T, \quad k = -N \to N \tag{4.91}$$

where the time instants t_k, $k = -N \to N$ as well as N need to be determined such that $t_{-N-1} = 0$ and $t_N = T$. Using (4.91) in (4.88), we obtain the Fourier coefficients to be

$$\begin{aligned} A_n &= \frac{A}{T}\sum_{k=-N}^{N} \int_{t_{k-1}}^{t_k} e^{j(\theta_k + k\omega_o t - n\omega_o t)}dt \\ &= \frac{A}{T}e^{j\theta_n}\int_{t_{n-1}}^{t_n} dt + \frac{A}{T}\sum_{\substack{k=-N \\ k\neq n}}^{N} e^{j\theta_k}\int_{t_{k-1}}^{t_k} e^{j(k-n)\omega_o t}dt \\ &= (t_n - t_{n-1})\frac{A}{T}e^{j\theta_n} + \frac{A}{T}\sum_{\substack{k=-N \\ k\neq n}}^{N} e^{j\theta_k}\left.\frac{e^{j(k-n)\omega_o t}}{j(k-n)\omega_o}\right|_{t_{k-1}}^{t_k}. \end{aligned} \tag{4.92}$$

Following Schroeder [28], suppose we choose the unknown time intervals in (4.91) so as to satisfy the identity[6]

$$(t_n - t_{n-1})\frac{A}{T} = P(n\omega_o) = p_n \tag{4.93}$$

then $t_{-N} = \frac{T}{A}P(-N\omega_o)$ and (4.92) reads

$$A_n = P(n\omega_o)e^{j\theta_n} + \underbrace{A\sum_{\substack{k=-N\\k\neq n}}^{N} e^{j\theta_k}\frac{\left(e^{j(k-n)\omega_o t_n} - e^{j(k-n)\omega_o t_{n-1}}\right)}{j(k-n)\omega_o}}_{\simeq 0,\ \text{as } N\to\infty}, \quad n = 0,\ \pm 1,\ \pm 2,\ \pm N. \tag{4.94}$$

As $N \to \infty$, the summation term in (4.94) tends to be small so that

$$A_n \to P(n\omega_o)\,e^{j\theta_n}, \quad n = 0,\ \pm 1,\ \ldots,\ \pm N \tag{4.95}$$

satisfying the requirement in (4.89). The desired phase components θ_n can be determined by requiring that $\psi(t)$ in (4.91) be continuous. This gives

$$\theta_{n-1} + (n-1)\omega_o t_{n-1} = \theta_n + n\omega_o t_{n-1} \tag{4.96}$$

or

$$\theta_n = \theta_{n-1} - \omega_o t_{n-1} = \theta_{-N} - \omega_o \sum_{k=-N}^{n-1} t_k, \quad -N \le n \le N. \tag{4.97}$$

But from (4.93)

$$t_k = t_{k-1} + \frac{T}{A}p_k = \frac{T}{A}\sum_{i=-N}^{k} p_i, \quad k = -N \to N \tag{4.98}$$

so that with (4.98) in (4.97) we obtain the unknown phase terms to be

$$\theta_n = \theta_{-N} - \frac{T\omega_o}{A}\sum_{k=-N}^{n-1}\sum_{i=-N}^{k} p_i = \theta_{-N} - \frac{2\pi}{A}\sum_{i=-N}^{n-1}(n-i)p_i. \tag{4.99}$$

[6]In Schroeder's approach [28], the time intervals in (4.93) are proportional to the relative power $|P(n\omega_o)|^2$.

It is easy to determine the unknown A from (4.98). Since $t_N = T$, we get

$$t_N = T = \frac{T}{A} \sum_{m=-N}^{N} p_m \tag{4.100}$$

or

$$A = \sum_{m=-N}^{N} p_m \tag{4.101}$$

and substituting (4.101) into (4.99), we get

$$\theta_n = \theta_{-N} - 2\pi \frac{\sum_{i=-N}^{n-1} (n-i) p_i}{\sum_{m=-N}^{N} p_m}, \quad n = -N \to N \tag{4.102}$$

where θ_{-N} is arbitrary. From (4.91) and (4.98), the desired phase function $\psi(t)$ is piece-wise linear and monotonically increasing, where the total time interval is split into N nonuniform segments $0 < t_{-N} < t_{-N+1} < \cdots < t_{-1} < t_o < t_1 < \cdots < t_N = T$ according to (4.98) as shown in Fig. 4.16(b).

From (4.98) and (4.101), since

$$t_k = t_{k-1} + \frac{T}{\sum_{m=-N}^{N} p_m} p_k = T \frac{\sum_{i=-N}^{k} p_i}{\sum_{m=-N}^{N} p_m} = T \frac{\sum_{i=-N}^{k} P(i\omega_o)}{\sum_{m=-N}^{N} P(m\omega_o)}, \tag{4.103}$$

$$k = -N, \cdots -1, 0, 1, \cdots N,$$

the last time segment is obtained by adding a scaled version of the given Fourier transform magnitude $P(k\omega_o)$ to the previous time segment (Figs. 4.16 and 4.17).

The $f(t)$ in (4.85) generated using (4.91) and (4.102) clearly has a constant envelope and its nth Fourier coefficient has the form

$$A_n \simeq P(n\omega_o) \exp\left(j \left(\theta_{-N} - 2\pi \frac{\sum_{i=-N}^{n-1} (n-i) P(i\omega_o)}{\sum_{i=-N}^{N} P(i\omega_o)} \right) \right), \tag{4.104}$$

$$i = 0, \pm 1, \pm 2, \cdots \pm N.$$

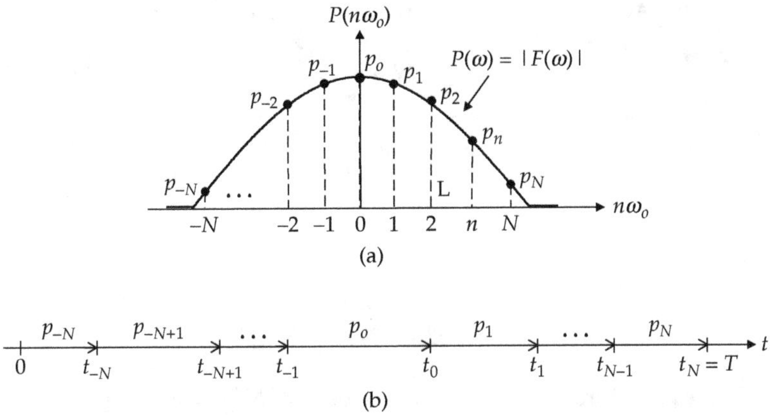

FIGURE 4.16 Given Fourier transform magnitude and various time segments.

Clearly these Fourier coefficients match the given magnitude transform $P(n\omega_o)$ approximately at discrete frequency points $\pm n\omega_o$, $n = 0, 1, 2, \cdots N$, as $N \rightarrow \infty$. The unknown number $2N+1$ needs to be sufficiently high so as to make the approximation in (4.104) reasonably accurate. The piece-wise linear phase function in Fig. 4.17 can be viewed as an approximation to a chirp signal (quadratic phase) while matching the given magnitude transform at the discrete frequency points.

Finally, the duration T of the desired signal itself can be estimated from the number of matches obtained in (4.104). Since there are $2N+1$ matches in (4.104) as shown in Fig. 4.15, if the transform of $f(t)$ is

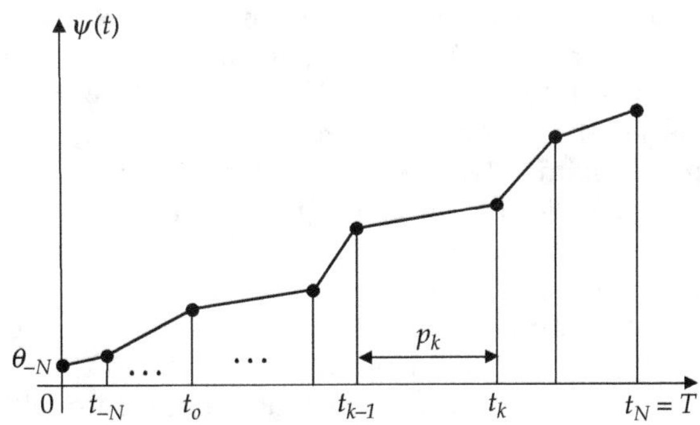

FIGURE 4.17 Phase function.

known to have a two-sided bandwidth $2B_o$, then we must have

$$2B_o \simeq (2N+1)\omega_o \tag{4.105}$$

or

$$\omega_o = \frac{2B_o}{2N+1} \simeq \frac{B_o}{N}. \tag{4.106}$$

Since

$$\omega_o = \frac{2\pi}{T}, \tag{4.107}$$

this gives the duration of the signal to be $T = N\frac{2\pi}{B_o} = N\Delta T$ where ΔT represents the sampling interval. Hence if $2N+1$ discrete frequency points matchings are desired in (4.104), then the constant envelope signal $f(t)$ in (4.85) needs to be at least $N\Delta T$ long.

In summary, given the Fourier transform magnitude transform $P(\omega)$ over $|\omega| < B_o$, the constant envelope signal

$$f(t) = Ae^{j\psi(t)}, \quad 0 < t < T \tag{4.108}$$

with monotonically increasing piece-wise linear phase function

$$\psi(t) = \theta_n + n\omega_o t, \quad t_{n-1} < t < t_n \leq T, \; n = 0, \; \pm1, \; \pm2, \; \cdots \pm N \tag{4.109}$$

where θ_n, t_n are given by (4.102) and (4.103), also satisfies the truncated Fourier series expansion

$$g(t) \simeq \sum_{n=-N}^{N} P(n\omega_o)\, e^{j(n\omega_o t+\theta_n)}, \quad 0 \leq t \leq T \tag{4.110}$$

for sufficiently large N [see (4.86)]. In that case $\omega_o = B_o/N$ and the signal duration $T = 2\pi/\omega_o = N2\pi/B_o$. Interestingly for any finite N, the function $f(t)$ in (4.108) and the function $g(t)$ in (4.110) are indeed different. While $f(t)$ has a constant envelope, the periodic function $g(t)$ has only an approximately constant envelope for finite N, and by design, its Fourier coefficients match the given Fourier transform magnitude values $P(n\omega_o)$, $n = 0, \pm1, \cdots \pm N$ exactly. The Fourier coefficients of $f(t)$ on the other hand, only approximately match the given transform coefficients. Figure 4.18 illustrates these effects for a given non-symmetric Fourier transform magnitude function $P(\omega)$ in Fig. 4.18(a) with a 63-point match in the frequency domain. Figure 4.18(c) shows the corresponding phase function $\psi(t)$, and Fig. 4.18(h) and (i) show the matched filter outputs corresponding to waveforms $f(t)$ and $g(t)$ in Fig. 4.18(d) and (g).

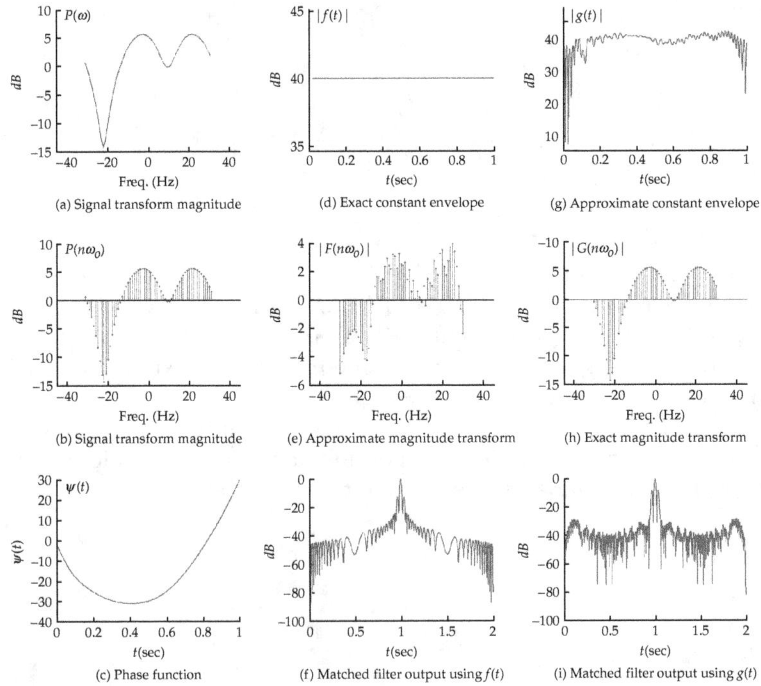

FIGURE 4.18 Nonsymmetric Fourier transform function and the corresponding constant envelope and approximately constant envelope signals.

In particular if $|F(\omega)|$ is symmetric, by selecting $\theta_{-n} = \theta_n$ in (4.109) and (4.110), we get

$$g(t) = p_o e^{j\theta_o} + 2\sum_{n=1}^{N} p_n \cos(n\omega_o t + \theta_n) \tag{4.111}$$

so that it is possible to write (4.108) as

$$f(t) = A\cos\psi(t), \quad 0 < t < T, \tag{4.112}$$

a real function. Proceeding similar to the complex case with

$$\psi(t) = \theta_n + n\omega_o t, \quad t_{n-1} \le t < t_n, \quad n = 0, 1, 2, \cdots N \tag{4.113}$$

it is easy to show that in this case

$$t_k = T\frac{\sum\limits_{i=0}^{k} P(i\omega_o)}{\sum\limits_{m=0}^{N} P(m\omega_o)} \tag{4.114}$$

and

$$\theta_n = \theta_o - 2\pi \frac{\sum_{i=0}^{n-1} (n-i)p_i}{\sum_{m=0}^{N} p_i}, \qquad n = 0, 1, 2 \cdots N. \tag{4.115}$$

From (4.111), as long as θ_o is either 0 or π, $g(t)$ there represents a real signal that matches the given Fourier transform coefficients. Similarly the signal $f(t)$ in (4.112) with $\psi(t)$ as in (4.113)–(4.115) refers to a "cosine-type" signal whose Fourier coefficients approximately match the given coefficients $P(n\omega_o)$, $n = 0, 1, 2, \cdots N$.

Figures 4.19 and 4.20 show two symmetric Fourier transform magnitude functions $P(\omega)$ with a 63-point match using $\theta_o = 0$ along with their phase functions, and respective matched filter outputs.

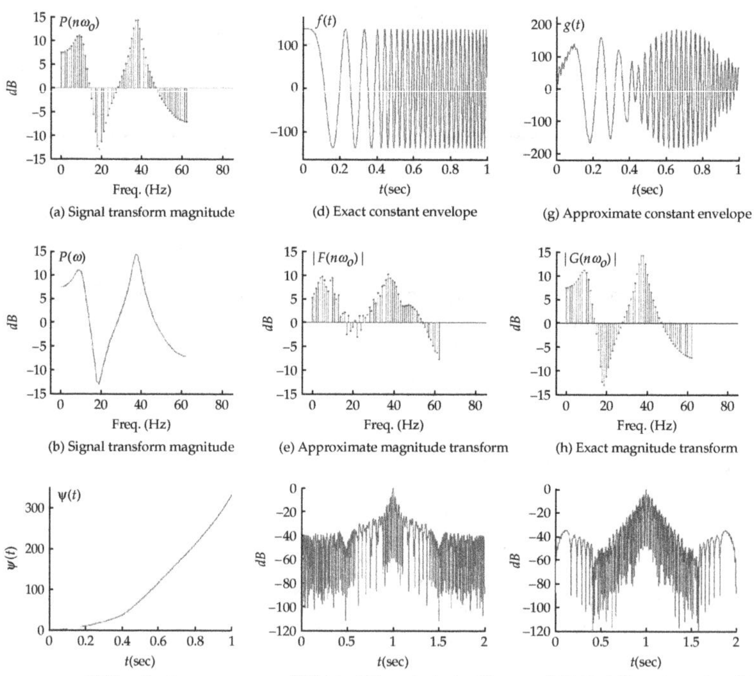

FIGURE 4.19 Symmetric Fourier transform function and corresponding constant envelope signal.

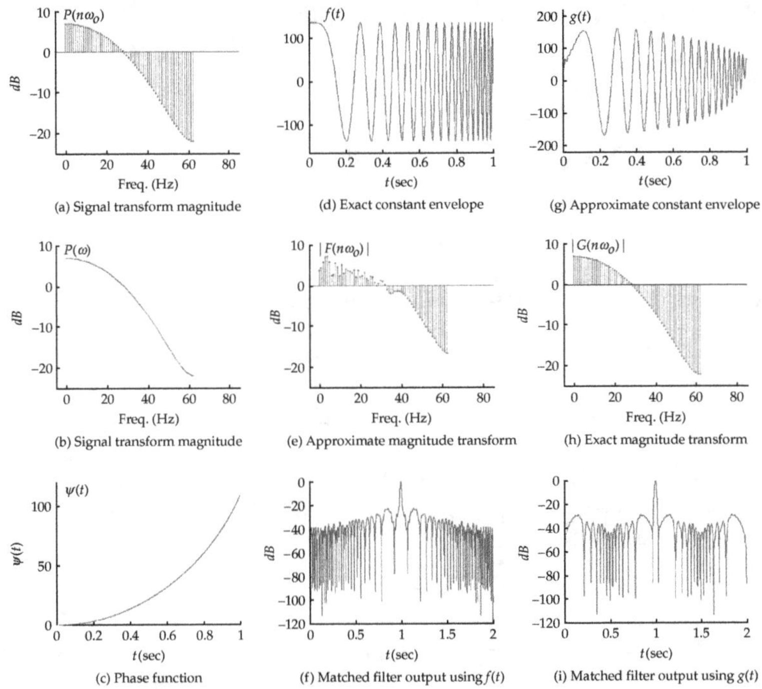

FIGURE 4.20 Symmetric Fourier transform function and corresponding constant envelope signal.

Lindenfeld's Method

Lindenfeld's approach to the design of finite-duration constant-modulus waveforms involves a set of N_b "forbidden bands," $(\omega_{k_1}, \omega_{k_2})$, with $1 \leq k \leq N_b$ and the waveform is designed so that its discrete time Fourier transform (DTFT) is concentrated in the "allowed" bands and not in the forbidden bands [30].

To design the waveform, it is proposed to find a phase function $\theta(n)$ that minimizes the cost function

$$J(\theta) = \sum_{k=1}^{N_b} \int_{\omega_{k_1}}^{\omega_{k_2}} |F(\omega)|^2 \, d\omega \tag{4.116}$$

where $F(\omega)$ represents the transfer function of the waveform. The cost function J can be written as

$$J = f^* \mathbf{R} f \tag{4.117}$$

where $\mathbf{R}$ is an $N \times N$ Toeplitz matrix, with

$$\mathbf{R}(m, n) = \begin{cases} \displaystyle\sum_{k=1}^{N_b} \frac{e^{j(m-n)\omega_{k2}} - e^{j(m-n)\omega_{k1}}}{j(m-n)}, & m \neq n \\ \omega_{k_2} - \omega_{k_1} & m = n. \end{cases} \tag{4.118}$$

An iterative gradient descent algorithm is given by

$$\theta_{i+1} = \theta_i - \mu \frac{\partial J}{\partial \theta_i} \tag{4.119}$$

where i is the iteration index and μ is a small positive number. The gradient is given by

$$\frac{\partial J}{\partial \theta} = 2\mathrm{Im}\{\bar{f} \odot \mathbf{R} f\} \tag{4.120}$$

where $\odot$ represents component-wise Schur multiplication and $\bar{f}$ is the complex conjugate of f. The update equation has the form

$$\theta_{i+1} = \theta_i - 2\mu\mathrm{Im}\{\bar{f} \odot \mathbf{R} f\}. \tag{4.121}$$

The above iterative algorithm can be initialized by setting $\theta(n)$ to any random value between 0 and 2π. Figure 4.21(a) shows the frequency magnitude transform of a constant modulus signal obtained using the Lindenfeld's method and for the purpose of comparison, the projection based signal g_k in (4.41) and (4.42) using the iterative method described there. Although the specifications are set for −80 dB notches, the constant modulus waveforms are able to achieve about −20 dB notches in the frequency domain in both designs. Notice that the projection-based waveform [Fig. 4.21(b)] exhibits somewhat better features over the passband. The matched filter output corresponding to the constant envelope waveform in each case is also shown in Fig. 4.21(c) and Fig. 4.21(d) respectively.

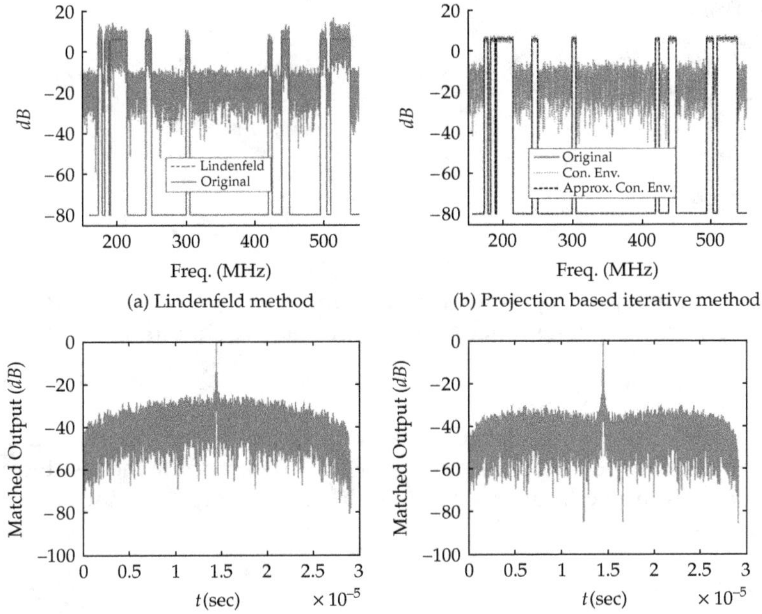

(a) Lindenfeld method (b) Projection based iterative method

(c) Matched filter output (d) Matched filter output

FIGURE 4.21 Constant envelope signals using Lindenfeld's method and the projection-based iterative method in (4.41) and (4.42).

4.4 Pulse Compression and Huffman Code

Consider the FIR sequence $\{h_k\}_{k=0}^{N}$ with transform

$$H(z) = \sum_{k=0}^{N} h_k z^{-k} \tag{4.122}$$

whose pulse compression properties are given by the matched filter output $R(z)$ in Fig. 4.22.

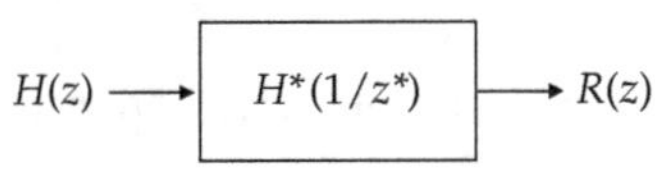

FIGURE 4.22 Matched filter output.

Thus

$$R(z) = H(z)H^*(1/z^*) \triangleq \sum_{k=-N}^{N} r_k z^{-k} \tag{4.123}$$

where

$$r_k = \sum_{i=0}^{N} h_{k+i}\, h_i^*, \quad k = 0, \pm 1, \cdots \pm N. \tag{4.124}$$

For optimum pulse compression, the output in Fig. 4.22 should ideally correspond to a delta function, in which case we should have

$$r_k = 0, \quad |k| \geq 1 \tag{4.125}$$

in (4.124). Since no nontrivial finite sequence $\{h_k\}_{k=0}^{N}$ can satisfy (4.125), the next best possibility is to require (4.125) except for the last term r_N, which should be as small as possible. Although a trivial sequence such as

$$H_o(z) = z^{-N} - a \tag{4.126}$$

will satisfy the above autocorrelation requirement, (4.126) on the other hand represents an undesirable transmit signal from a practical point of view, since all of its energy is concentrated at the two extreme positions. In this context, the desirable signal should distribute energy equally among its samples, approximating the constant envelope property, while satisfying the pulse compression property.

Huffman showed that it is indeed possible to select nontrivial sequences $\{h_k\}_{k=0}^{N}$ such that its autocorrelation sequence has nonzero terms only at the origin as well as at the last term, provided the zeros of its z-transform $H(z)$ are placed at equal angular intervals of $\frac{2\pi}{N}$ with each zero located either on a circle of radius r or a circle of radius $1/r$ with $r < 1$ as shown in Fig. 4.23 [31,32].

To see this, let

$$\theta_k = \frac{2\pi k}{N}, \quad k = 1, 2, \cdots N \tag{4.127}$$

represent the N equidistant zeros in the z-domain and select two radii r and $1/r$ with $r < 1$. Select *any* K zeros from (4.127) to associate with radius r and the remaining zeros correspond to radius $1/r$, and consider the transfer function $H(z)$ generated from these zeros. Thus

$$H(z) = z^{-N} \prod_{k=1}^{K} (z - re^{j\theta_{i_k}}) \prod_{m>K} \left(z - \frac{1}{r} e^{j\theta_{i_m}} \right). \tag{4.128}$$

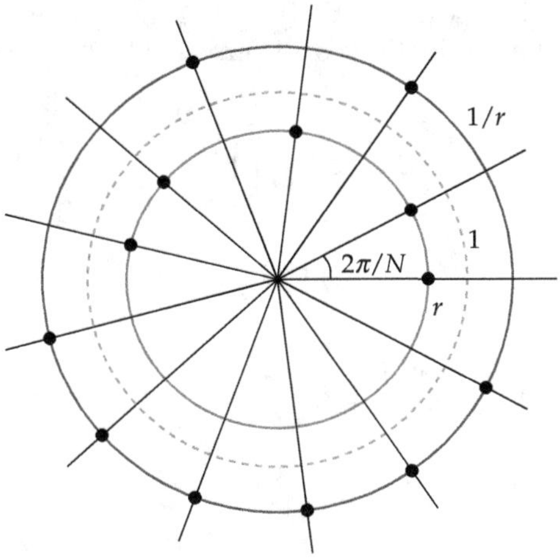

FIGURE 4.23 Zeros of the Huffman signals lie at equal intervals either on a circle of radius r or a circle of radius $1/r$.

Thus

$$
\begin{aligned}
H_*(z) &\triangleq H^*(1/z^*) = z^N \prod_{k=1}^{K} (z^{-1} - re^{-j\theta_{i_k}}) \prod_{m>K} \left(z^{-1} - \frac{1}{r} e^{-j\theta_{i_m}} \right) \\
&= \prod_{k=1}^{K} (1 - rze^{-j\theta_{i_k}}) \prod_{m>K} \left(1 - \frac{z}{r} e^{-j\theta_{i_m}} \right) \\
&= \left(\frac{1}{r} \right)^{N-K} r^K e^{-j\sum_{k=1}^{N} \theta_{i_k}} \prod_{k=1}^{K} \left(\frac{1}{r} e^{j\theta_{i_k}} - z \right) \prod_{m>K} (re^{j\theta_{i_m}} - z).
\end{aligned}
\tag{4.129}
$$

But

$$
\sum_{k=1}^{N} \theta_{i_k} = \frac{2\pi}{N} (1 + 2 + \cdots + N) = \pi(N+1) \tag{4.130}
$$

so that

$$
e^{-j\sum_{k=1}^{N} \theta_{i_k}} = e^{-j\pi(N+1)} = (-1)^{N+1} \tag{4.131}
$$

and substituting this into (4.129) we get

$$
H_*(z) = -\left(\frac{1}{r} \right)^{N-2K} \prod_{k>K} (z - re^{j\theta_{i_k}}) \prod_{m=1}^{K} \left(z - \frac{1}{r} e^{j\theta_{i_m}} \right). \tag{4.132}
$$

Multiplying (4.128)–(4.132) and observing that the $\{\theta_{i_k}\}$ set as well as the $\{\theta_{i_m}\}$ set in the product gives all zeros in (4.127), we get

$$
\begin{aligned}
R(z) = H(z)H_*(z) &= -\frac{1}{r^{N-2K}} z^{-N} \prod_{k=1}^{N} (z - re^{j\theta_{i_k}}) \prod_{k=1}^{N} \left(z - \frac{1}{r} e^{j\theta_{i_k}}\right) \\
&= -\frac{1}{r^{N-2K}} z^{-N} (z^N - r^N)\left(z^N - \frac{1}{r^N}\right) \\
&= \frac{1}{r^{N-2K}} (z^N - r^N)\left(\frac{z^{-N}}{r^N} - 1\right) \\
&= \frac{1}{r^{N-2K}} \left(r^N + \frac{1}{r^N} - z^N - z^{-N}\right) \\
&= \frac{1+r^{2N}}{r^{2(N-K)}} \left(1 - \frac{r^N}{1+r^{2N}}(z^N + z^{-N})\right) \\
&= r_o + r_N(z^N + z^{-N})
\end{aligned} \tag{4.133}
$$

where

$$r_o = \frac{1+r^{2N}}{r^{2(N-K)}} \tag{4.134}$$

and

$$r_N = \frac{-r^N}{1+r^{2N}}, \quad 0 < r < 1. \tag{4.135}$$

Notice that $r_N \to 0$ as $N \to \infty$. From (4.133), the power spectrum of any Huffman signal has the form

$$S(\omega) = r_o + 2r_N \cos N\omega T \tag{4.136}$$

and it is essentially flat with a small ripple term due to the second term. Expanding (4.128), the Huffman signal $H(z) = \sum_{k=0}^{N} h_k z^{-k}$ in general has nonzero terms at every delay; however, their time domain envelope can fluctuate wildly as shown in Table 4.1 for a sequence of length 13.

Interestingly, starting with the Huffman signal, we can make use of the projection-like substitution algorithms discussed in Sec. 4.1 to generate a closely related new signal f_k having constant envelope or an approximately constant envelope signal g_k.

The original signal in Table 4.1, the result of iterations f_k and g_k after 2000 iterations using (4.39)–(4.42), are plotted in Fig. 4.24(a)–(c).

Amplitude	Phase
1.0000	0
1.0872	−120
0.5911	120
0.7846	0
0.6784	−120
0.9029	−60
1.0748	180
0.9029	−120
0.6784	120
0.7846	180
0.5911	−120
1.0872	−60
1.0000	0

TABLE 4.1 Huffman Sequence of Length 13

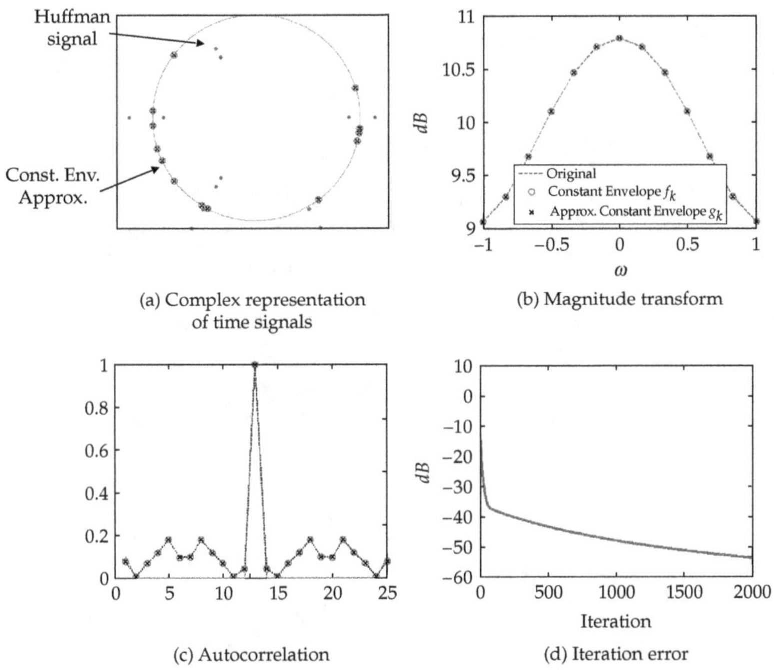

FIGURE 4.24 The Huffman signal of length 13 and its approximations using (4.41). (a) Complex representation of time signals; (b) magnitude transform; (c) autocorrelation; (d) iteration error.

Figure 4.24(a) shows the time domain representation of these signals in the complex domain with real and imaginary parts of each instant plotted as x-axis and y-axis. The scattered dots there represent the original Huffman signal which is obviously not of constant envelope. However, both f_k and g_k maintain the constant envelope property after a sufficient number of iterations as shown by circles and cross markings on the unit circle. Similar behavior is shown by the magnitude transform in Fig. 4.24(b) as well. Notice that the magnitude transform of f_k is close to the original, whereas the approximation g_k has exactly the same magnitude transform as the original signal. Figure 4.24(c) shows the autocorrelation function of the original and the two approximations. Notice that the approximations f_k and g_k possess off-center (non-zero lag) autocorrelations and hence, compared to the original Huffman signal, their pulse compressed outputs will have sidelobes. Finally, Fig. 4.24(d) shows the convergence rate of the iteration where the error d_k in (4.43) is plotted.

Figure 4.25 shows similar details for a 32 length Huffman sequence with details indicated there. From these figures, the original Huffman

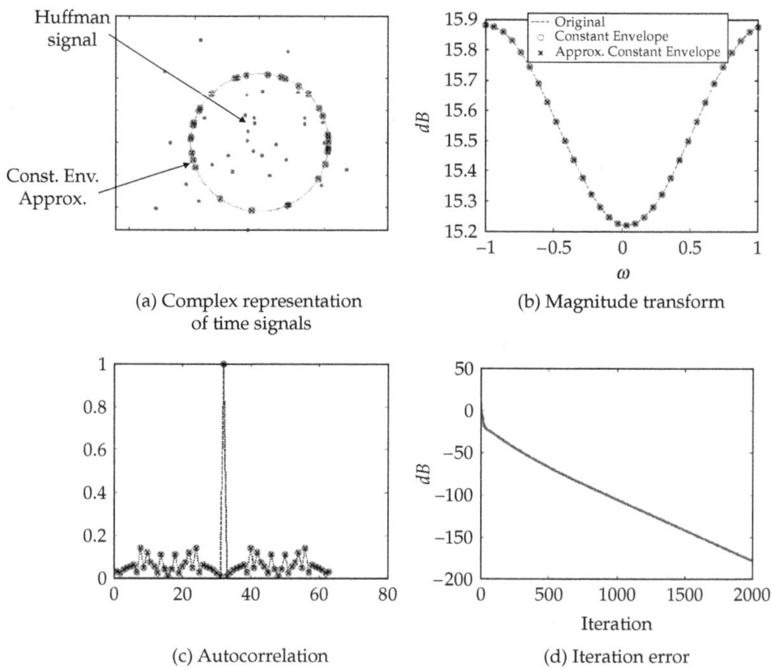

FIGURE 4.25 The Huffman signal of length 32 and its approximations using (4.41). (a) Complex representation of time signals; (b) magnitude transform; (c) autocorrelation; (d) iteration error.

sequence [scattered dots in Fig. 4.25(a)] has poor constant envelope property, but has excellent pulse compression property, where the two approximations f_k and g_k have both excellent constant envelope property but poorer pulse compression because of the spurious sidelobes present in their autocorrelation function.

References

[1] D.C.Youla, "Mathematical Theory of Image Restoration by the Method of Convex Projections," *Chapter 2, Theorem 2.4-1, in Image Recovery: Theory and Application*, edited by H. Stark, pp. 29–77, New York: Academic press, Inc.

[2] A. Levi, H. Stark, "Image Restoration by the Method of Generalized Projections with Applications to Restoration from Magnitude," *IEEE International Conference on Acoustics, Speech and Signal Processing*, pp. 88–91, March 1984.

[3] V. Velasco, "Some Engineering Applications of the Theory of Nonexpansive Operators," *PhD Dissertation, Polytechnic Institute of NYU*, Farmindale, NY, Jan. 1986.

[4] H.H. Bauschka, J.M. Borwein, "On Projection Algorithms for Solving Convex Feasibility Problems," *SIAM Review*, vol. 38, no. 3, pp. 367–426, 1996.

[5] D.C. Youla, "Generalized Image Restoration by the Method of Alternating Projections," *IEEE Transactions on Circuits and Systems*, vol. 25, pp. 694–702, Sep. 1978.

[6] L.G. Gubin, B.T. Polyak, E.V. Raik, "The Method of Projections for Finding the Common Point of Convex Sets," *U.S.S.R. Computational Mathematics and Mathematical Physics*, vol. 7, no. 6, pp. 1–24, 1967.

[7] Z. Opial, "Weak Convergence of the Sequence of Successive Approximations for Nonexpansive Mappings," *Bull. Amer. Soc.*, vol. 73, pp. 591–597, 1967.

[8] J.V. Neumann, "Functional Operators, Vol. II," *Annals of Mathematics Studies*, no. 22, Theorem 13.7, p. 55, Princeton, NJ, 1950.

[9] R.W. Gerchberg, W. Saxton, "A Practical Algorithm for the Determination of Phase from Image and Diffraction Plane Pictures," *Optik*, vol. 35, no. 237, 1972.

[10] J.R. Fienup, "Iterative Method Applied to Image Reconstruction and to Computer-Generated Holograms," *Opt. Eng.*, vol. 19, pp. 297–305, May–June 1980.

[11] P. Stoica, H. He, J. Li, "New Algorithms for Designing Unimodular Sequences with Good Correlation Properties," *IEEE Transactions on Signal Processing*, vol. 57, no. 4, April 2009.

[12] H. He, P. Stoica, J. Li, "Unimodular Sequence Sets with Good Correlations for MIMO Radar," *2009 IEEE Radar Conference*, Pasadena, CA, May 4–8, 2009.

[13] S.U. Pillai, B. Himed, "Unimodular Sequences Using Cyclic Algorithms and Their Relationship to the Gerchberg-Saxton Algorithm," *DASP Workshop*, Lihue, HI, Sep. 26–Oct. 1, 2009.

[14] T. Svensson, A. Svensson, "Design and Performance of Constrained Envelope Continuous Phase Modulation," *IEEE International Waveform Diversity & Design Conference*, Orlando, FL, Feb. 8–13, 2009.

[15] S. Thompson, J. Stralka, "Constant Envelope OFDM for Power-Efficient Radar and Data Communications," *IEEE International Waveform Diversity & Design Conference*, Orlando, FL, Feb. 8–13, 2009.

[16] C.D. Chung, S.M. Cho, "Constant-Envelope Orthogonal Frequency Division Multiplexing Modulation," *Proc. APCC/OECC*, vol. 1, Beijing, Chaina, pp. 629–632, Oct. 1999.

[17] J. Tan, G.L. Stuber, "Constant Envelope Multi-Carrier Modulation," *Proc. IEEE Milcom*, vol. 1, Anaheim, pp. 607–611, Oct. 2002.

[18] F. Harris, C. Dick, "Non Iterative Techniques for Low Peak-to Average Power Ratio OFDM Signals," *IEEE International Waveform Diversity & Design Conference*, Orlando, FL, Feb. 8–13, 2009.

[19] D. Fuhrmann, P. Browning, M. Rangaswamy, "Constant-Modulus Partially Correlated Signal Design for Uniform Linear and Rectangular MIMO Radar Array," *IEEE International Waveform Diversity & Design Conference*, Orlando, FL, Feb. 8–13, 2009.

[20] L.K. Patton, B.D. Rigling, "Modulus Constraints in Adaptive Radar Waveform Design," *IEEE International Waveform Diversity & Design Conference*, Orlando, FL, Feb. 8–13, 2008.
[21] Gerchberg-Saxton Algorithm, *Wikipedia,* http://en.wikipedia.org/wiki/Gerchberg%E2%80%93Saxton_algorithm
[22] L.K. Patton, B.D. Rigling, "Modulus Constraints in Adaptive Radar Waveform Design," *2008 IEEE Radar Conference*, Rome, Italy, May 26–30, 2008.
[23] P. Stoica, J. Li and X. Zhu, "Waveform Synthesis for Diversity-Based Transmit Beampattern Design," *IEEE Transactions on Signal Processing*, vol. 56, no. 6, June 2008.
[24] M. Schroeder, "Synthesis of Low-Peak-Factor Signals and Binary Sequences with Low Autocorrelation," *IEEE Transactions on Information Theory*, vol. 16, no. 1, pp. 85–89, Jan. 1970.
[25] C.D. Groot, D. Wurtz, K.H. Hoffmann, "Low Autocorrelation Binary Sequences: Exact Enumeration and Optimization by Evolutionary Strategies," *Optimization*, vol. 23, no. 4, pp. 369–384, 1992.
[26] H. Deng, B. Himed, "Virtual Radar Antenna Beamforming Using Orthogonal Coding Waveforms," *IEEE International Waveform Diversity & Design Conference*, Orlando, FL, Feb. 8–13, 2009.
[27] A. Brodzik, R. Tolimieri, "Polyphase Sequence Design with Good Correlation Properties," *IEEE International Waveform Diversity & Design Conference*, Orlando, FL, Feb. 8–13, 2009.
[28] M. Schroeder, "Synthesis of Low-Peak-Factor Signals and Binary Sequences with Low Autocorrelation," *IEEE Transactions on Information Theory*, vol. 16, no. 1, pp. 85–89, Jan. 1970.
[29] P. Stoica, H. He, J. Li, "New Algorithms for Designing Unimodular Sequences with Good Correlation Properties," *IEEE Transactions on Signal Processing*, vol. 57, no. 4, April 2009.
[30] M.J. Lindenfeld, "Sparse Frequency Transmit and Receive Waveform Design," *IEEE Trans. Aerospace and Electronic Systems*, vol. 40, no. 3, pp. 851–861, July 2004.
[31] D.A. Huffman, "The Generation of Impulse-Equivalent Pulse Trains," *IRE Trans. Informa. Theory*, vol. IT-8, pp. 10–16, 1962.
[32] M.H. Ackroyd, "Synthesis of Efficient Huffman Sequences," *IEE Transactions on Aerospace and Electronic Systems*, vol. AES-8, no. 1, Jan. 1972.

CHAPTER 5

Optimum Waveform Design

This chapter presents optimum transmit waveform design in the general case—in presence of clutter and noise.

5.1 Point Target in White Noise and Arbitrary Clutter

Consider an impulse target in white noise and arbitrary clutter. Thus [1]

$$Q(\omega) \equiv 1 \qquad \text{and} \qquad G_n(\omega) = \sigma_n^2 \tag{5.1}$$

and from (3.27), λ is given by

$$\lambda = \frac{2\pi E}{\sigma_n \int\limits_{\Omega_+} \frac{1}{G_c(\omega)} d\omega} + \sigma_n \tag{5.2}$$

and is greater than $\max\limits_{\omega \in \Omega_+} \sqrt{G_n(\omega)} \,/\, |Q(\omega)| \;=\; \sigma_n$, and therefore satisfies the inequality requirement (3.25). Also from (3.30) we get

$$SINR_{\max} = \frac{E \int\limits_{\Omega_+} \frac{1}{G_c(\omega)} d\omega}{2\pi E + \sigma_n^2 \int\limits_{\Omega_+} \frac{1}{G_c(\omega)} d\omega} = \frac{1}{\frac{\sigma_n^2}{E} + \frac{1}{\frac{1}{2\pi} \int\limits_{\Omega_+} \frac{1}{G_c(\omega)} d\omega}} = \frac{1}{\frac{\sigma_n^2}{E} + \eta} \tag{5.3}$$

where

$$\eta = \frac{1}{\frac{1}{2\pi}\int\limits_{-\infty}^{+\infty} \frac{1}{G_c(\omega)} d\omega}. \tag{5.4}$$

Thus to maximize *SINR*, the quantity η in (5.3) should be minimized or from (5.4) $\int\limits_{-\infty}^{+\infty} \frac{1}{G_c(\omega)} d\omega$ should be maximized, and to achieve that the operating band Ω_+ in (5.4) should be the entire interval $(-\infty < \omega < +\infty)$. In that case, $\lambda = \frac{E\eta}{\sigma_n} + \sigma_n$, and

$$SINR_{\max} = \frac{1}{\frac{\sigma_n^2}{E} + \eta} \tag{5.5}$$

and from (3.38) and (5.2), the optimum transmit signal transform satisfies

$$|F(\omega)|^2 = \frac{\sigma_n(\lambda - \sigma_n)}{G_c(\omega)} = \frac{E\,\eta}{G_c(\omega)}, \tag{5.6}$$

the classic radar solution [2]. From (3.12), the optimum receiver in that case equals

$$H_{opt}(\omega) = \frac{F^*(\omega)e^{-j\omega t_o}}{E\,\eta + \sigma_n^2} \tag{5.7}$$

or

$$h_{opt}(t) = c_o f(t_o - t), \tag{5.8}$$

where $c_o = 1/(E\eta + \sigma_n^2)$. Interestingly, (5.8) represents the classic matched filter solution *irrespective* of the presence of clutter, provided the noise and target have *flat* spectral characteristics.

From (5.6), in the point target and the white noise case, the optimum transmit signal transform must be inversely proportional to the clutter spectrum. This of course assumes that there is no restriction on the transmit signal bandwidth, and that *infinite* bandwidth is available at the transmitter to accommodate the requirement in (5.6). This requirement is illustrated next for the flat-clutter spectrum [3,4].

5.1.1 Point Target, White Noise, White Clutter (Flat-Flat-Flat Case)

In this case we have

$$Q(\omega) \equiv 1, \qquad G_c(\omega) = \sigma_c^2, \qquad \text{and} \qquad G_n(\omega) = \sigma_n^2 \tag{5.9}$$

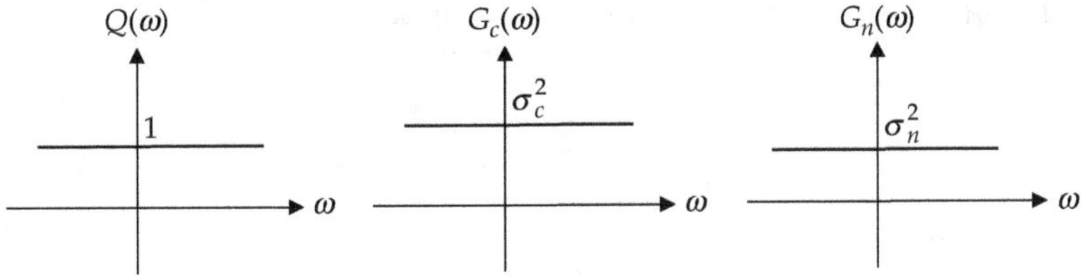

FIGURE 5.1 Flat-flat-flat case.

and let $2B_o$ represent the available bandwidth. From (5.4) we obtain

$$\eta = \frac{1}{\frac{1}{2\pi}\int_{-B_o}^{B_o} \frac{1}{\sigma_c^2} d\omega} = \frac{\pi\sigma_c^2}{B_o} \tag{5.10}$$

so that from (5.6)

$$|F(\omega)|^2 = \frac{\pi E}{B_o}. \tag{5.11}$$

Thus the transmit signal has flat spectral characteristics as well, the classic solution due to Manasse [2]. In that case from (5.5)

$$SINR_{\max} = \frac{1}{\frac{\sigma_n^2}{E} + \frac{\pi\sigma_c^2}{B_o}}. \tag{5.12}$$

From (5.12)

$$SINR_{\max} \to \frac{E}{\sigma_n^2}, \quad \text{as } B_o \to \infty. \tag{5.13}$$

Thus by increasing the transmit bandwidth *indefinitely* it is possible to eliminate flat-clutter completely and the *SINR* is only noise-limited as in (5.13) [5].

Interestingly the chirp signal satisfies the flat-spectral characteristics feature, and hence in the flat-flat-flat case (Fig. 5.1) so long as there is no restriction on the transmit signal bandwidth, an appropriate chirp signal (see Sec. 2.6) is optimum and by increasing its bandwidth, the effect of clutter can be eliminated [6].

5.1.2 Finite Transmit Bandwidth

Next, we proceed to examine the consequences of imposing a finite bandwidth at the transmitter output [7].

To start with, the $SINR_{max}$ in (3.14) is upper-bounded by

$$\begin{aligned} SINR_{max} &= \frac{1}{2\pi}\int_{-\infty}^{+\infty} \frac{|Q(\omega)|^2\,|F(\omega)|^2}{G_c(\omega)|F(\omega)|^2 + G_n(\omega)}\,d\omega \\ &\le \frac{1}{2\pi}\int_{-\infty}^{+\infty} \frac{|Q(\omega)|^2\,|F(\omega)|^2}{G_n(\omega)}\,d\omega \end{aligned} \tag{5.14}$$

which for the point target and white noise case reduces to

$$SINR_{max} \le \frac{1}{2\pi\sigma_n^2}\int_{-\infty}^{+\infty} |F(\omega)|^2\,d\omega = \frac{E}{\sigma_n^2}. \tag{5.15}$$

We will show that if there is no bandwidth restriction at the transmitter output, an appropriate chirp waveform with sufficiently large bandwidth can be made to achieve the upper bound in (5.15). To see this, consider the chirp signal

$$f(t) = \begin{cases} \cos(\omega_o t + \beta t^2), & 0 < t < T \\ 0, & \text{otherwise.} \end{cases} \tag{5.16}$$

Then

$$\begin{aligned} F(\omega) &= \frac{1}{2}\text{Re}\left(e^{-j\left(\frac{\omega-\omega_o}{2\sqrt{\beta}}\right)^2}\sqrt{\frac{\pi}{2\beta}}\left\{K\left(\frac{\omega-\omega_o}{2\sqrt{\beta}}\right) - \left(\frac{\omega-\omega_o}{2\sqrt{\beta}} - \sqrt{B}\,T\right)\right\}\right) \\ &\approx \begin{cases} \frac{1}{2}\sqrt{\pi/\beta}, & \omega_o < |\omega| < \omega_o + 2\beta T \\ 0, & \text{otherwise.} \end{cases} \end{aligned} \tag{5.17}$$

Thus for a chirp signal as $\beta \to \infty$, its bandwidth $2\beta T \to \infty$. However from (5.17) its transform amplitude $|F(\omega)| \sim \frac{1}{\sqrt{\beta}} \to 0$ so that its energy remains constant. Thus as the transmit bandwidth increases, the clutter contribution

$$G_c(\omega)|F(\omega)|^2 = \frac{G_c(\omega)}{\beta} \to 0 \tag{5.18}$$

in (5.14), and the inequalities in (5.14) and (5.15) become equalities. Thus in the absence of any transmitter output bandwidth restriction, by increasing the bandwidth of the transmit chirp signal the effect of

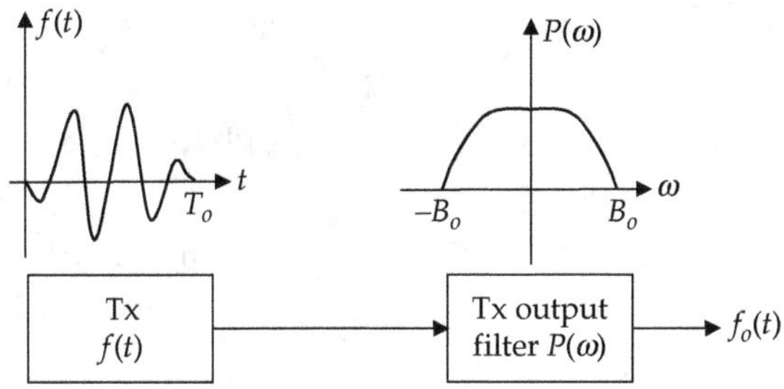

FIGURE 5.2 Transmitter output filter $P(\omega)$.

clutter can be driven to zero, and the chirp realizes the noise-limited maximum achievable $SINR = E/\sigma_n^2$ asymptotically [8].

However, the situation is quite different if a transmitter output filter $P(\omega)$ is present as shown in Fig. 5.2.

We will require the final transmit output waveform $f_o(t)$ to be of finite energy, and this is guaranteed if the transmitter output filter $p(t) \leftrightarrow P(\omega)$ satisfies the *integrability* and *square-integrability* condition. Thus absolute integrability on $p(t)$ gives

$$\int_0^{\infty} |p(t)|\, dt = M_1 < \infty \tag{5.19}$$

that is guaranteed by the bounded-input, bounded-output (BIBO) stability condition, and the square integrability on $p(t)$ gives

$$\int_0^{\infty} |p(t)|^2\, dt = \frac{1}{2\pi} \int_{-\infty}^{+\infty} |P(\omega)|^2\, d\omega = M_2 < \infty. \tag{5.20}$$

In that case, the transmitter final output signal energy is given by

$$\begin{aligned} \int_0^{\infty} |f_o(t)|^2\, dt &= \frac{1}{2\pi} \int_{-\infty}^{+\infty} |F_o(\omega)|^2\, d\omega = \frac{1}{2\pi} \int_{-\infty}^{+\infty} |F(\omega)|^2\, |P(\omega)|^2\, d\omega \\ &\leq M_1^2 \int_{-\infty}^{+\infty} |F(\omega)|^2 d\omega = M_1^2 E < \infty, \end{aligned} \tag{5.21}$$

since from (5.19)

$$|P(\omega)| = \left| \int_0^\infty p(t)e^{-j\omega t}dt \right| \leq \int_0^\infty |p(t)|\, dt < M_1. \tag{5.22}$$

With transmit output filter $P(\omega)$ present, the $SINR_{\max}$ in (5.14) becomes [6]

$$\begin{aligned} SINR_{\max} &= \frac{1}{2\pi} \int_{-\infty}^{+\infty} \frac{|Q(\omega)|^2\, |P(\omega)|^2\, |F(\omega)|^2}{G_c(\omega)\, |P(\omega)|^2\, |F(\omega)|^2 + G_n(\omega)} d\omega \\ &\leq \frac{1}{2\pi} \int_{-\infty}^{+\infty} \frac{|Q(\omega)|^2\, |P(\omega)|^2\, |F(\omega)|^2}{G_n(\omega)} d\omega \end{aligned} \tag{5.23}$$

which for point target and white noise case reduces to

$$\begin{aligned} SINR_{\max} &\leq \frac{1}{2\pi\sigma_n^2} \int_{-\infty}^{+\infty} |P(\omega)|^2\, |F(\omega)|^2\, d\omega \leq \frac{|F_{\max}(\omega)|^2}{2\pi\sigma_n^2} \int_{-\infty}^{+\infty} |P(\omega)|^2\, d\omega \\ &= \frac{M_2\, |F_{\max}(\omega)|^2}{2\pi\sigma_n^2} = \frac{M_2}{8\sigma_n^2} \frac{1}{\beta} \quad \rightarrow \; 0 \end{aligned} \tag{5.24}$$

since $p(t)$ is square integrable, and from (5.17), $|F_{\max}|^2 = \frac{\pi}{4\beta}$. Hence as $\beta \to \infty$, clutter gets eliminated as in (5.18); however, from (5.24) $SINR_{\max}$ also tends to zero.

In other words, increasing the chirp bandwidth so as to eliminate clutter and maximize the output *SINR* will work only if the transmitter output filter has infinite bandwidth. With finite bandwidth imposed through a square integrable transmit output filter, increasing chirp bandwidth eliminates clutter; however, it is no longer possible to achieve maximum *SINR* in that manner even in the white noise case since $SINR \to 0$, and selective transmitter-receiver design must be carried out to maximize the output *SINR* [1].

5.1.3 Optimality of Chirp Waveforms

From the above discussion, in the case of a point target (flat spectrum) present in white noise and any type of clutter, the chirp waveform is asymptotically optimum, provided there is no restriction on its bandwidth. In that case by increasing the bandwidth of the chirp signal, the clutter can be completely cancelled out and it is possible to realize the maximum achievable noise-limited *SINR* using a chirp waveform. The bandwidth required for realizing this goal can be unacceptably large [9].

However, the chirp transmit signal is no longer optimum if any one of the above conditions does not hold; that is,

- Frequency selective target (nonpoint target) is present
- Colored noise
- Finite bandwidth restrictions at the transmitter output

In particular, with finite bandwidth restrictions usually imposed through a transmitter output filter (Fig. 5.2), as (5.24) shows, chirp cannot be made optimum by simply increasing its bandwidth, and selective transmitter-receiver design must be carried out to maximize the output *SINR* for target detection and identification.

In this context, the noncausal receiver in the general case is given by (3.12), and Secs. 3.2–3.4 discuss two transmitter design strategies. Interestingly, both these methods lead to specific solutions for the transmit signal transform magnitude $|F(\omega)|$ as given by (3.38) and (3.86). In general, it may not be possible to realize phase functions $\psi(\omega)$ so that the inverse transforms of $|F(\omega)|e^{j\psi(\omega)}$ correspond to causal pulse-like waveforms.

Next, we address this problem with the restriction that both the transmitter and the receiver are of the causal type.

5.2 Optimum Tx-Rx: Causal Case[1]

Consider the general situation in (3.1)–(3.4), where a finite energy transmit signal $f(t)$ interacts with a target with impulse response $q(t)$ to generate the signal $s(t)$ [10, 11]. Let $w_c(t)$ represent the stochastic (clutter generating) process that interacts with $f(t)$ to generate the clutter waveform $w_c(t) * f(t)$, and $n(t)$ the noise. In that case, the receiver input signal $r(t)$ is given by (3.3), and the total input interference plus noise power spectrum is as given in (3.4).

The received signal $r(t)$ is presented to a causal receiver $h_c(t)$. The causal receiver in the case of a signal $s(t)$ contaminated by white or colored noise is well known (Sec. 2.2.1). In the case of colored noise the optimum causal receiver is given by (2.69) and in the case of white noise it is given by (2.75) [12]. As Fig. 2.14 shows, a whitening operation can be performed to convert the colored noise case to the white noise case. A similar whitening procedure can be applied in the present situation as well, where the total interference plus noise

[1] © 2000 IEEE. Reprinted with permission, from "Optimum Transmit-Receiver Design in the Presence of Signal-Dependent Interference and Channel Noise" by S. U. Pillai, H. S. Oh, D. C. Youla, and J. R. Guerci, *IEEE Transactions on Information Theory*, Vol. 46, No. 2, pps. 577–584, March 2000.

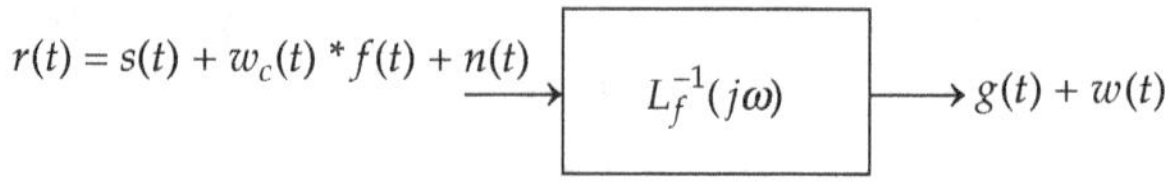

FIGURE 5.3 Whitening filter.

power spectrum is given by (3.4). Assume that $G_I(\omega)$ satisfies the Paley-Wiener condition in (2.48) [with $G(\omega)$ replaced by $G_I(\omega)$]. In that case, knowing $|F(\omega)|$, the total interference spectrum $G_I(\omega)$ can be factorized as in (2.49). Thus, [1]

$$G_I(\omega) = G_c(\omega)|F(\omega)|^2 + G_n(\omega) = |L_f(j\omega)|^2 \tag{5.25}$$

where $L_f(s)$ is *analytic* together with its *inverse* $L_f^{-1}(s)$ in the right half-complex plane. As in Fig. 2.14, $L_f^{-1}(j\omega)$ represents the whitening filter in this case. Observe that $L_f(j\omega)$ depends on the transmit signal transform magnitude $|F(\omega)|$. Let $g(t) + w(t)$ represent the whitening filter output to the input $r(t)$ (Fig. 5.3).

Clearly, the signal part $g(t)$ is given by

$$g(t) \leftrightarrow L_f^{-1}(j\omega)S(\omega) = L_f^{-1}(j\omega)Q(\omega)F(\omega) \tag{5.26}$$

and $w(t)$ represents the white noise output of unit spectral density. Hence from (2.75), the causal receiver in this case is given by $g^*(t_o - t)u(t)$, where t_o represents the decision instant. Combining the above two steps (whitening followed by white noise matched filtering), we obtain the setup shown in Fig. 5.4, where we define the transform $K(\omega)$ through the relation [1]

$$g^*(t_o - t)\, u(t) \leftrightarrow K(\omega) \tag{5.27}$$

as in (2.65).

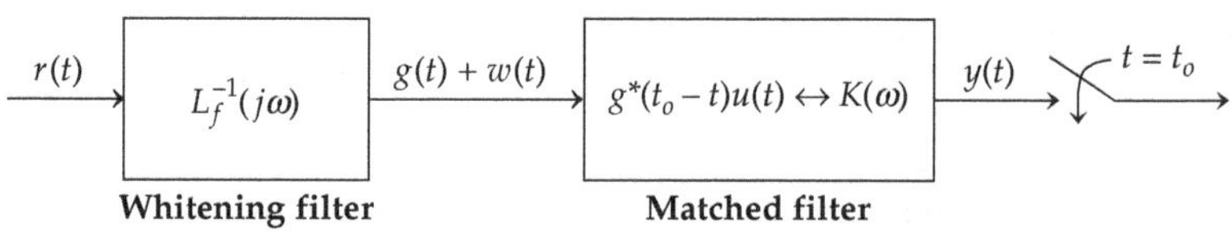

FIGURE 5.4 Optimum causal receiver.

From Fig. 5.4, the optimum causal receiver transform $H_c(\omega)$ is given by

$$H_c(\omega) = L_f^{-1}(j\omega)K(\omega) \tag{5.28}$$

or in time domain

$$h_c(t) = l_{inv}(t) * g^*(t_o - t)u(t) \tag{5.29}$$

as in (2.69) and (2.70). Notice that unlike (2.69), the whitening filter $L_f^{-1}(j\omega)$ in the present case makes use of the transmit signal transform magnitude as well. Thus *knowing* $F(\omega)$, the optimum causal receiver design is given by (5.28).[2] In that case, following (2.71), the maximum realizable output *SINR* is given by

$$\rho(t_o) = \frac{1}{2\pi}\int_{-\infty}^{+\infty} |K(\omega)|^2\, d\omega = \int_{-\infty}^{+\infty} \left|g^*(t_o - t)u(t)\right|^2 dt = \int_{0}^{t_o} |g(t)|^2\, dt. \tag{5.30}$$

Observe that $\rho(t_o)$ in (5.30) is a function of the transmit signal $f(t)$, and hence the final part of the optimum causal Tx-Rx design is to maximize the output *SINR* in (5.30) by varying the transmit signal subject to the energy condition in (3.1) [13].

Next, we establish a key property of the desired optimum transmit pulse when clutter and noise are present [1].

Minimum Phase Optimum Transmit Signal

Section 2.4 outlines the optimum causal Tx-Rx design when the total interference is of the noise type. In that case, the optimum transmit pulse solution that maximizes the output *SNR* in (2.103) has been shown to be a minimum phase signal [see (2.108)–(2.116)]. Interestingly, the same proof holds in the current situation also since the output *SINR* in (5.30) has the same functional form as the white noise only case in (2.116), and $g(t)$ there can be expressed as [see (5.26)] [14, 15]

$$g(t) = \Phi_f(t) * f(t) \tag{5.31}$$

[2] If $F(\omega)$ is *unknown*, this design procedure obviously cannot be completed.

where

$$\Phi_f(t) \leftrightarrow L_f^{-1}(j\omega)Q(\omega). \tag{5.32}$$

Hence proceeding as in (2.108)–(2.116), define the transmit pulse $f_o(t)$ that is generated from $f(t)$ as in (2.106). Notice that the transform of $f_o(t)$ has fewer zeros in the right half plane compared to that of $f(t)$. In that case since

$$|F(\omega)|^2 = |F_o(\omega)|^2, \tag{5.33}$$

from (5.25) and (5.33), we obtain $L_f^{-1}(j\omega)$ and hence $\Phi_f(t)$ associated with *both* $f(t)$, and $f_o(t)$ to be the *same* [see (5.32)]. Hence the output *SINR* due to the transmit pulse $f_o(t)$ can be expressed as

$$\rho_o(t) = \int_0^{t_o} |g_o(t)|^2\, dt \tag{5.34}$$

where

$$g_o(t) = \Phi_f(t) * f_o(t) \tag{5.35}$$

and following the same arguments that lead to (2.116) we obtain

$$\int_0^{t_o} |g_o(t)|^2\, dt \geq \int_0^{t_o} |g(t)|^2\, dt \tag{5.36}$$

proving the claim that an input pulse with "fewer" zeros on the right half plane in its transform attains a larger output *SINR* [1, 16]. Hence in the causal case, the optimum transmit signal that maximizes the output *SINR* is necessarily minimum phase in the general case (clutter/interference plus colored noise) as well.

The minimum phase nature of $f(t)$ demands that energy of the optimum transmit pulse must be front-loaded in time. Hence front-loading the excitation with more energy also can set the target response (if any) to contain more energy. This makes sense since only the early part of the output (up to t_o) is used in decision making, and to maximize *SINR*, it should contain as much target information as possible.

To maximize the output *SINR* $\rho(t_o)$ in (5.30) subject to the transmit energy constraint, one may proceed as in (2.117)–(2.126). The procedure there involves the eigensolution of the integral equation in (2.124), where the kernel $K(\tau_1, \tau_2)$ depends only on the target characteristics as in (2.118) [1, 15–17]. However in the present case, the kernel depends also on the unknown transmit signal $f(t)$ and hence this procedure leads to a highly nonlinear optimization problem. To see this, using (5.28) and (5.29) we get

$$\begin{aligned}\rho(t_o) &= \int_0^{t_o} |g(t)|^2\,dt = \int_0^{t_o}\left|\int_0^{t_o} |\Phi_f(t-\tau)f(\tau)|^2 d\tau\right|^2 dt\\ &= \int_0^{t_o} f^*(\tau_1)d\tau_1 \int_0^{t_o} \underbrace{\left(\int_0^{t_o} \Phi_f^*(t-\tau_1)\Phi_f(t-\tau_2)dt\right)}_{\Omega(\tau_1,\,\tau_2)} f(\tau_2)d\tau_2 \\ &= \int_0^{t_o} f^*(\tau_1)d\tau_1 \int_0^{t_o} \Omega(\tau_1,\,\tau_2)f(\tau_2)d\tau_2\end{aligned} \tag{5.37}$$

where the kernel

$$\Omega(\tau_1,\,\tau_2) = \int_0^{t_o} \Phi_f^*(t-\tau_1)\Phi_f(t-\tau_2)dt \tag{5.38}$$

with $\Phi_f(t)$ as in (5.32). From (5.25), the causal function $\Phi_f(t)$ depends on the unknown transmit signal $f(t)$ to be optimized and hence the kernel in (5.38) is *unknown*.

The integral equation in (5.37) formally defines an integral operator

$$\psi(\tau_1) = \int_0^{t_o} \Omega(\tau_1,\,\tau_2)f(\tau_2)d\tau_2 \tag{5.39}$$

and using this, (5.37) can be expressed as

$$\rho(t_o) = \int_0^{t_o} f^*(\tau_1)\psi(\tau_1)d\tau_1 \le \left(\int_0^{t_o} |f(\tau_1)|^2\,d\tau_1\right)^{1/2}\left(\int_0^{t_o} |\psi(\tau_1)|^2\,d\tau_1\right)^{1/2} \tag{5.40}$$

where we have used Schwarz' inequality.

Clearly (5.40) is maximized if and only if $\psi(\tau_1)$ is chosen proportional to $f(\tau_1)$ for $0 < \tau_1 < t_o$. Thus for maximum *SINR*, $f(\tau_1)$ must be a solution of the integral equation

$$\psi(\tau_1) = \int_0^{t_o} \Omega(\tau_1, \tau_2) f(\tau_2) d\tau_2 = \lambda f(\tau_1), \quad 0 < \tau_1 < t_o \tag{5.41}$$

and substituting (3.1) and (5.41) into (5.40), we obtain

$$\rho(t_o) = \sqrt{E}\,\lambda \leq \sqrt{E}\,\lambda_1 \tag{5.42}$$

where λ_1 is the largest eigenvalue of the integral equation in (5.41). Hence the output *SINR* $\rho(t_o)$ is maximized by selecting the transmit signal $f(t)$ to be the eigenvector associated with the largest eigenvalue of the kernel $\Omega(\tau_1, \tau_2)$ in (5.41). Unfortunately the kernel in (5.41) itself involves the *unknown* $f(t)$ in a nonlinear fashion through $L_f^{-1}(j\omega)$, and hence direct solution as in the white noise case is not possible in the general case [18].

Before addressing the general case involving both clutter and noise, we consider two special situations:

- Noise only case (clutter is insignificant)
- High clutter case (noise is low or clutter dominated case)

5.2.1 Noise Only Case

This situation corresponds to high noise or insignificant clutter case. Thus in the absence of clutter we have $G_c(\omega) \equiv 0$, and (5.25) leads to the Wiener factor $L_n(s)$ given by

$$G_n(\omega) = |L_n(j\omega)|^2. \tag{5.43}$$

Notice that given $G_n(\omega)$, the Wiener factor $L_n(j\omega)$ is known and substituting this into (5.32) we obtain

$$\Phi(t) = L_n^{-1}(j\omega) Q(\omega) \tag{5.44}$$

and with $\Phi_f(t) = \Phi(t)$ in (5.38) the kernel $K(\tau_1, \tau_2)$ can be computed. Knowing the kernel, the optimum transmit signal turns out to be proportional to the eigenvector $\psi_1(t)$ associated with the largest eigenvalue of the integral equation (5.41). Thus

$$f_{opt}(t) = \sqrt{E}\,\psi_1(t), \quad 0 \le t \le t_o \tag{5.45}$$

and from (5.28) the corresponding optimum causal receiver is given by

$$H_c(\omega) = L_n^{-1}(j\omega)K(\omega) \tag{5.46}$$

where from (5.26) and (5.27)

$$g^*(t_o - t)u(t) \;\leftrightarrow\; K(\omega) \tag{5.47}$$

and

$$g(t) \;\leftrightarrow\; L_n^{-1}(j\omega)Q(\omega)F_{opt}(\omega). \tag{5.48}$$

Using (5.47) and (5.48), (5.46) can be rewritten as

$$H_c(\omega) = \frac{e^{-j\omega t_o}\int_0^{t_o} g(t)e^{j\omega t}dt}{L_n(j\omega)}. \tag{5.49}$$

In the absence of clutter, equations (5.45) and (5.46) are exact solutions for the optimum transmitter and receiver in presence of colored noise [1, 19].

Figures 5.5 and 5.6 show the optimum transmitter-receiver design in the presence of noise for two different target and noise scenes. Notice that in Fig. 5.5, the target is resonant at two dominant frequencies ω_1 and ω_2 with ω_1 being the stronger component. Since the noise has flat spectral characteristics, the optimum transmit waveform in this case concentrates most of its energy around the ω_1 region that results in an output $SNR = 12.0$ dB. In Fig. 5.6, the target has flat features in the frequency domain and the noise is low pass. As a result, the optimum transmitter puts out most of its energy toward the high-frequency region of the target where the noise is less dominant. Thus, the target is excited mostly toward "the edge" of its frequency spectrum and this results in an output $SNR = 2.4$ dB; an ordinary chirp waveform under these conditions will generate output $SNR = 2.1$ dB and -1.5 dB for Figs. 5.5 and 5.6, respectively. The improvement in terms of output SNR here can be attributed to the joint optimum transmitter-receiver design [20–23].

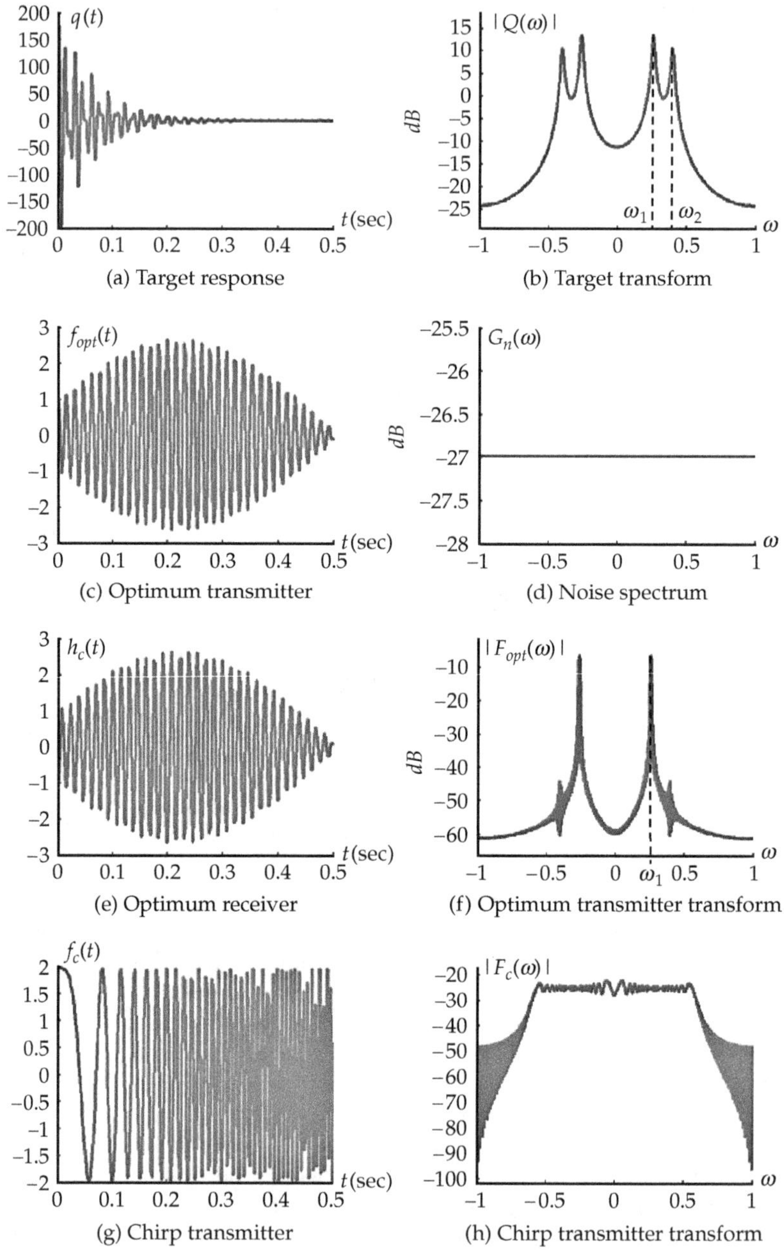

FIGURE 5.5 Optimum transmitter-receiver in presence of noise. Optimum transmitter latches on to the dominant resonant mode of the target in the flat noise case. Input target to noise ratio (*TNR*) is 0 dB and output *SNR* is 12.0 dB. A comparable input chirp generates an output *SNR* = 2.1 dB.

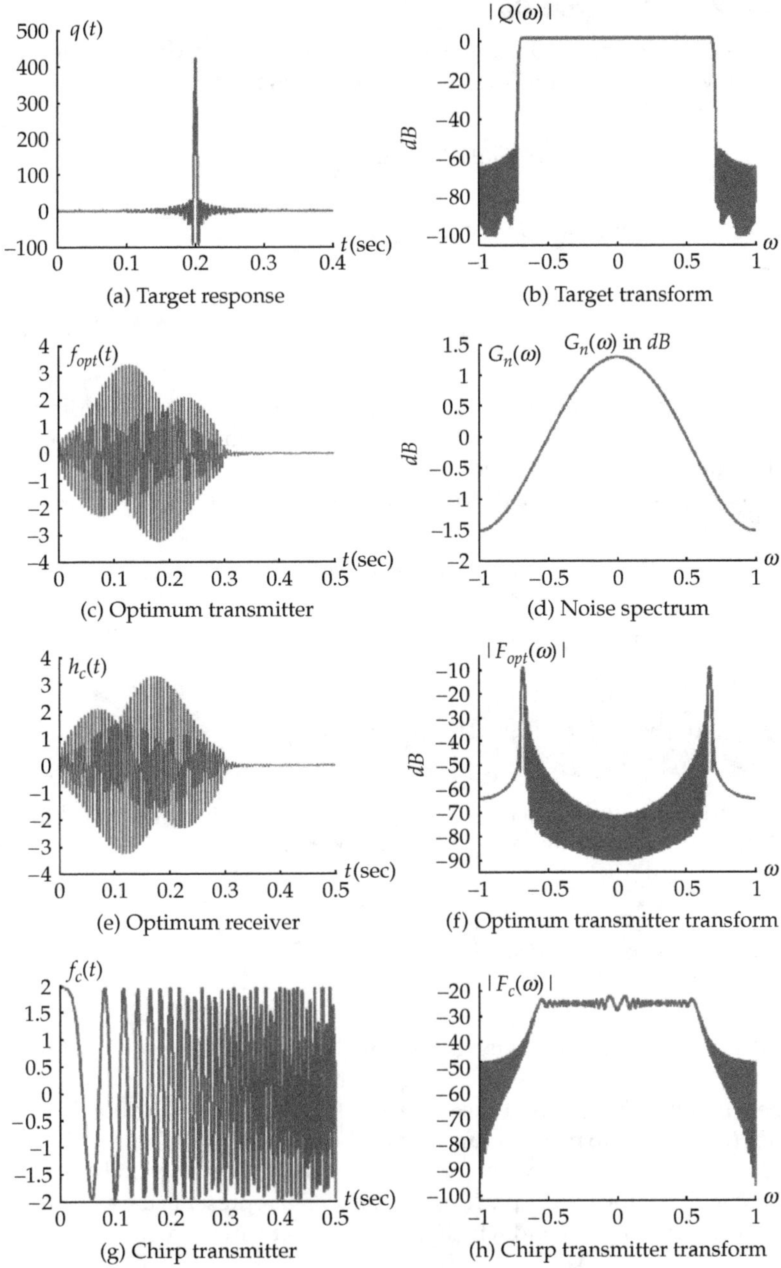

FIGURE 5.6 Optimum transmitter-receiver in presence of colored noise. Optimum transmitter puts out most of its energy in the low-noise region. Input $TNR = 0$ dB and output $SNR = 2.4$ dB. A comparable input chirp generates an output $SNR = -1.5$ dB.

5.2.2 High Clutter Case [1]

At the other extreme, if the noise is negligible compared to the clutter,[3] we have $G_n(\omega) \equiv 0$ and the Wiener factor equation in (5.25) gives

$$G_c(\omega)|F(\omega)|^2 = \left|L_f(j\omega)\right|^2. \tag{5.50}$$

Let $L_c(s)$ represent the Wiener factor associated with the clutter spectrum $G_c(\omega)$. Thus

$$G_c(\omega) = |L_c(j\omega)|^2 \tag{5.51}$$

where $L_c(s)$ is minimum phase, and considering that the optimum (unknown) transmit signal is also minimum phase, $F(\omega)$ in (5.50) though *unknown* can be *chosen* to be minimum phase. Using this together with (5.51) in (5.50) we get

$$L_f(j\omega) = L_c(j\omega)F(\omega) \tag{5.52}$$

and from (5.26) this gives

$$g_{sat}(t) \;\leftrightarrow\; L_f^{-1}(j\omega)Q(\omega)F(\omega) = \frac{Q(\omega)F(\omega)}{L_c(j\omega)F(\omega)} = \frac{Q(\omega)}{L_c(j\omega)}. \tag{5.53}$$

In this case, unlike (5.26) $g(t)$ is *not* a function of the transmit signal $f(t)$, and substituting (5.53) into (5.30) we obtain

$$\rho_{\max}(t_o) = \int_0^{t_o} |g_{sat}(t)|^2\, dt. \tag{5.54}$$

Thus when clutter is truly dominant, the shape of the transmit signal is *irrelevant* to signal design so long as the transmit signal is chosen to be *minimum phase*. In that case thresholding occurs and the maximum *SINR* turns out to be *not* a function of the transmit signal shape. From (5.53), $g(t)$ is only a function of the target and clutter spectra [through its Wiener factor $L_c(j\omega)$], and $\rho_{\max}(t_o)$ in (5.54) represents the saturated maximum value beyond which it is *impossible* to improve performance

[3]In the underwater sonar scene, this situation corresponds to reverberation-limited case. Usually the range bins close to the transmitter exhibit this property since transmit signal power level is high at low range points.

either by transmitting more energy or a different transmit signal.[4] As in (5.49), the optimum receiver in this case is given by

$$H_c(\omega) = \frac{e^{-j\omega t_o}\int_0^{t_o} g_{sat}(t)e^{j\omega t}dt}{L_c(j\omega)F(\omega)} = \frac{V(\omega)}{F(\omega)}, \tag{5.55}$$

and it is inversely proportional to the transmit signal transform [23, 24]. In summary, in high clutter situations the shape of the transmit signal is *irrelevant* so long as it is chosen to be *minimum phase,* and the saturated performance in terms of maximum achievable *SINR* is given by (5.54).

High Energy Case

Interestingly from (3.2) and (3.5)–(3.10), the high clutter scene can be simulated by allowing the transmitter energy E to increase. In that case the target return and the clutter returns are significant (compared to the noise) and we obtain the saturated performance situation in (5.53) and (5.54). As a result by systematically increasing the transmitter energy, the system can be driven to the saturation mode in terms of achievable output *SINR* where the actual transmit signal shape is irrelevant. Figure 5.7 shows the maximum *SINR* as a function of the transmitter energy for the optimum transmitter and a minimum phase chirp transmitter. The target power to noise ratio (*TNR*) is set at 0 dB and *CNR* is set at 50 dB (The optimum transmitter curve is designed using the procedure developed in the next section.) From Fig. 5.7(d), in the present situation, increasing the transmit energy beyond 20 dB serves no purpose in terms of improving output *SINR*.[5]

Thus there exists a threshold value for the transmit energy beyond which it serves no purpose to increase it any further. Under such a saturated operating condition the shape of the transmit signal is *irrelevant* so long as it is chosen to be minimum phase, as in the high

[4]From (5.54), *SINR* can be improved by increasing the integration time t_o. However, even this has an upper bound, since if $g_{sat}(t)$ in (5.53) has most of its energy concentrated in some interval $(0, T_s)$, then T_s acts as an upper limit for the integration time t_o. The actual value of T_s will of course depend on the target and clutter characteristics [see (5.53)].

[5]In practice, especially in underwater sonar scene, there is always an incentive to transmit more power since range covered is directly proportional to the transmit power. Because of the severe attenuation at sonar frequencies, to cover a certain range a fixed amount of power must be transmitted. The natural tendency then is to transmit as much power as possible hoping to detect targets as far away as possible. However, in range-limited littoral situations, an optimum transmitter-receiver design can be traded off for transmit power.

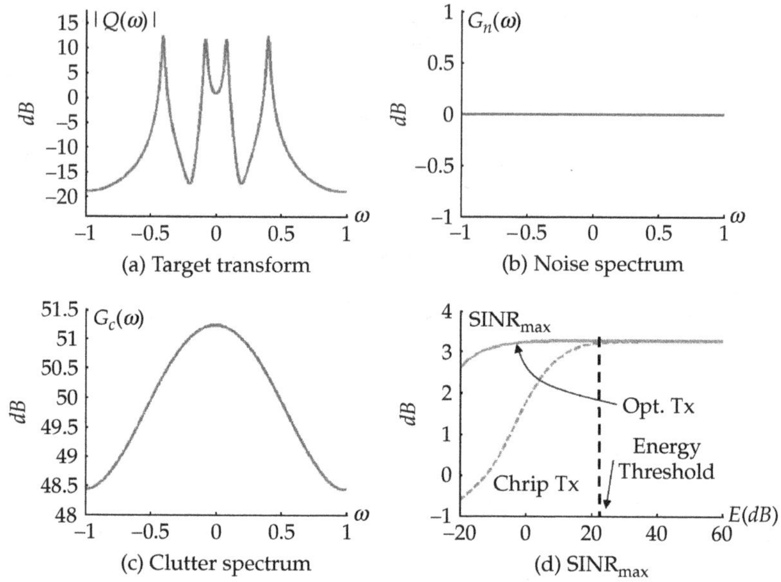

FIGURE 5.7 Maximum *SINR* as function of transmitter energy in high clutter case.

clutter case above. In summary, the transmit signal design can be traded off at the expense of more transmit energy. By transmitting very high energy, the shape of the transmit signal plays no role *so long as* it is chosen to be a minimum phase signal [1].

However, if the transmit energy is at premium, then the optimum transmitter-receiver design pays off. In that case, the general nonlinear problem in (5.35)–(5.42) must be solved.

5.2.3 General Case: Clutter and Noise

In the most general situation, neither channel noise nor clutter is negligible and equation

$$|L_f(j\omega)|^2 = G_n(\omega) + G_c(\omega)|F(\omega)|^2 \tag{5.56}$$

for the Wiener-Hopf factor $L_f(j\omega)$ permits no simplification. Nonetheless, it is possible to construct a successful iterative procedure for the determination of an optimal pair $f(t)$, $h_c(t)$ that extends the integral equation approach developed above and seems to converge robustly over a wide range of input *SINR*. To start with, given the target impulse response $q(t)$, or its transform $Q(\omega)$, and the spectral densities

$G_n(\omega)$ and $G_c(\omega)$, select t_o and the energy $\|f(t)\|^2 = E$ of the desired optimum pulse $f(t)$.

Let any causal pulse $f_o(t)$ of duration t_o and energy E act as the initialization. At stage k, assume $f_k(t) \leftrightarrow F_k(\omega)$ and find the Wiener-Hopf solution of the equation

$$|L_k(j\omega)|^2 = G_n(\omega) + G_c(\omega)\,|F_k(\omega)|^2 \tag{5.57}$$

by writing $L_k(j\omega) = |L_k(j\omega)|\,e^{j\theta_k(\omega)}$ and observing that $\theta_k(\omega)$ is the Hilbert transform of

$$\ln|L_k(j\omega)| = \frac{1}{2}\ln\left\{G_n(\omega) + G_c(\omega)\,|F_k(\omega)|^2\right\}. \tag{5.58}$$

Let

$$\Phi_k(t) \;\leftrightarrow\; L_k^{-1}(j\omega)Q(\omega) \tag{5.59}$$

and compute

$$\Omega_k(\tau_1,\,\tau_2) = \int_0^{t_o} \Phi_k^*(t-\tau_1)\Phi_k(t-\tau_2)dt, \quad 0 \le \tau_1,\ \tau_2 \le t_o. \tag{5.60}$$

Ideally, referring back to (5.41) and (5.42), the largest eigenvalue $\lambda_1^{(k)}$ of the integral equation

$$\int_0^{t_o} \Omega_k(\tau_1,\,\tau_2)\,\psi_1^{(k)}(\tau_2)d\tau_2 = \lambda_1^{(k)}\psi_1^{(k)}(\tau_1), \quad 0 < \tau_1 < t_o, \tag{5.61}$$

and corresponding normalized eigenfunction $\psi_1^{(k)}(t)$ should act as the solution pair since

$$\max\rho(t_o) = \int_0^{t_o}\int_0^{t_o} \Omega_k(\tau_1,\,\tau_2)\,f_k^*(\tau_1)\,f_k(\tau_2)d\tau_1 d\tau_2. \tag{5.62}$$

Thus for some k, ideally we should have the identity

$$f_k(t) = \sqrt{E}\,\psi_1^{(k)}(t), \tag{5.63}$$

where $f_k(t)$ is the kth iteration value used in (5.57). However, that may not be the case, and their difference given by

$$e_k(t) = f_k(t) - \sqrt{E}\,\psi_1^{(k)}(t) \tag{5.64}$$

is a meaningful error signal at that stage. By direct computation

$$\begin{aligned} \varepsilon_k^2 &= \|e_k(t)\|^2 = \left\| f_k(t) - \sqrt{E}\,\psi_1^{(k)}(t) \right\|^2 \\ &= \left\| f_k(t) \right\|^2 + E \left\| \psi_1^{(k)}(t) \right\|^2 - 2\sqrt{E} \int_0^{t_o} f_k^*(t)\psi_1^{(k)}(t)dt \\ &= 2E - 2\sqrt{E}\,c_1^{(k)} = 2\sqrt{E}\left(\sqrt{E} - c_1^{(k)}\right) \end{aligned} \tag{5.65}$$

where we have defined

$$c_1^{(k)} \triangleq \int_0^{t_o} f_k^*(t)\psi_1^{(k)}(t)dt \tag{5.66}$$

and the magnitude of the error signal in (5.64) can be used to update $f_k(t)$ by suitably combining it with $\psi_1^{(k)}(t)$. Toward this, define the update rule

$$f_{k+1}(t) = \mu\left(f_k(t) + \varepsilon_k \psi_1^{(k)}(t) \right). \tag{5.67}$$

The constant μ should be selected so that $\| f_{k+1}(t)\|^2 = E$, and this gives

$$f_{k+1}(t) = \frac{f_k(t) + \varepsilon_k \psi_1^{(k)}(t)}{\sqrt{\left(1 + \frac{\varepsilon_k}{\sqrt{E}}\right)^2 - \left(\frac{\varepsilon_k}{\sqrt{E}}\right)^3}}. \tag{5.68}$$

The updated pulse $f_{k+1}(t)$ should be used to update $L_k(j\omega)$ to $L_{k+1}(j\omega)$ in the iteration scheme (5.57) and the entire algorithm should be repeated until ε_k becomes acceptably small. Then

$$f_{opt}(t) = \underset{k\to\infty}{\text{limit}}\; f_k(t). \tag{5.69}$$

Knowing $f_{opt}(t)$, the optimum receiver impulse response $h(t)$ is computed accordingly, and

$$\max_f \max_h \rho(t_o) = E\lambda_1, \tag{5.70}$$

where $\lambda_1 = \lim_{k\to\infty} \lambda_1^{(k)}$.

Figure 5.8 shows the optimum transmitter-receiver design in the presence of clutter and noise. The *TNR* is set at 0 dB and *CNR* is set at 20 dB. The final output *SINR* for the optimum transmitter-receiver in this case is 11.5 dB, and it represents a 13 dB gain compared to a chirp transmit signal. From Fig. 5.8(b), (f), and (h), the optimum design puts most of the transmitter energy over the frequency region where the target is dominant and clutter is weak [25]. In this case the optimum design favors the region around the resonant frequency ω_2, although to a casual observer the region around ω_3 is preferable because of the absence of clutter there. This final tradeoff is not easy to predict, since the design should also turn out to be a minimum phase causal pulse waveform [26,27].

Figure 5.9 shows the optimum transmitter for four different types of noise and clutter spectra. The target has two dominant resonant features at ω_1 and ω_2. For the flat noise and low pass clutter situation in (b), the optimum transmitter $F_1(\omega)$ distributes energy at both these resonant frequencies equally; however, when clutter and noise are both low pass as in (c), the higher target resonant frequency ω_2 is preferred (see $|F_2(\omega)|$) by the optimum design; similarly when both clutter and noise are high pass as in (d), as $|F_3(\omega)|$ shows, the lower resonant frequency ω_1 is preferred by the optimum transmitter. Finally, when clutter and noise are mutually complementary as in (e), the optimum transmitter strategy depends on the relative bandwidth, relative power, and spectral shapes and is difficult to predict. In this case, as $|F_4(\omega)|$ shows, the transmitter shares energy between the two resonant modes. In all these cases, optimum strategy will outperform all other designs in terms of realizable output *SINR* gain. The actual *SINR* gain will depend on the available distinguishable features of the target, clutter, and noise spectra. For example, if a resonant target is present, the optimum transmitter tries to latch onto its dominant resonant modes in the frequency domain where the clutter and noise are low. If the target has no resonant features, then differences in available bandwidths or spectral features of the clutter and noise spectra are exploited. In summary, the optimum transmitter exploits the differences among the target, clutter, and noise spectral shapes and synthesizes a finite duration waveform that further amplifies their difference, thus achieving a better output *SINR* than any other waveform. It follows

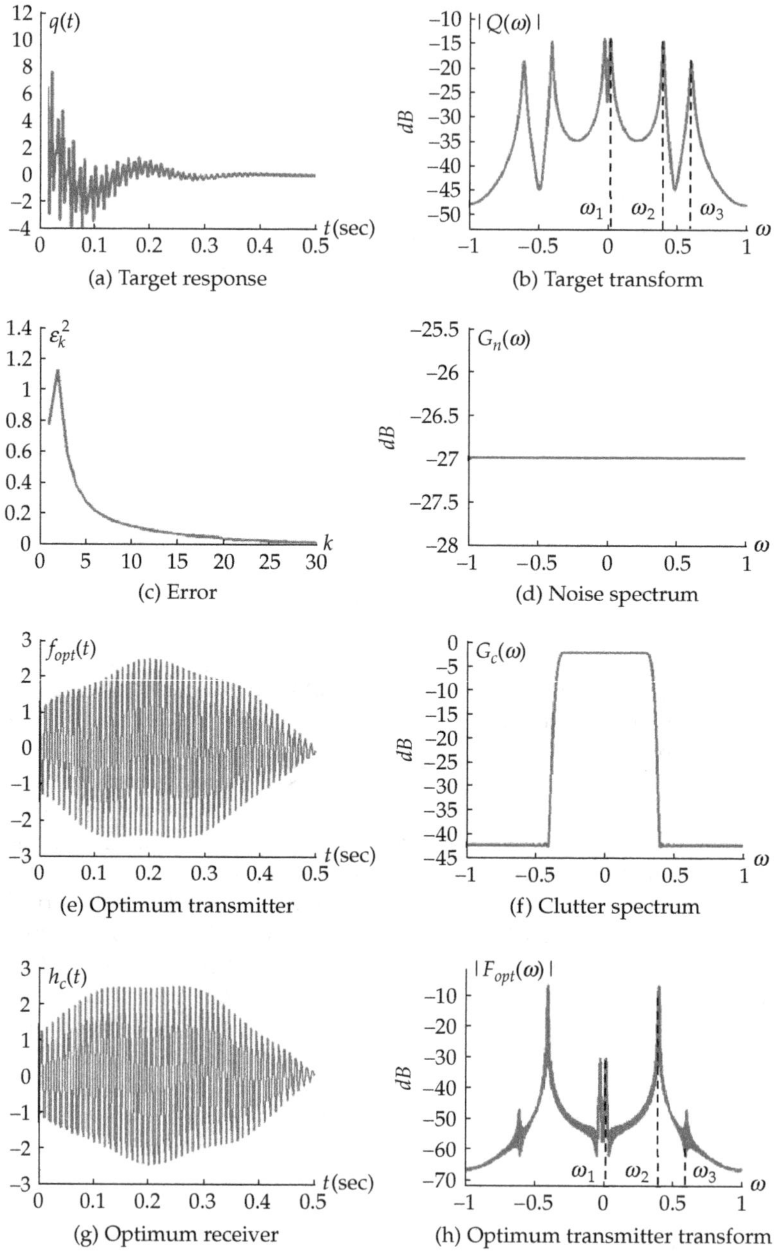

FIGURE 5.8 Optimum transmitter-receiver design in presence of clutter and noise. Optimum transmitter latches onto dominant resonant frequency in the clutter free region. Input $TNR = 0$ dB and $CNR = 20$ dB. Output $SINR = 11.5$ dB. A comparable input chirp generates an output $SINR = -1.5$ dB.

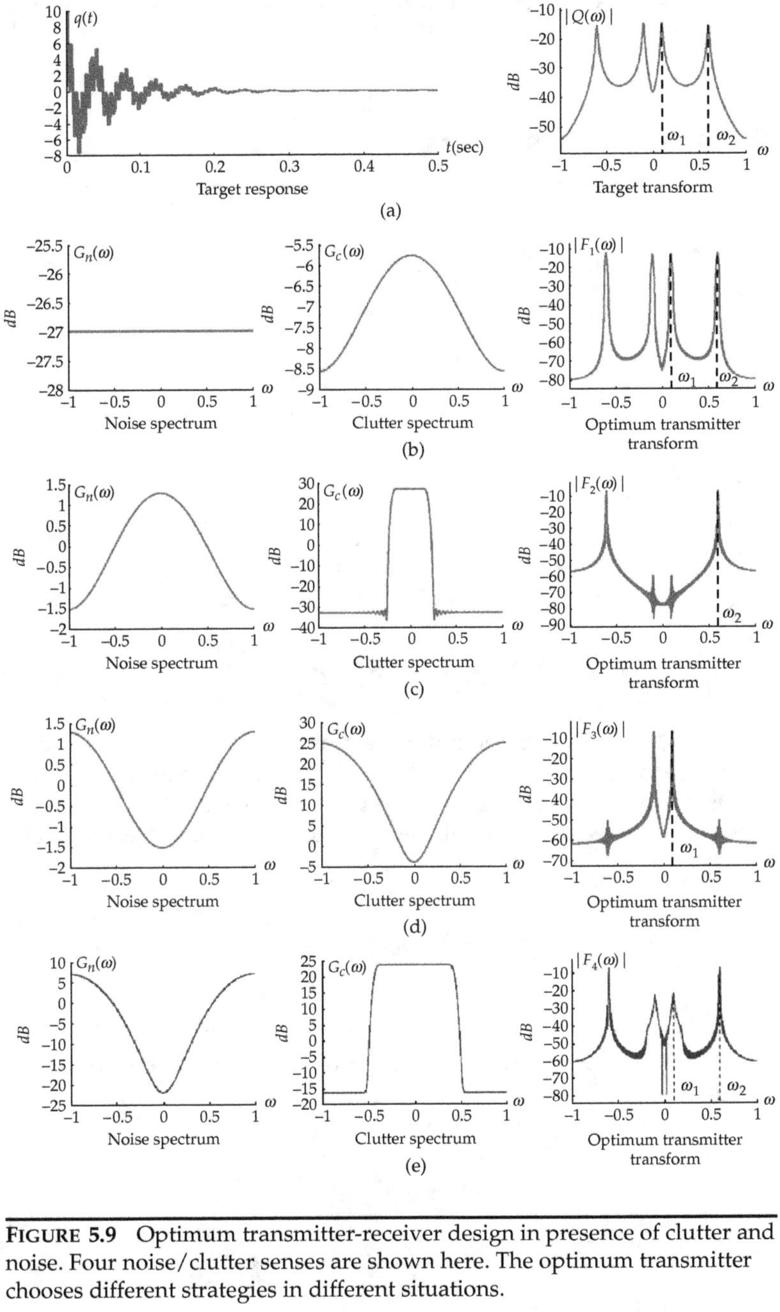

FIGURE 5.9 Optimum transmitter-receiver design in presence of clutter and noise. Four noise/clutter senses are shown here. The optimum transmitter chooses different strategies in different situations.

that if the distinguishing features among the target, clutter, and noise are not significant, then the optimum transmitter design payoff will not be very significant as well [28, 29].

5.3 Rational All-Pass Filters

A rational all-pass filter $H(z)$ of degree n has the form

$$H(z) = \frac{a_n^* + a_n^* z^{-1} + \cdots + a_o^* z^{-n}}{a_o + a_1 z^{-1} + \cdots + a_n z^{-n}} = \frac{\tilde{A}(z)}{A(z)} \tag{5.71}$$

where

$$A(z) = a_o + a_1 z^{-1} + \cdots + a_n z^{-n} \tag{5.72}$$

and

$$\tilde{A}(z) = z^n A_*(z) \tag{5.73}$$

with

$$A_*(z) = A^*(1/z^*). \tag{5.74}$$

Here $\tilde{A}(z)$ represents the polynomial reciprocal to $A(z)$, and $A_*(z)$ is the paraconjugate of $A(z)$. If $H(z)$ is also stable, then the polynomial $A(z)$ is free of zeros outside the closed unit circle ($|z| \leq 1$) and it is said to be strict-Hurwitz, that is, all zeros of $A(z)$ are inside the unit circle. Let

$$H(z) = \sum_{k=0}^{\infty} b_k z^{-k}, \qquad |z| \geq 1 \tag{5.75}$$

represent its power series expansion that is valid in $|z| \geq 1$. Equating (5.71) and (5.75), cross multiplying and comparing coefficients, we obtain

$$\begin{aligned} a_n^* &= b_o a_o \\ a_{n-1}^* &= b_1 a_o + b_o a_1 \\ &\;\vdots \\ a_o^* &= b_n a_o + b_{n-1} a_1 + \cdots + b_o a_n \end{aligned} \tag{5.76}$$

and in general

$$b_k a_o + b_{k-1} a_1 + \cdots + b_{k-n} a_n = 0, \qquad k > n. \tag{5.77}$$

By direct substitution, we have

$$H_*(z)H(z) = H(z)H_*(z) = \frac{z^n A_*(z)}{A(z)} \frac{z^{-n} A(z)}{A_*(z)} = 1 \tag{5.78}$$

which $z = e^{j\omega}$ translates into

$$|H(e^{j\omega})|^2 = 1 \tag{5.79}$$

that exhibits the all-pass nature of (5.71). Substituting (5.75) into (5.78) and comparing coefficients of like powers, we get

$$\sum_{k=0}^{\infty} |b_k|^2 = 1 \tag{5.80}$$

and

$$\sum_{k=0}^{\infty} b_k b_{k+m}^* = 0, \qquad m \geq 1. \tag{5.81}$$

To represent equations (5.80) and (5.81) compactly in matrix form, define the lower-triangular infinite dimensional Toeplitz matrix **H** as

$$\mathbf{H} = \left(\begin{array}{cccc|ccc} b_o & 0 & \cdots & 0 & & & \\ b_1 & b_o & \ddots & 0 & & 0 & \\ \vdots & \vdots & \ddots & 0 & & & \\ b_n & b_{n-1} & \cdots & b_o & & & \\ \hline b_{n+1} & b_n & \cdots & b_1 & b_o & 0 & \cdots \\ \vdots & \vdots & \vdots & \vdots & \vdots & \vdots & \vdots \\ b_m & b_{m-1} & \cdots & b_{m-n} & b_{m-n-1} & \cdots & \cdots \\ \vdots & \vdots & \vdots & \vdots & \vdots & \vdots & \vdots \end{array}\right) = \left(\begin{array}{c|c} \mathbf{H}_n & \mathbf{0} \\ \hline \mathbf{X}_n & \mathbf{Y}_n \end{array}\right). \tag{5.82}$$

Using (5.82), equations (5.78)–(5.81) can be written as

$$\mathbf{H}^*\mathbf{H} = \mathbf{I} \tag{5.83}$$

or

$$\left(\begin{array}{c|c} \mathbf{H}_n^* & \mathbf{X}_n^* \\ \hline \mathbf{0} & \mathbf{Y}_n^* \end{array}\right)\left(\begin{array}{c|c} \mathbf{H}_n & \mathbf{0} \\ \hline \mathbf{X}_n & \mathbf{Y}_n \end{array}\right) = \left(\begin{array}{c|c} \mathbf{I} & \mathbf{0} \\ \hline \mathbf{0} & \mathbf{I} \end{array}\right) \tag{5.84}$$

which gives

$$\mathbf{H}_n^*\mathbf{H}_n + \mathbf{X}_n^*\mathbf{X}_n = \mathbf{I}_n, \tag{5.85}$$

$$\mathbf{X}_n^*\mathbf{Y}_n = 0, \tag{5.86}$$

and

$$\mathbf{Y}_n^*\mathbf{Y}_n = \mathbf{I}. \tag{5.87}$$

From (5.85),

$$\mathbf{I}_n - \mathbf{H}_n^*\mathbf{H}_n = \mathbf{X}_n^*\mathbf{X}_n \geq 0, \tag{5.88}$$

that is, all eigenvalues of $\mathbf{H}_n^*\mathbf{H}_n$ are bounded by unity.

If we define

$$\underline{a} = \left(a_o, a_1, \cdots a_n\right)^T, \tag{5.89}$$

then (5.77) can be expressed as

$$\mathbf{X}_n \underline{a} = \underline{0}. \tag{5.90}$$

Hence from (5.88)

$$\underline{a}^* \left(\mathbf{I} - \mathbf{H}_n^*\mathbf{H}_n\right) \underline{a} = \underline{a}^*\mathbf{X}_n^*\mathbf{X}_n\underline{a} = 0 \tag{5.91}$$

$$\Rightarrow \left(\mathbf{I}_n - \mathbf{H}_n^*\mathbf{H}_n\right) \underline{a} = 0 \tag{5.92}$$

$$\mathbf{H}_n^*\mathbf{H}_n\underline{a} = \underline{a} \tag{5.93}$$

that is, $\mathbf{I}_n - \mathbf{H}_n^*\mathbf{H}_n$ is singular and $\mathbf{H}_n^*\mathbf{H}_n$ has its largest eigenvalue equal to unity, and the corresponding eigenvector is given by $\underline{a}$, the vector generated by the coefficients of the all-pass function.

Since (5.77) represents the simplest such representation, we also have

$$rank(\mathbf{X}_n) = n \tag{5.94}$$

and using (5.88), this gives

$$rank\left(\mathbf{I}_n - \mathbf{H}_n^*\mathbf{H}_n\right) = n. \tag{5.95}$$

Hence the largest eigenvalue of $\mathbf{H}_n^*\mathbf{H}_n$ is simple, and the associated eigenvector $\underline{a}$ is also unique.

Proceeding in a similar manner, it follows that for any $k \geq n$, with $\mathbf{H}_k$ representing the lower triangular Toeplitz matrix generated from $b_o, b_1, \cdots b_k$ as in (5.82), we have

$$\mathbf{I}_n - \mathbf{H}_k^*\mathbf{H}_k \geq 0, \tag{5.96}$$

and

$$rank\left(\mathbf{I}_n - \mathbf{H}_k^*\mathbf{H}_k\right) = rank\,(\mathbf{X}_k) = n, \quad k \geq n. \tag{5.97}$$

From (5.96), since the eigenvalues of $\mathbf{H}_k^*\mathbf{H}_k$ are bounded by unity, it follows that for $k \geq n$, the largest $k - n + 1$ eigenvalues of $\mathbf{H}_k^*\mathbf{H}_k$ are equal to unity.

Equations (5.92) and (5.93) also follow from (5.76). To see this, observe that (5.76) can be expressed as

$$\mathbf{H}_n\,\underline{a} = \mathbf{J}\,\underline{\bar{a}} \tag{5.98}$$

where

$$\mathbf{J} = \begin{pmatrix} 0 & \cdots & 0 & 1 \\ 0 & \cdots & 1 & 0 \\ \vdots & \iddots & \vdots & \vdots \\ 1 & 0 & \cdots & 0 \end{pmatrix}. \tag{5.99}$$

This gives

$$\underline{a}^*\mathbf{H}_n^*\mathbf{H}_n\,\underline{a} = (\mathbf{J}\,\underline{\bar{a}})^*(\mathbf{J}\,\underline{\bar{a}}) = \overline{(\underline{a}^*\underline{a})} = \underline{a}^*\underline{a} \tag{5.100}$$

since $\mathbf{J}^2 = \mathbf{I}_n$. Thus,

$$\underline{a}^*(\mathbf{I}_n - \mathbf{H}_n^*\mathbf{H}_n)\underline{a} = 0 \tag{5.101}$$

which is the same as (5.91).

In summary, for all all-pass functions

$$rank\,(\mathbf{H}_k) = k, \tag{5.102}$$

$$rank\,(\mathbf{I}_k - \mathbf{H}_k^*\mathbf{H}_k) = \begin{cases} k, & k \leq n \\ n, & k > n \end{cases} \tag{5.103}$$

and the coefficient vector $\underline{a}$ satisfies

$$\mathbf{H}_n^*\mathbf{H}_n\underline{a} = \underline{a}. \tag{5.104}$$

References

[1] S.U. Pillai, H.S. Oh, D.C. Youla, and J.R. Guerci, "Optimum Transmit-Receiver Design in the Presence of Signal-Dependent Interference and Channel Noise", *IEEE Transactions on Information Theory*, Vol. 46, No. 2, pps. 577–584, March 2000.

[2] R. Manasse, The Use of Pulse Coding to Discriminate Against Clutter, Technical Report 312-12, *MIT Lincoln Laboratory*, Lexington, MA, Aug. 1957.

[3] K.W. Forsythe, D.W. Bliss, "Waveform Correlation and Optimization Issues for MIMO Radar," *Proc. 39th Asilomar Conf. Signals, Systems, Computer*, Pacific Grove, CA, pp. 1306–1310, Nov. 2005.

[4] B. Friedlander, "Waveform Design for MIMO Radars," *IEEE Transactions on Aerospace and Electronic Systems*, vol. 43, no. 3, pp. 1227–1238, Oct. 2007.

[5] Y. Chi, A. Pezeshki, R. Calderbank, S. Howard, "Range Sidelobe Suppression in a Desired Doppler Interval," *IEEE International Waveform Diversity & Design Conference*, Orlando, FL, Feb. 8–13, 2009.

[6] Y. Li, W. Moran, S. Sira, A. Papandreou-Suppappola, "Adaptive Waveform Design in Rapidly Varying Radar Scenes," *IEEE International Waveform Diversity & Design Conference*, Orlando, FL, Feb. 8–13, 2009.

[7] M. Inggs, M. Brooker, "Extensible Simulator for Waveform Diversity Testing," *IEEE International Waveform Diversity & Design Conference*, Orlando, FL, Feb. 8–13, 2009.

[8] Y. Paichard, M. Brooker, M. Inggs, "A Signal Level Simulator for Netted Radar Waveforms Evaluation," *IEEE International Waveform Diversity & Design Conference*, Orlando, FL, Feb. 8–13, 2009.

[9] M. Inggs, "Passive Coherent Location as Cognitive Radar," *IEEE International Waveform Diversity & Design Conference*, Orlando, FL, Feb. 8–13, 2009.

[10] H. Mendelson, P. Zulch, "Mitigation of Radar Range Ambiguities via Waveform Diversity," *IEEE International Waveform Diversity & Design Conference*, Orlando, FL, Feb. 8–13, 2009.

[11] J. Goodman, B. Miller, M. Herman, M. Wai, P. Monticciolo, "Extending the Dynamic Range of RF Receiver with Nonlinear Equalization," *IEEE International Waveform Diversity & Design Conference*, Orlando, FL, Feb. 8–13, 2009.

[12] J. Zhang, A. Papandreou-Suppappola, "MIMO Radar with Frequency Diversity," *IEEE International Waveform Diversity & Design Conference*, Orlando, FL, Feb. 8–13, 2009.

[13] J. Li, J.R. Guerci, L. Xu, "Signal Waveform's Optimal-Under-Restriction Design for Active Sensing," *IEEE Signal Process. Lett.*, vol. 13, no. 9, pp. 565–568, Sep. 2006.

[14] P.F. Sammartino, C. Baker, "The Frequency Diverse Bistatic System," *IEEE International Waveform Diversity & Design Conference*, Orlando, FL, Feb. 8–13, 2009.

[15] T. Sturman, "Enabling Chaotic Waveform Diversity," *IEEE International Waveform Diversity & Design Conference*, Orlando, FL, Feb. 8–13, 2009.

[16] Y. Paichard, "Orghogonal Multicarrier Phased Coded Signal for Netted Radar Systems," *IEEE International Waveform Diversity & Design Conference*, Orlando, FL, Feb. 8–13, 2009.

[17] J. Stiles, J. Jenshak, "Sparse Array Construction Using Marginal Fisher's Information," *IEEE International Waveform Diversity & Design Conference*, Orlando, FL, Feb. 8–13, 2009.

[18] H.B. Manjunath, J. Zhang, A. Papandreou-Suppappola, D. Morrell, "Waveform-Agile Sensing for Range and DoA Estimation in MIMO Radars," *IEEE International Waveform Diversity & Design Conference*, Orlando, FL, Feb. 8–13, 2009.

[19] Y. Yang, R. Blum, "Waveform Design for MIMO Radar Based on Mutual Information and Minimum Mean-Square Error Estimation,"*Proc. 40th Annual Conference on Information Sciences and Systems*, pp. 111–116, March 2006.

[20] S. Kay, "Optimal Signal Design for Detection of Gaussian Point Targets in Stationary Gaussian Clutter/Reverberation," *IEEE J. Sel. Topics in Sig. Proc. Mag.*, vol. 1, no. 1, pp. 31–41, June 2007.

[21] D.A. Garren, M.K. Osborn, A.C. Odom, J.S. Goldstein, S.U. Pillai, J.R. Guerci, "Enhanced Target Detection and Identification via Optimized Radar Transmission Pulse Shape," *Proc. IEEE*, vol. 148, no. 3, pp. 130–138, June 2001.

[22] D.T. Gjessing, *Target Adaptive Matched Illumination Radar: Principles and Applications*, London, England: Peter Peregrinus, 1986.

[23] F. Ahmad, M. Amin, H. Estephan, "Multistatic Waveform Design for Seeing Through the Wall," *IEEE International Waveform Diversity & Design Conference*, Orlando, FL, Feb. 8–13, 2009.

[24] L.L. Monte, R. Ansari, D. Erricolo, M. Wicks, "The Use of Geometric Diversity for Spectral Dominance in Underground Imaging," *IEEE International Waveform Diversity & Design Conference*, Orlando, FL, Feb. 8–13, 2009.

[25] H. Deng, "Polyphase Coding Signal Design for Orthogonal Netted Radar Systems," *IEEE Transactions on Signal Processing*, vol. 52, no. 11, Nov. 2004.

[26] D.D. Weiner, M.C. Wicks, G.T. Capraro, "Waveform Diversity and Sensors as Robots in Advanced Military Systems," *1st International Waveform Diversity and Design Conference*, Edinburgh, UK, Nov. 2004.

[27] R. Bonneau, "A Coherence Gradient Method in Waveform Design for Apertures," *IEEE International Waveform Diversity & Design Conference*, Orlando, FL, Feb. 8–13, 2009.

[28] R. Romero, N. Goodman, "Improved Waveform Design for Target Recognition with Multiple Transmissions," *IEEE International Waveform Diversity & Design Conference*, Orlando, FL, Feb. 8–13, 2009.

[29] M.R. Bell, "Information Theory and Radar Waveform Design," *IEEE Trans. Inform. Theory*, vol. 39, no. 5, pp. 1578–1597, Sep. 1993.

CHAPTER 6

Discrete-Time Waveform Design

The discrete-time domain implementation of the concepts and algorithms described in earlier chapters with regard to both transmitter design and receiver implementation is described in this chapter. Most of the algorithms follow from their continuous counterpart, and the discrete case affords efficient matrix representation whereby matrix theory results can be used to make similar conclusions.

6.1 Matched Filter

In the discrete version of the problem [see (2.3)], the received signal $r(n)$ is the sum of an incoming signal $s(n)$ and noise $w(n)$. Thus,

$$r(n) = s(n) + w(n), \quad n = 0,\ 1,\ 2, \ldots N, \tag{6.1}$$

where $(0 \rightarrow N)$ represents the finite observation interval. The incoming signal $s(n)$ is usually the output of a channel/target with impulse response $q(n)$. If we assume $f(n)$ to be the transmit signal, in a linear time-invariant model we have[1] (Fig. 6.1)

$$s(n) = f(n) * q(n) = \sum_{k=0}^{N} q(n-k) f(k), \quad n = 0,\ 1,\ 2, \ldots N \tag{6.2}$$

[1] In (6.2), we are assuming the channel response and the transmit signal to be causal. The noncausal case follows in a similar manner.

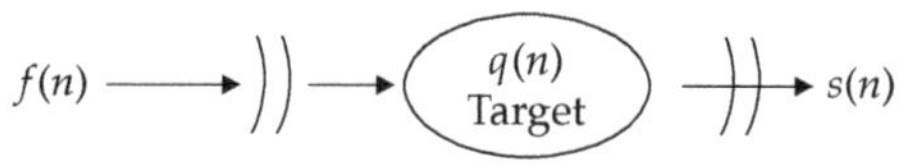

FIGURE 6.1 Transmit signal and target response.

which can be represented in matrix form as

$$\underline{s} = \begin{pmatrix} s_o \\ s_1 \\ s_2 \\ \vdots \\ s_N \end{pmatrix} = \begin{pmatrix} q_o & 0 & 0 & \cdots & 0 \\ q_1 & q_o & 0 & \cdots & 0 \\ q_2 & q_1 & q_o & \ddots & 0 \\ \vdots & \vdots & \vdots & \ddots & \vdots \\ q_N & q_{N-1} & q_{N-1} & \cdots & q_o \end{pmatrix} \begin{pmatrix} f_o \\ f_1 \\ f_2 \\ \vdots \\ f_N \end{pmatrix} = \mathbf{Q}\underline{f} \quad (6.3)$$

where we define

$$s_k = s(k), \quad f_k = f(k), \quad q_k = q(k). \quad (6.4)$$

In (6.3), **Q** represents the lower triangular Toeplitz matrix generated from the target response $q_o, q_1, \ldots, q_N$ shown there, and

$$\underline{f} = (f_o, \;\; f_1, \;\; \cdots \;\; f_N)^T \quad (6.5)$$

a column vector of transmit signal values. In general, if there are no restrictions on the transmit signal length, then we can assume the transmit signal to have the same duration as the number of observations in (6.3). The transmit signal can be shorter in duration compared to the observations, in which case, the transmit signal vector in (6.5) has the form

$$\underline{f} = (f_1, \;\; f_2, \;\; \cdots \;\; f_M, \;\; 0 \;\; \cdots \;\; 0)^T = \left(\underline{f}_1^T, \;\; \underline{0}\right)^T \quad (6.6)$$

and using this in (6.3), we can rewrite the data vector in (6.3) as

$$\underline{s} = \begin{pmatrix} s_o \\ s_1 \\ \vdots \\ s_M \\ \vdots \\ s_N \end{pmatrix} = \begin{pmatrix} q_o & 0 & \cdots & 0 \\ q_1 & q_o & \ddots & \vdots \\ \vdots & \vdots & \ddots & 0 \\ q_M & q_{M-1} & \cdots & q_o \\ \vdots & \vdots & \ddots & \vdots \\ q_N & q_{N-1} & \cdots & q_{N-M} \end{pmatrix} \begin{pmatrix} f_o \\ f_1 \\ f_2 \\ \vdots \\ f_M \end{pmatrix} = \mathbf{Q}_1 \underline{f}_1, \quad (6.7)$$

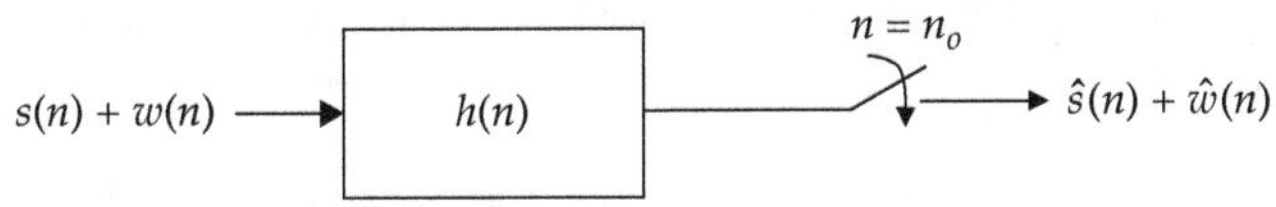

FIGURE 6.2 Receiver design.

where $\underline{f}_1$ is the truncated version of $\underline{f}$ consisting of its nonzero part. Notice that the $\mathbf{Q}_1$ in (6.7) is also a truncated version of $\mathbf{Q}$ where only the first $(M+1)$ columns of $\mathbf{Q}$ are retained. Thus

$$\mathbf{Q}_1 = \begin{pmatrix} q_o & 0 & \cdots & 0 \\ q_1 & q_o & \ddots & \vdots \\ \vdots & \vdots & \ddots & 0 \\ q_M & q_{M-1} & \cdots & q_o \\ \vdots & \vdots & \ddots & \vdots \\ q_N & q_{N-1} & \cdots & q_{N-M} \end{pmatrix} \tag{6.8}$$

and

$$\underline{f}_1 = (f_o, \quad f_1, \quad \cdots \quad f_M)^T. \tag{6.9}$$

Proceeding as in Fig. 2.4, with $h(n) \leftrightarrow H(e^{j\omega})$ representing the receiver filter impulse response and transfer function to be designed, let

$$\hat{s}(n) = s(n) * h(n) \tag{6.10}$$

and

$$\hat{w}(n) = w(n) * h(n) \tag{6.11}$$

represent the receiver output signal and output noise components as shown in Fig. 6.2.

To improve the output signal at decision instant $n = n_o$, the output signal to noise ratio (SNR_o) given by

$$SNR_o = \frac{|\hat{s}(n_o)|^2}{E\{|\hat{w}(n)|^2\}} \tag{6.12}$$

can be maximized as in (2.9). Following (2.15), the average output noise power is given by

$$E\{|\tilde{w}(n)|^2\} = \frac{1}{2\pi}\int_{-\pi}^{\pi} G_w(\omega)|H(e^{j\omega})|^2 d\omega \tag{6.13}$$

where $G_w(\omega)$ represents the input noise power spectrum, and $H(z)$ the unknown receiver transfer function. If we assume the input noise is white, then $G_w(\omega) = \sigma^2$, and following the procedure in (2.9)–(2.28), we obtain the optimum receiver to be the matched filter given by

$$H(e^{j\omega}) = S^*(e^{j\omega})e^{j\omega n_o} \tag{6.14}$$

or the receiver impulse response is given by

$$h(n) = s^*(n_o - n). \tag{6.15}$$

In this case, the maximum attainable *SNR* corresponds to

$$SNR = \frac{\sum_n |s(n)|^2}{\sigma^2}. \tag{6.16}$$

In general, $s(n)$ can be nonzero beyond $n = n_o$, and in those cases, the receiver in (6.15) will be noncausal. In situations where it is desirable to have a causal receiver filter, we can proceed as in Sec. 2.2.2 to derive the optimum causal matched filter. This is illustrated next where the input noise is assumed to be nonwhite with power spectral density $G_w(\omega)$.

Let $L(z)$ represent the minimum phase spectral factorization filter associated with the power spectrum $G_w(\omega)$. Then

$$G_w(\omega) = |L(e^{j\omega})|^2 \tag{6.17}$$

and $L(z)$ is stable together with its inverse $L^{-1}(z)$. In that case, the average output noise power is given by

$$E\{|\tilde{w}(n)|^2\} = \frac{1}{2\pi}\int_{-\pi}^{\pi} |L(e^{j\omega})|^2|H(e^{j\omega})|^2 d\omega. \tag{6.18}$$

We can use these expressions to derive the optimum causal matched filter. Notice that the above spectral factorization is always possible provided the power spectrum $G_w(\omega)$ satisfies the condition

$$\frac{1}{2\pi}\int_{-\pi}^{\pi} \ln G_w(\omega)\, d\omega > -\infty. \tag{6.19}$$

6.1.1 Optimum Causal Matched Filter

To obtain the causal matched filter, define the signal $g(n)$ as [see (2.59)]

$$g(n) \leftrightarrow L^{-1}(e^{j\omega})S(e^{j\omega}) = L^{-1}(e^{j\omega})Q(e^{j\omega})F(e^{j\omega}), \tag{6.20}$$

where

$$S(e^{j\omega}) = \sum_n s(n)\, e^{-jn\omega} \tag{6.21}$$

$$F(e^{j\omega}) = \sum_n f(n)\, e^{-jn\omega} \tag{6.22}$$

so that

$$g^*(n_o - n) \leftrightarrow L^{*-1}(e^{j\omega})S^*(e^{j\omega})e^{-jn_o\omega}. \tag{6.23}$$

From Fig. 6.2, the output signal component at the decision instant $n = n_o$ can be expressed as

$$\begin{aligned} \hat{s}(n_o) &= \frac{1}{2\pi}\int_{-\pi}^{\pi} H(e^{j\omega})S(e^{j\omega})e^{jn_o\omega}d\omega \\ &= \frac{1}{2\pi}\int_{-\pi}^{\pi} \{L(e^{j\omega})H(e^{j\omega})\}\{L^{*-1}(e^{j\omega})S^*(e^{j\omega})e^{-jn_o\omega}\}^* d\omega \\ &= \sum_{n=-\infty}^{\infty} v(n)g(n_o - n) \end{aligned} \tag{6.24}$$

where we have made use of (6.23) and Parsevel's theorem. Here

$$v(n) \leftrightarrow L(e^{j\omega})H(e^{j\omega}). \tag{6.25}$$

Observe that $v(n)$ in (6.25) represents a causal signal since the spectral factor $L(z)$ represents a causal filter, and the receiver $H(z)$ is by definition also causal. Hence the right side of (6.24) can be expressed as

$$\begin{aligned}\hat{s}(n_o) &= \sum_{n=-\infty}^{\infty} v(n)g(n_o-n) = \sum_{n=0}^{\infty} v(n)g(n_o-n)\\ &= \sum_{n=-\infty}^{+\infty} v(n)g(n_o-n)u(n)\\ &= \sum_{n=-\infty}^{+\infty} v(n)\{g^*(n_o-n)u(n)\}^*\\ &= \frac{1}{2\pi}\int_{-\pi}^{\pi}\{L(e^{j\omega})H(e^{j\omega})\}K^*(j\omega)d\omega\end{aligned} \tag{6.26}$$

where $u(n)$ represents the unit step function and $K(j\omega)$ represents the transform of the causal signal $g^*(n_o-n)u(n)$. Thus

$$g^*(n_o-n)u(n) \leftrightarrow K(e^{j\omega}). \tag{6.27}$$

Using (6.26) in (6.24) we get the output signal component to be

$$\hat{s}(n_o) = \frac{1}{2\pi}\int_{-\pi}^{\pi} L(e^{j\omega})H(e^{j\omega})K^*(e^{j\omega})\,d\omega. \tag{6.28}$$

Substituting (6.28) and (6.18) into (6.12) and applying Schwartz' inequality, we obtain

$$\begin{aligned}SNR_o &= \frac{\left|\frac{1}{2\pi}\int_{-\pi}^{\pi} L(e^{j\omega})H(e^{j\omega})K^*(e^{j\omega})\,d\omega\right|^2}{\frac{1}{2\pi}\int_{-\pi}^{\pi}|L(e^{j\omega})H(e^{j\omega})|^2 d\omega}\\ &\leq \frac{1}{2\pi}\int_{-\pi}^{\pi}|K(e^{j\omega})|^2 d\omega = \sum_{n=-\infty}^{+\infty}|g^*(n_o-n)|^2 u(n)\\ &= \sum_{n=0}^{n_0}|g(n)|^2 = SNR_{\max}.\end{aligned} \tag{6.29}$$

Hence the maximum SNR is realized when

$$L(e^{j\omega})H_c(e^{j\omega}) = K(e^{j\omega}) \tag{6.30}$$

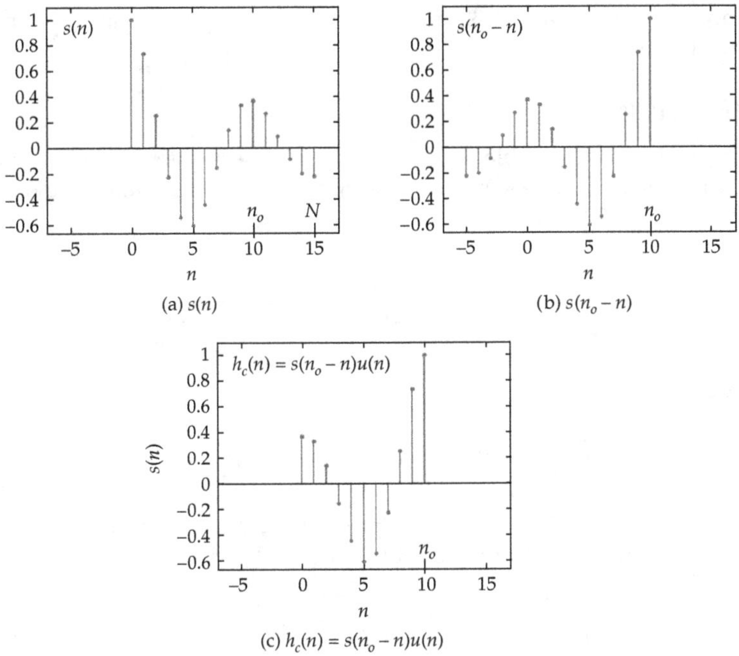

FIGURE 6.3 Optimum Receiver.

or the transform associated with the optimum causal receiver is given by

$$H_c(e^{j\omega}) = L^{-1}(e^{j\omega})K(e^{j\omega}). \tag{6.31}$$

In the time domain, this gives the optimum causal receiver to be

$$h_c(n) = l_{inv}(n) * g^*(n_o - n)u(n). \tag{6.32}$$

Here $l_{inv}(n)$ represents the impulse response of the causal filter $L^{-1}(e^{j\omega})$. In the white noise case

$$L(e^{j\omega}) = \sigma \tag{6.33}$$

so that $g(n) = s(n)$ in (6.20) and

$$h_c(n) = g^*(n_o - n)u(n) = s^*(n_o - n)u(n). \tag{6.34}$$

Thus, if the input noise is white, as Fig. 6.3(b)–(c) shows, the optimum causal matched filter is the truncated impulse response of the non causal filter in (6.15). However, if the noise is not white, then this is no longer true, and (6.32) must be used to determine the optimum causal receiver.

Maximum *SNR*

The maximum realizable *SNR* in the causal case is given by (6.29), and it can be used to make some interesting conclusions about the transmit signal in Fig. 6.1. From (6.29)

$$SNR_{\max} = \sum_{n=0}^{n_o} |g(n)|^2 = \underline{g}^* \underline{g} \tag{6.35}$$

where

$$\underline{g} = [\, g(0), \quad g(1), \quad \cdots \quad g(n_o)\,]^T. \tag{6.36}$$

From (6.20), define $P(e^{j\omega})$ to represent the combined inverse filter $L^{-1}(e^{j\omega})$ together with the target transform $Q(e^{j\omega})$. Thus

$$P(e^{j\omega}) = L^{-1}(e^{j\omega}) Q(e^{j\omega}) \triangleq \sum_{k=0}^{\infty} p_k e^{-jk\omega} \tag{6.37}$$

so that

$$p(n) = l_{inv}(n) * q(n) \tag{6.38}$$

and from (6.20)

$$g(n) = p(n) * f(n) = \sum_{k=0}^{n} p_{n-k} f_k. \tag{6.39}$$

Using (6.39), the vector in (6.36) can be expressed as

$$\underline{g} = \begin{pmatrix} g(0) \\ g(1) \\ \vdots \\ g(n_o) \end{pmatrix} = \begin{pmatrix} p_o & 0 & \cdots & 0 \\ p_1 & p_o & \ddots & \vdots \\ \vdots & \vdots & \ddots & 0 \\ p_{n_o} & p_{n_o-1} & \cdots & p_o \end{pmatrix} \begin{pmatrix} f_o \\ f_1 \\ \vdots \\ f_{n_o} \end{pmatrix} = \mathbf{P} \underline{f} \tag{6.40}$$

where $\underline{f}$ is as defined in (6.5) with N replace by n_o. Hence

$$SNR_{\max} = \underline{g}^* \underline{g} = \underline{f}^* \mathbf{P}^* \mathbf{P} \, \underline{f} \tag{6.41}$$

where **P** represents the lower triangular Toeplitz matrix

$$\mathbf{P} = \begin{pmatrix} p_o & 0 & \cdots & 0 \\ p_1 & p_o & \ddots & \vdots \\ \vdots & \vdots & \ddots & 0 \\ p_{n_o} & p_{n_o-1} & \cdots & p_o \end{pmatrix} = \begin{pmatrix} l_o & 0 & \cdots & 0 \\ l_1 & l_o & \ddots & \vdots \\ \vdots & \vdots & \ddots & 0 \\ l_{n_o} & l_{n_o-1} & \cdots & l_o \end{pmatrix}$$
$$\times \begin{pmatrix} q_o & 0 & \cdots & 0 \\ q_1 & q_o & \ddots & \vdots \\ \vdots & \vdots & \ddots & 0 \\ q_{n_o} & q_{n_o-1} & \cdots & q_o \end{pmatrix} = \mathbf{L}\mathbf{Q} \tag{6.42}$$

with p_k defined as in (6.38). Notice that if the additive noise is white, then $p_k = q_k$ represents the target response characteristics and $\mathbf{P} = \mathbf{Q}$.

Optimum Transmitter

The design of an optimum transmit waveform can be given a meaningful formulation using this approach. We can use (6.41) and (6.42) to design an optimum transmit waveform $\underline{f}$. From (6.41), to further maximize $SNR_{\max}$, the transmit waveform $\underline{f}$ must be chosen as the eigenvector associated with the largest eigenvalue of the positive-definite matrix

$$\mathbf{\Omega} = \mathbf{P}^*\mathbf{P} > 0. \tag{6.43}$$

Notice that $\mathbf{\Omega}$ depends both on the channel/target as well as the noise spectral characteristics through (6.37) and (6.38). With n_o representing the decision instant, the normalized optimum transmit waveform $\underline{f}$ is given by the eigenvector $\underline{e}_1$ associated with the largest eigenvalue $\lambda_{\max}$ of the positive definite matrix $\mathbf{\Omega}$ in (6.43). If the largest eigenvalue $\lambda_{\max}$ does not repeat, then the corresponding eigenvector is unique. Thus the optimum transmit waveform is given by this unique eigenvector. Hence

$$\underline{f} = \underline{e}_1 \tag{6.44}$$

where

$$\mathbf{\Omega}\,\underline{e}_1 = \lambda_{\max}\,\underline{e}_1. \tag{6.45}$$

It is easy to show that the above optimum solution always represents a minimum phase solution,[2] provided the largest eigenvalue is simple (multiplicity is one).[3]

To prove the minimum phase nature of the optimum transmit signal $\underline{f} = \underline{e}_1$, let

$$\underline{e}_1 = \begin{pmatrix} e_o, & e_1, & \cdots & e_{n_o} \end{pmatrix}^T \tag{6.46}$$

and define the polynomial

$$F(z) = e_o + e_1 z^{-1} + \cdots + e_{n_o} z^{-n_o} \tag{6.47}$$

constructed from the eigenvector $\underline{e}_1$. Let $z_k, k = 1, 2, \cdots n_o$ represent the zeros of $F_1(z)$. Thus

$$F(z) = z^{-n_o} \prod_{k=1}^{n_o} (z - z_k). \tag{6.48}$$

If $F(z)$ is not minimum phase, then it has K zeros, say at $z = z_k, k = 1 \to K$, outside the unit circle, and hence it can be expressed as

$$F(z) = F_o(z) \prod_{k=1}^{K} \frac{z - z_k}{1 - z\, z_k} = F_o(z) B(z), \qquad |z_k| > 1 \tag{6.49}$$

where

$$B(z) = \prod_{k=1}^{K} \frac{z - z_k}{1 - z\, z_k} = \sum_{m=0}^{\infty} b_m z^{-m}, \quad |z| > 1, \quad |z_k| > 1 \tag{6.50}$$

represents a degree K stable all-pass function, and using the representation in (6.48), the new polynomial

$$F_o(z) = z^{-n_o} \prod_{k=1}^{K} (1 - z\, z_k) \prod_{k=K+1}^{n_o} (z - z_k) \tag{6.51}$$

is minimum phase with degree n_o and with all zeros of it inside the unit circle. Moreover, from (6.49) and (6.50), both $F_o(z)$ and $F(z)$ have the same energy ($\equiv 1$). Let

$$F_o(z) = \sum_{k=0}^{n_o} a_k z^{-k} \tag{6.52}$$

---

[2]A minimum phase polynomial has all its zeros inside the unit circle ($|z| < 1$).

[3]If $\Omega = \mathbf{P}^*\mathbf{P}$ is constructed from an all-pass function of degree n, then the largest eigenvalue has multiplicity equal to $k - n + 1$, where $k + 1$ represents the size of $\mathbf{P}$ [See (5.103)].

represent the new polynomial and define the vector

$$\underline{a} = (a_o, \quad a_1, \quad \cdots \quad a_{n_o})^T \tag{6.53}$$

associated with $F_o(z)$. We will show that the normalized vector $\underline{a}$ when used as the transmit vector satisfies

$$\underline{a}^*\Omega\underline{a} \geq \underline{e}_1^*\Omega\underline{e}_1 = \lambda_{\max} \tag{6.54}$$

violating the maximizing property of the unique eigenvector $\underline{e}_1$ in (6.45). To show (6.54), observe that using (6.44)–(6.46), (6.50), (6.52) and (6.53) in (6.49), we get

$$\underline{e}_1 = \begin{pmatrix} e_o \\ e_1 \\ \vdots \\ e_{n_o} \end{pmatrix} = \begin{pmatrix} b_o & 0 & \cdots & 0 \\ b_1 & b_o & \ddots & \vdots \\ \vdots & \vdots & \ddots & 0 \\ b_{n_o} & b_{n_o-1} & \cdots & b_o \end{pmatrix} \begin{pmatrix} a_o \\ a_1 \\ \vdots \\ a_{n_o} \end{pmatrix} = \mathbf{B}\underline{a} \tag{6.55}$$

where $\mathbf{B}$ is a lower triangular Toeplitz matrix generated from the all-pass function response in (6.50), and using (6.55) we get

$$\underline{e}_1^*\Omega\underline{e}_1 = \underline{a}^*\mathbf{B}^*\Omega\,\mathbf{B}\underline{a}. \tag{6.56}$$

Thus

$$\underline{a}^*\Omega\underline{a} - \underline{e}_1^*\Omega\underline{e}_1 = \underline{a}^*\mathbf{P}^*\mathbf{P}\underline{a} - \underline{a}^*\mathbf{B}^*\mathbf{P}^*\mathbf{P}\mathbf{B}\,\underline{a} = \underline{a}^*(\mathbf{P}^*\mathbf{P} - \mathbf{B}^*\mathbf{P}^*\mathbf{P}\mathbf{B})\,\underline{a} \tag{6.57}$$

where we have used (6.43). Using the classic result that lower triangular Toeplitz matrices commute, we have

$$\mathbf{PB} = \mathbf{BP} \tag{6.58}$$

and hence (6.57) simplifies to

$$\underline{a}^*\Omega\underline{a} - \underline{e}_1^*\Omega\underline{e}_1 = \underline{a}^*\mathbf{P}^*(\mathbf{I} - \mathbf{B}^*\mathbf{B})\mathbf{P}\underline{a} \geq 0 \tag{6.59}$$

since (see (5.96), Sec. 5.3)

$$\mathbf{I} - \mathbf{B}^*\mathbf{B} \geq 0 \tag{6.60}$$

which follows from $\mathbf{B}$ being a lower triangular Toeplitz matrix corresponding to an all-pass function as in (6.50). Equation (6.59) shows that the normalized minimum phase vector $\underline{a}$ is not an inferior transmit vector in terms of maximizing output *SNR*, and it contradicts the unique minimum phase nature of $\underline{e}_1$ in terms of maximizing the output *SNR*. Hence to start with, the original waveform $F(z)$ in (6.47) must be minimum phase. In summary, in the case of causal systems, when decisions are made based on a finite segment of the system output, the optimum transmit excitation waveform is always minimum phase irrespective of the nature of the additive noise at the receiver input. In the case of white noise, from (6.38)

$$p_k = q_k, \qquad k = 0 \to \infty \tag{6.61}$$

and the lower triangular Toeplitz matrix **P** in (6.42) can be generated using the target impulse response only. The following examples illustrate the minimum phase nature of the optimum transmit signal.

Example 6.1 Consider the impulse response sequences corresponding to the three targets given below (Fig. 6.4). Find the optimum transmit waveform in each case for the decision instant (a) $n_o = 10$ and (b) $n_o = 100$. Assume the additive noise

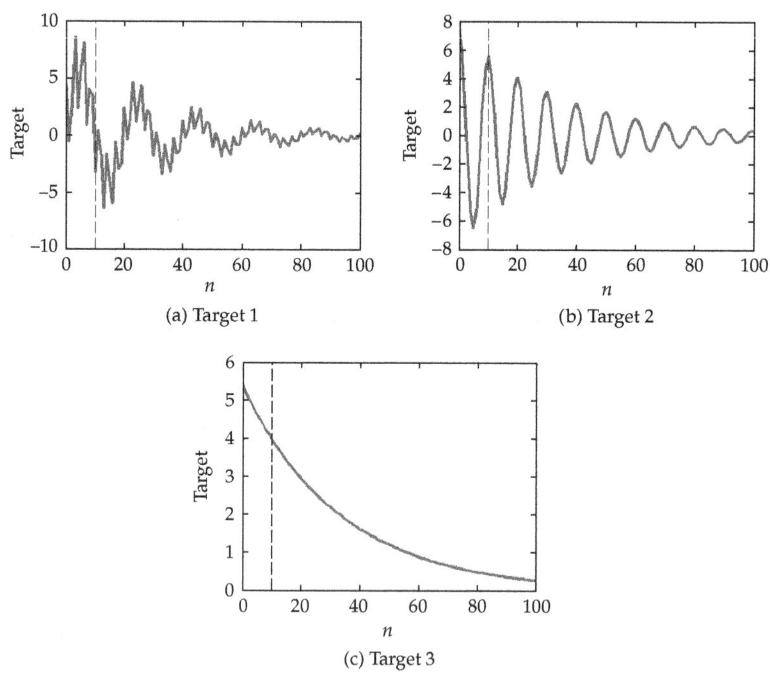

FIGURE 6.4 Three target responses.

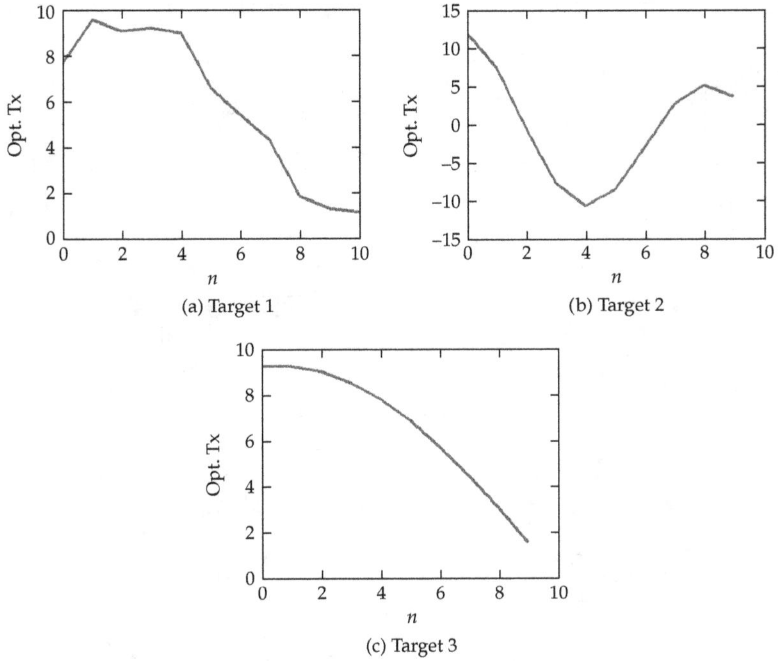

FIGURE 6.5 Optimum transmit signal of length $n_o = 10$ for targets 1, 2, and 3.

to be normalized white noise with $\sigma^2 = 1$.

$$\text{Target 1: } q_1(n) = 3e^{-0.03n}\cos(0.4\pi n) + 10e^{-0.32n}\sin(0.4\pi n), \tag{6.62}$$

$$\text{Target 2: } q_2(n) = e^{-0.03n}\cos(0.2\pi n) \tag{6.63}$$

and

$$\text{Target 3: } q_3(n) = e^{-0.03n} \tag{6.64}$$

Solution The optimum transmit signal $\underline{f}_1$ is given by the solution of (6.45). The optimum transmit signal responses $\{f_n\}$ with $n_o = 10$ for targets 1, 2, and 3 are shown in Fig. 6.5. Let

$$F(z) = \sum_{n=0}^{10} f_n z^{-n} \tag{6.65}$$

represent the polynomial associated with the optimum solution. The zeros of these three polynomials are plotted in Fig. 6.6(a)–(c). Notice that all solutions are minimum phase. Figure 6.7 shows the target spectra along with the transform of the optimum transmit signals generated using the transmit signal samples in

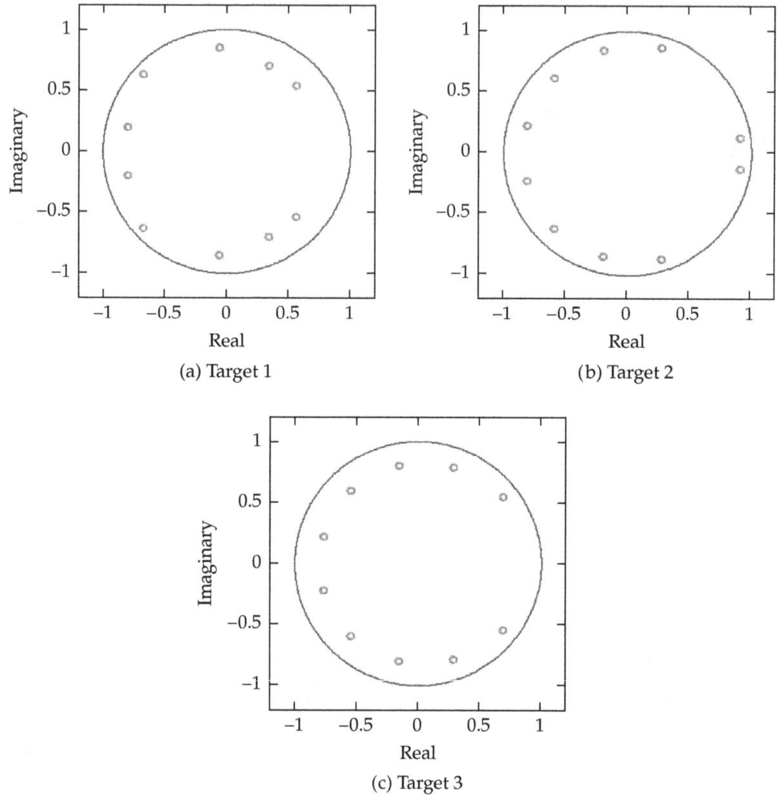

FIGURE 6.6 Zeros of the optimum transmit signals generated using (6.65) for targets 1, 2, and 3 for $n_o = 10$.

Fig. 6.5. Figure 6.7 shows that the transmit spectral peaks coincide with the target spectral peaks, implying that the finite duration optimum transmit signal tries to pump out energy corresponding to those frequencies where the target also has pronounced responses. This results in sympathetic resonating echo returns from the target that have superior amplitudes compared to arbitrary transmit signals.

Figure 6.8 shows the optimum transmit waveforms for the three targets shown in Fig. 6.4 with $n_o = 100$. Figure 6.9 shows the 100 zeros of these waveforms and Fig. 6.10 shows the frequency transform of these optimum transmit waveforms. From there, the transmit signals deliver energy highly selectively in the frequency domain, and this is clearly highlighted for transmit signal associated with target 1. Notice that both Figs. 6.7(d) and 6.10(a) refer to the transform of optimum

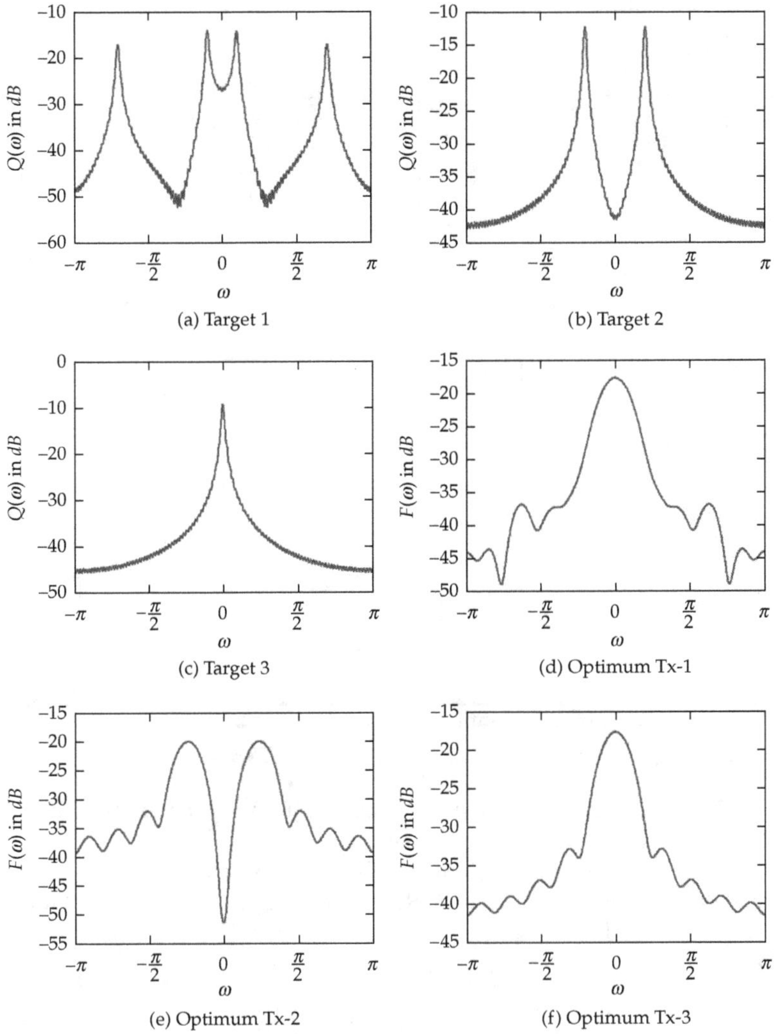

FIGURE 6.7 Target and optimum transmit signal frequency characteristics for $n_o = 10$.

transmit signal associated with target 1 of durations 11 and 101, respectively. However, the spectral peak in Fig. 6.10(a) is sharper compared to that in Fig. 6.7(d). This is realized by the longer duration of the transmit signal (101 vs. 11). Figure 6.11 shows the output *SNR* as a function of n_o for the three targets in Fig. 6.4. Notice that the *SNR* increases as a function of the transmit signal length n_o. However, if

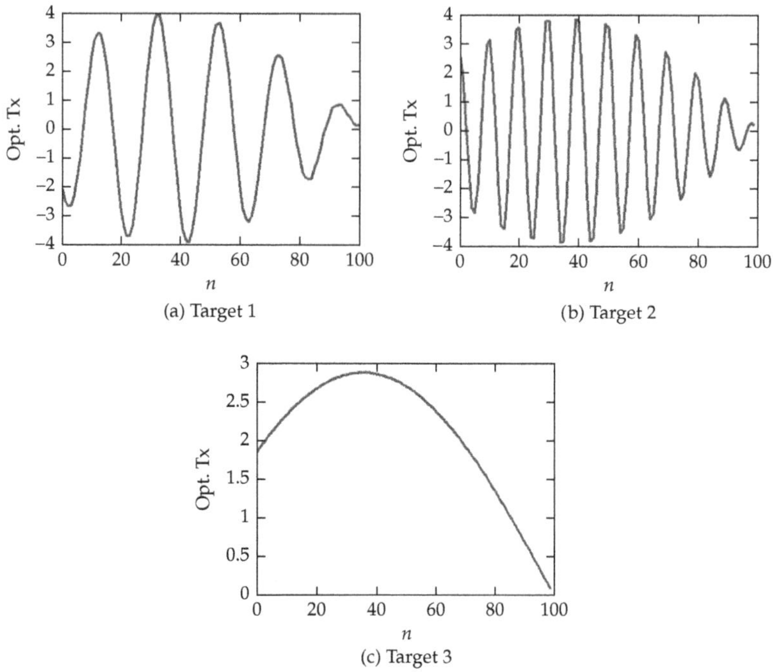

FIGURE 6.8 Optimum transmit signal for targets 1, 2, and 3 for $n_o = 100$.

the *SNR* saturates beyond a certain n_o, then that value may be used as the optimum duration of the transmit signal.

Finally, if the transmit signal length m is constrained to be less than the decision instant n_o as in (6.9), this corresponds to the situation shown in (6.7). Proceeding as in (6.7), (6.40)–(6.41) in this case simplify to

$$SNR_{\max}(m) = \underline{g}_1^* \underline{g}_1 = \underline{f}_1^* \mathbf{P}_1^* \mathbf{P}_1 \underline{f}_1 \tag{6.66}$$

where $\underline{f}_1$ is as in (6.9) and

$$\mathbf{P}_1 = \begin{pmatrix} p_o & 0 & \cdots & 0 \\ p_1 & p_o & \ddots & 0 \\ \vdots & \vdots & \ddots & \vdots \\ p_m & p_{m-1} & \cdots & p_o \\ \vdots & \vdots & \vdots & \vdots \\ p_{n_o} & p_{n_o-1} & \cdots & p_{n_o-m} \end{pmatrix} \tag{6.67}$$

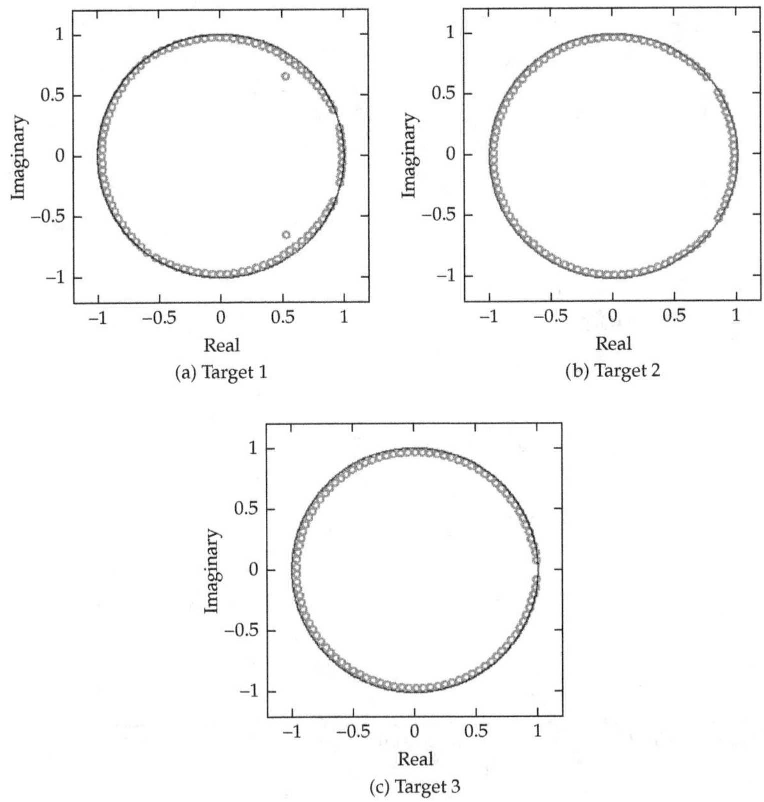

FIGURE 6.9 Zeros of the optimum transmit signal for targets 1, 2, and 3, for $n_o = 100$.

and the optimum transmit signal in this case is given by the eigenvector $\underline{u}_1$ associated with the largest eigenvalue of

$$\mathbf{\Omega}_1 = \mathbf{P}_1^* \mathbf{P}_1. \tag{6.68}$$

Thus

$$\mathbf{\Omega}_1 \underline{u}_1 = \lambda_{\max}(m, n_o) \underline{u}_1 \tag{6.69}$$

where $\lambda_{\max}(m, n_o)$ is the largest eigenvalue of $\mathbf{\Omega}_1$, and it represents the maximum achievable *SNR* for a normalized transmit signal of length m. In this case also it is easy to show that the optimum solution $\underline{u}_1$ generates a minimum phase polynomial. Figure 6.12 shows the *SNR* as a function of the transmit signal duration m with the decision instant held constant at $n_o = 100$.

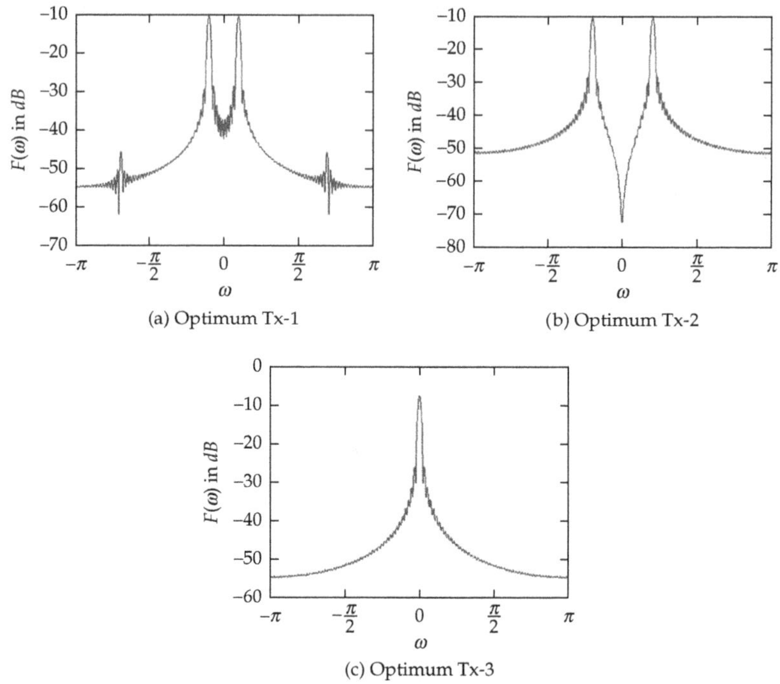

FIGURE 6.10 Optimum transmit signal frequency transforms for $n_o = 100$.

The transmit waveform solutions in Figs. 6.5 and 6.8 are not constant modulus waveforms. To realize constant modulus waveforms, their transform magnitudes in Fig. 6.7 or 6.10 can be used either in (4.112)–(4.115) or in the iterative projection-based methods (4.41). The constant modulus waveforms generated from Fig. 6.10 using (4.112)–(4.115) are shown in Fig. 6.13 along with their spectra in Fig. 6.14 where the optimum spectra from Fig. 6.10 are also shown. Figure 6.15 shows the spectra of the constant modulus waveform generated using the iterative projection method along with the corresponding optimum spectra from Fig. 6.10. Observe that in all these cases, the actual spectra associated with constant modulus waveforms have inferior sidelobes compared to the optimum spectra. This may be considered as a penalty for introducing the constant modulus property in the time domain.

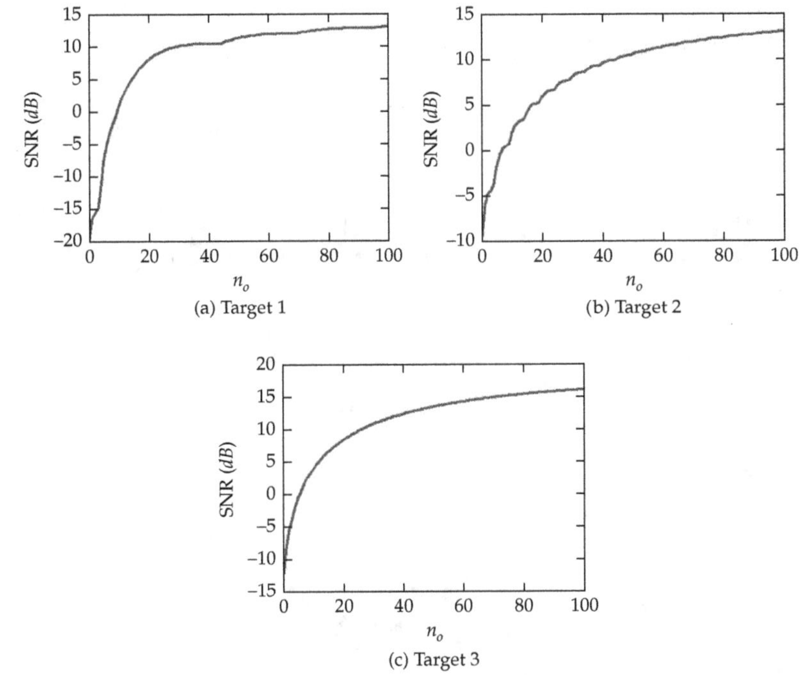

FIGURE 6.11 *SNR* as function of n_o.

6.2 Tx-Rx Design in Colored Noise and Interference

In the general case, a finite energy finite duration transmit signal excites both the desired target with impulse response $q(n)$ and interferences. If we model the interferences as a stochastic process $w_c(n)$, then the received signal $r(n)$ can be expressed as in (3.3)

$$r(n) = s(n) + w_c(n) * f(n) + w(n) \tag{6.70}$$

where

$$s(n) = f(n) * q(n). \tag{6.71}$$

Here $w(n)$ represents the additive colored noise process with spectral density $G_n(\omega)$ and $w_c(n)$ represents clutter/interference process

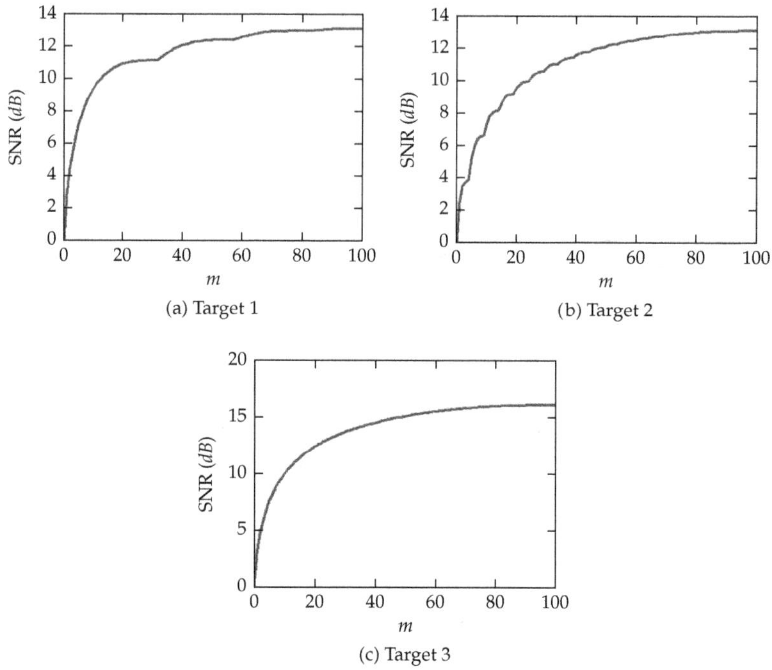

FIGURE 6.12 *SNR* as function of transmit signal duration m when $n_o = 100$.

with power spectrum $G_c(\omega)$. Thus

$$G_I(\omega) = G_c(\omega)|F(e^{j\omega})|^2 + G_n(\omega) \tag{6.72}$$

represents the total interference plus noise power spectral density. The received signal $r(n)$ is passed through a receiver $h(n) \leftrightarrow H(e^{j\omega})$ as in Fig. 6.2 to generate an output signal component

$$\hat{s}(n) = s(n) * h(n) = \frac{1}{2\pi}\int_{-\pi}^{\pi} S(e^{j\omega})H(e^{j\omega})\,e^{j\omega n_o}\,d\omega, \tag{6.73}$$

and the output interference plus noise component $w_o(n)$with power spectral density

$$G_o(\omega) = \left(G_c(\omega)|F(e^{j\omega})|^2 + G_n(\omega)\right)|H(e^{j\omega})|^2. \tag{6.74}$$

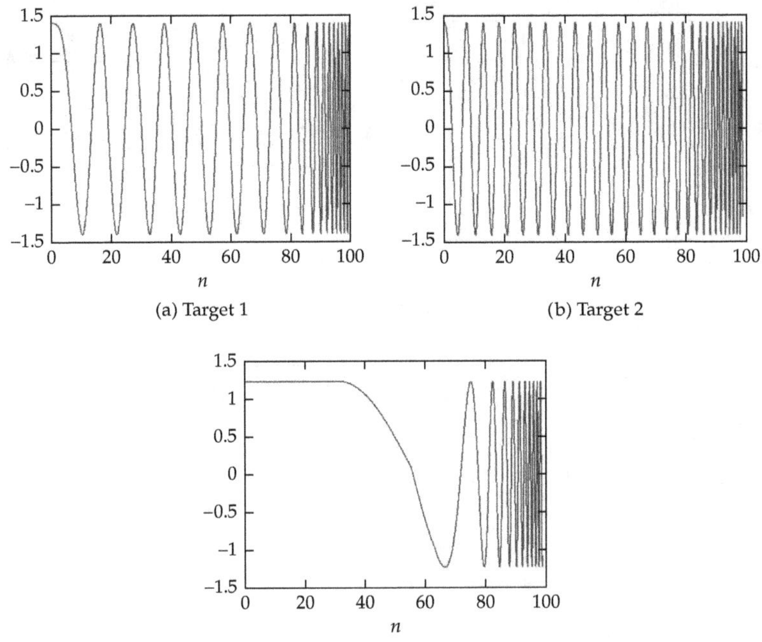

FIGURE 6.13 Constant modulus waveforms obtained using (4.112)–(4.115).

This gives the output *SINR* to be [see (3.10)]

$$SINR_o = \frac{\left| \frac{1}{2\pi} \int_{-\pi}^{\pi} S(e^{j\omega}) H(e^{j\omega}) e^{j\omega n_o} \, d\omega \right|^2}{\frac{1}{2\pi} \int_{-\pi}^{\pi} G_I(\omega) |H(e^{j\omega})|^2 d\omega}. \tag{6.75}$$

Let $L_f(z)$ represent the minimum phase spectral factor associated with the total interference input power spectrum $G_I(\omega)$ as in (6.17). Then

$$\left| L_f(e^{j\omega}) \right|^2 = G_I(\omega) = G_c(\omega) |F(e^{j\omega})|^2 + G_n(\omega), \tag{6.76}$$

and with (6.76) in (6.75) and proceeding as in (6.20)–(6.34), we get the optimum causal receiver to be

$$H_c(\omega) = L_f^{-1}(e^{j\omega}) K(e^{j\omega}) \tag{6.77}$$

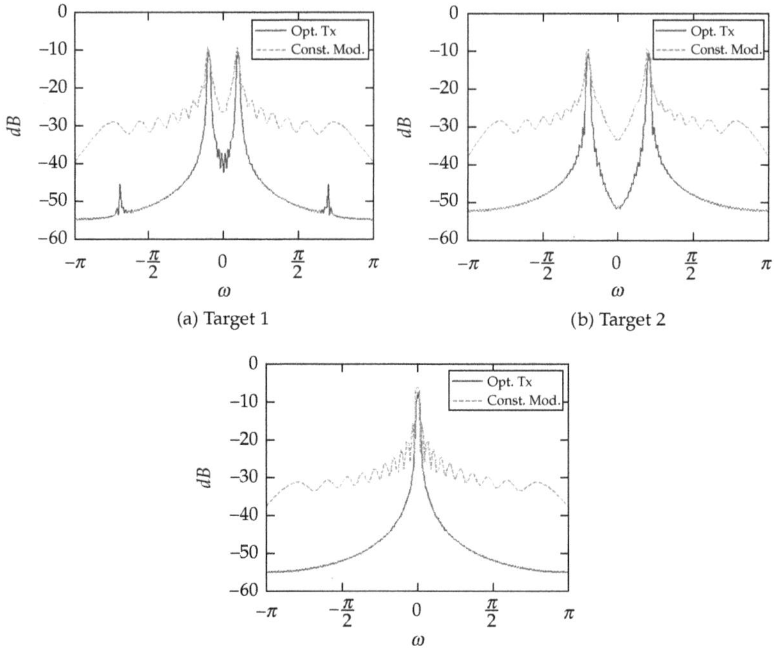

FIGURE 6.14 Comparison of spectra: optimum transmitters versus constant modulus transmit signals, obtained using (4.112)–(4.115).

where

$$g^*(n_o - n)u(n) \leftrightarrow K(e^{j\omega}) \tag{6.78}$$

and $g(n)$ is given by

$$g(n) \leftrightarrow L_f^{-1}(e^{j\omega})Q(e^{j\omega})F(e^{j\omega}). \tag{6.79}$$

In time domain, (6.77) translates to

$$h_c(n) = l_{inv}(n) * g(n_o - n)u(n) \tag{6.80}$$

as in (6.32). Here $l_{inv}(n)$ is the causal response associated with the minimum phase factor $L_f(e^{j\omega})$. Notice that unlike the noise only case in (6.17), the minimum phase spectral factor $L_f(z)$ in (6.76) depends on the transmit signal $f(n)$ through its transform $F(e^{j\omega})$. This nonlinear dependence of $L_f(z)$ on $f(n)$ makes the associated transmit

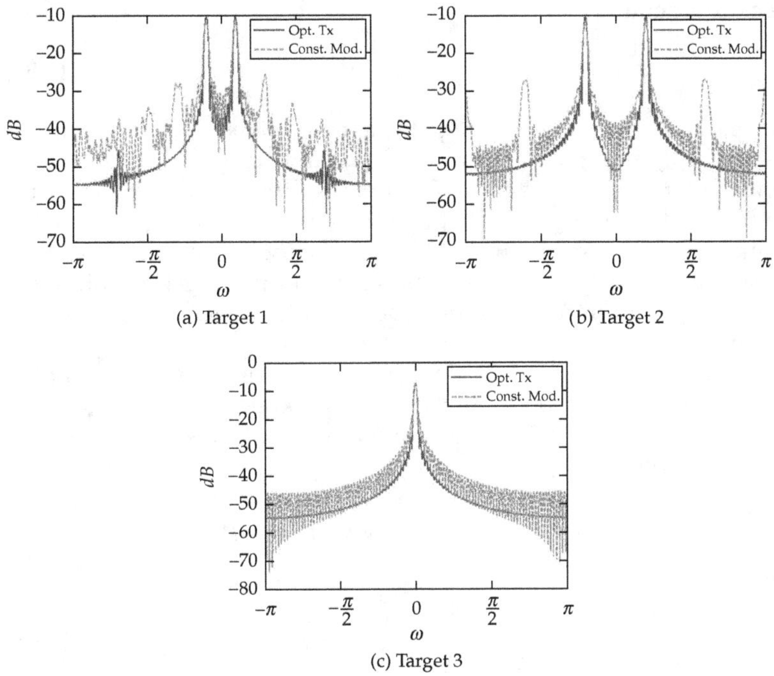

FIGURE 6.15 Spectra of constant modulus waveforms using projection-based method in (4.41) and the optimum transmit signals.

signal design problem much more complicated. The transmit design problem in this general case can be solved by proceeding iteratively as in Sec. 5.2.3.

To start with, notice that the maximum output *SINR* can be expressed as

$$SINR_{\max} = \frac{1}{2\pi}\int_{-\pi}^{\pi} |K(e^{j\omega})|^2 d\omega = \sum_{n=0}^{n_o} |g(n)|^2 = \underline{g}^* \underline{g} = \underline{f}^* \mathbf{P}^* \mathbf{P}\, \underline{f} \quad (6.81)$$

where

$$\mathbf{P} = \mathbf{L}\,\mathbf{Q} \quad (6.82)$$

is as defined in (6.42). Once again, the optimum transmit signal is given by the eigenvector associated with the largest eigenvalue of the positive definite matrix

$$\mathbf{\Omega} = \mathbf{P}^*\mathbf{P} \tag{6.83}$$

in (6.81). Observe that although $\mathbf{\Omega}$ depends on the unknown transmit signal $f(n)$ through the interference spectral factor $L_f^{-1}(z)$, from (6.76) only the transform magnitude $|F(e^{j\omega})|$ of the transmit signal plays a role in determining the shape of the minimum phase factor $L_f(z)$. Since $|F(e^{j\omega})|$ is insensitive to the location of the zeros of $F(z)$, proceeding as in (6.41)–(6.54), it follows that the optimum transmit signal $f(n)$ is minimum phase in the general clutter plus noise situation as well.

Finally, to determine the desired transmit signal, we can proceed iteratively as in (5.58)–(5.70) . Thus at stage k, since $f_k(n) \leftrightarrow F_k(e^{j\omega})$ is assumed to be known, find the minimum phase factor $L_k(z)$ associated with the equation

$$\left|L_k(e^{j\omega})\right|^2 = G_c(\omega)\left|F_k(e^{j\omega})\right|^2 + G_n(\omega)\ . \tag{6.84}$$

Define

$$p_k(n) \leftrightarrow L_k^{-1}(e^{j\omega})Q(e^{j\omega}) \tag{6.85}$$

and compute

$$\mathbf{\Omega}_k = \mathbf{P}_k^*\mathbf{P}_k \tag{6.86}$$

where

$$\mathbf{P}_k = \begin{pmatrix} p_k(0) & 0 & \cdots & 0 \\ p_k(1) & p_k(0) & \ddots & 0 \\ \vdots & \vdots & \ddots & \vdots \\ p_k(n_o) & p_k(n_o-1) & \cdots & p_k(0) \end{pmatrix}. \tag{6.87}$$

Ideally, the normalized eigenvector $\underline{u}_1^{(k)}$ associated with the largest eigenvalue $\lambda_1^{(k)}$ of the equation

$$\mathbf{\Omega}_k \underline{u}_1^{(k)} = \lambda_1^{(k)} \underline{u}_1^{(k)} \tag{6.88}$$

should correspond to the desired solution at some stake k, so that the identity

$$\underline{f}_k = \sqrt{E}\,\underline{u}_1^{(k)} \tag{6.89}$$

holds. Here E represents the transmit signal energy. However, (6.89) may not hold and in that case, the error

$$\underline{\varepsilon}_k = \underline{f}_k - \sqrt{E}\,\underline{u}_1^{(k)} \tag{6.90}$$

can be used to update $\underline{f}_k$ as in (5.65)–(5.68). Proceeding in that manner, since

$$\|\underline{\varepsilon}_k\|^2 = \underline{\varepsilon}_k^* \underline{\varepsilon}_k = 2E - 2\sqrt{E}\,c_1^{(k)} \tag{6.91}$$

where

$$c_1^{(k)} = \underline{f}_k^* \underline{u}_1^{(k)}, \tag{6.92}$$

the update rule

$$\underline{f}_{k+1} = \mu\left(\underline{f}_k + \|\underline{\varepsilon}_k\|\,\underline{u}_1^{(k)}\right) \tag{6.93}$$

after renormalization is given by

$$\underline{f}_{k+1} = \frac{\underline{f}_k + \|\underline{\varepsilon}_k\|\,\underline{u}_1^{(k)}}{\sqrt{\left(1 + \frac{\|\underline{\varepsilon}_k\|}{\sqrt{E}}\right)^2 - \left(\frac{|\underline{\varepsilon}_k|}{\sqrt{E}}\right)^3}} \tag{6.94}$$

and this can be used in (6.84) to update $L_k(z)$ to $L_{k+1}(z)$. This algorithm can be repeated until the error $\|\underline{\varepsilon}_k\|$ is acceptably small. In summary, the optimum transmit signal is given by

$$\underline{f} = \lim_{k\to\infty} \underline{f}_k \tag{6.95}$$

and the maximum output $SINR$ is given by

$$SINR_{\max} = \lim_{k\to\infty} \lambda_1^{(k)}. \tag{6.96}$$

Figure 6.16 shows the optimum transmitter design in presence of clutter and noise. The input target to noise ratio (TNR) and clutter to noise ratio (CNR) are set at 0 dB and 20 dB, respectively. The final output $SINR$ for the optimum transmitter in this case is 16.6 dB. The target here is similar to the one in Figs. 6.4(a) and 6.7(a). In Fig. 6.10, the optimum transmitter puts all of its energy at the dominant spectral peak of the target, whereas in Fig. 6.16(h), the transmitter energy is

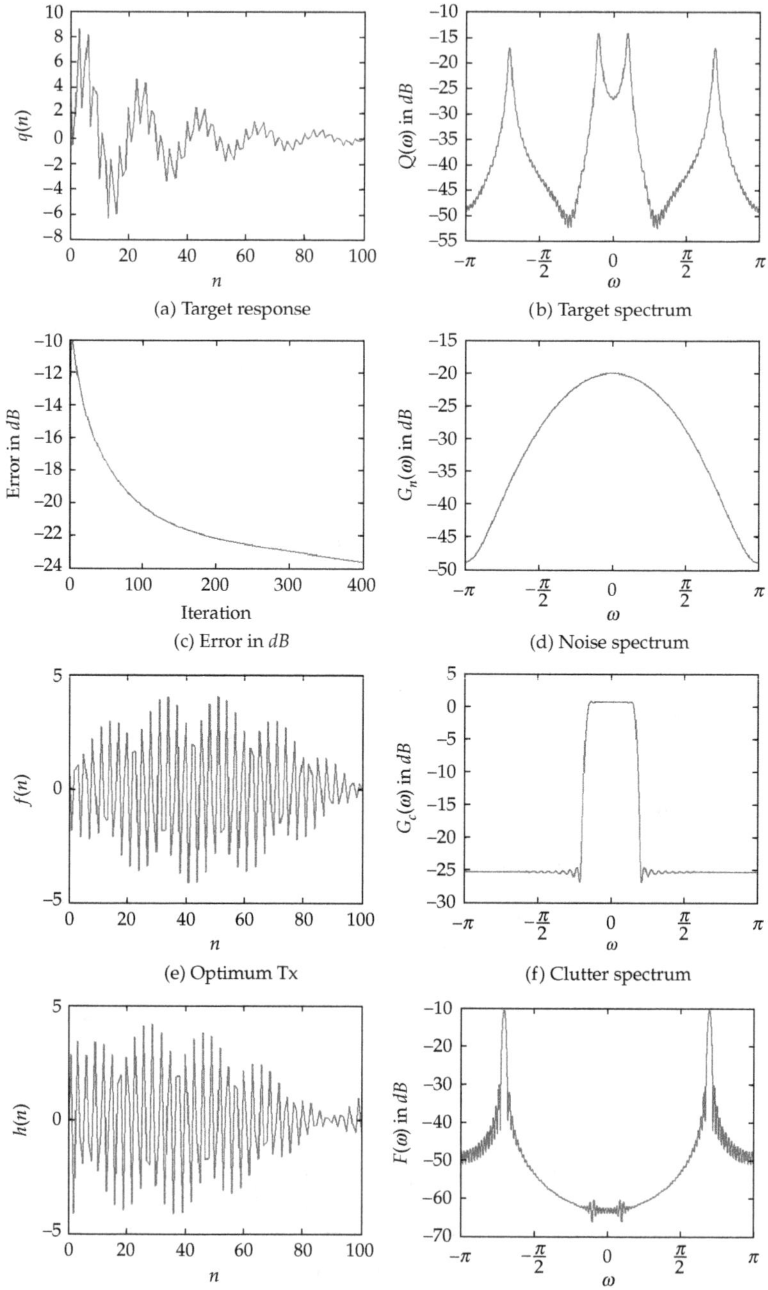

(a) Target response (b) Target spectrum

(c) Error in *dB* (d) Noise spectrum

(e) Optimum Tx (f) Clutter spectrum

(g) Optimum Rx (h) Optimum Tx spectrum

FIGURE 6.16 Optimum transmitter design in presence of clutter and noise. Input $TNR = 0$ dB and $CNR = 20$ dB. Output $SINR = 16.6$ dB.

concentrated around the second peak to suppress the low-pass clutter spectral components in Fig. 6.16(f).

Figure 6.17 shows the optimum transmitter design in presence of different clutter and noise for the same target scene as in Fig. 6.16. The input *TNR* and *CNR* are set at 0 dB and 20 dB, respectively. The final output *SINR* for the optimum transmitter in this case is 6.5 dB.

Notice that the clutter and noise generate conflicting demands in the spectral domain, and the best transmit strategy is unclear here. Nevertheless, the optimum transmitter has the spectral shape shown in Fig. 6.17(h) that corresponds to maximum tradeoff in terms of clutter rejection and noise suppression along with target detection.

6.3 Multichannel Transmit Signal Design

In the multichannel version of the transmitter-receiver design problem, multiple transmit signals $f_1(n),\ f_2(n),\ \cdots f_m(n)$ located at m distinct spatial locations are transmitted toward the target complex, and the m receiver locations receive $x_1(n),\ x_2(n),\ \cdots x_m(n)$. In the general case, the number of receivers can be different from the number of transmitters. Also, depending on the transmitter and receiver locations, the target can exhibit distinct characteristics. As Fig. 6.18 shows, let $q_{i\,j}(n)$ represent the target response between the transmitter at jth location and the receiver at the ith location. Thus

$$\underline{q}_j(n) = \begin{pmatrix} q_{1\,j}(n) \\ q_{2\,j}(n) \\ \vdots \\ q_{m\,j}(n) \end{pmatrix} \tag{6.97}$$

represents the target response vector between the transmitter at the jth location and the various receiver locations. This gives the receiver output vector from all transmit signals to be

$$\underline{x}(n) = \sum_{j=1}^{m} \underline{q}_j(n) * f_j(n), \quad j = 1 \rightarrow m. \tag{6.98}$$

Here $*$ represents the discrete convolution operation.

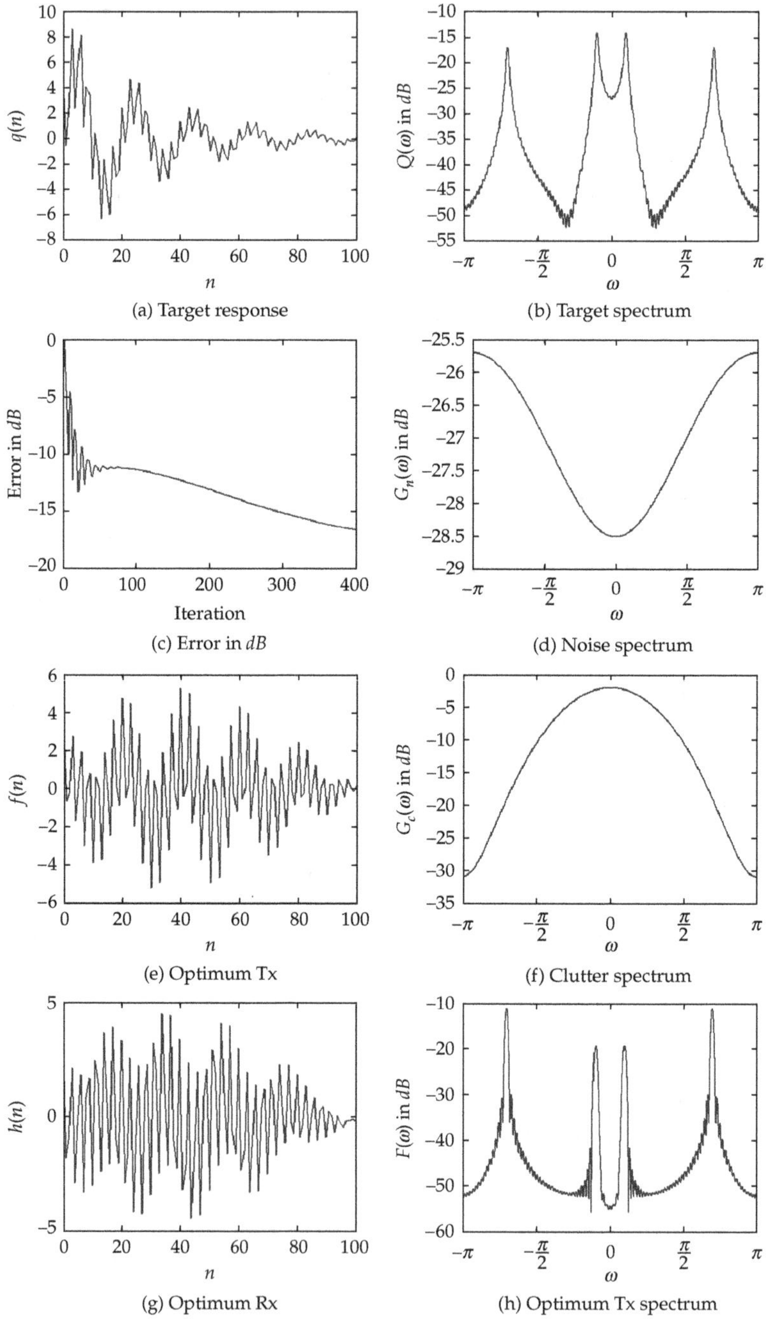

FIGURE 6.17 Optimum transmitter design in presence of clutter and noise. Input $TNR = 0$ dB and $CNR = 20$ dB. Output $SINR = 6.5$ dB.

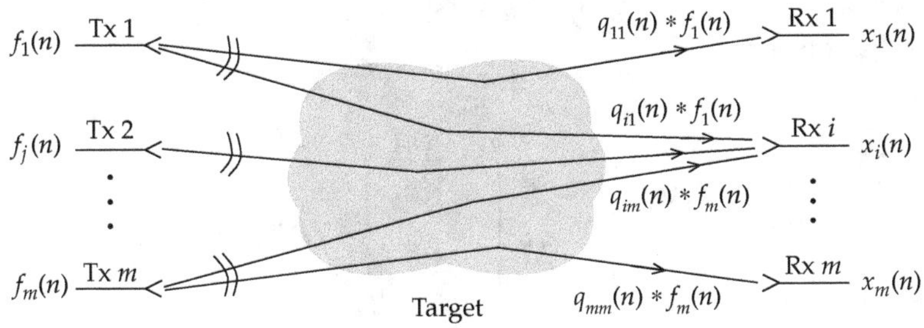

FIGURE 6.18 MIMO target scene.

Similarly, as Fig. 6.19 shows, let $w_{ij}(n)$ represent the clutter/interference process that is excited by the jth transmit signal $f_j(n)$ and received at the ith receiver. This gives the transmit signal-dependent clutter/interference component at the ith receiver to be

$$\sum_{j=1}^{m} w_{ij}(n) * f_j(n). \tag{6.99}$$

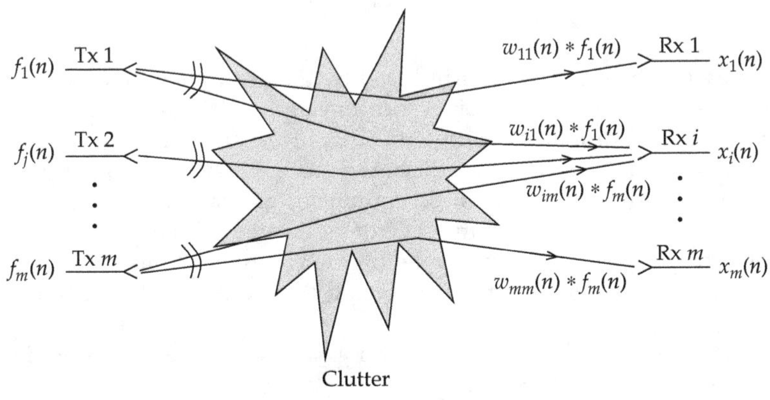

FIGURE 6.19 MIMO clutter scene.

Using (6.98) and (6.99), the total received vector $\underline{x}(n)$ can be expressed as

$$\underline{x}(n) = \begin{pmatrix} x_1(n) \\ x_2(n) \\ \vdots \\ x_m(n) \end{pmatrix} = \sum_{j=1}^{m} \begin{pmatrix} q_{1\,j}(n) \\ q_{2\,j}(n) \\ \vdots \\ q_{m\,j}(n) \end{pmatrix} * f_j(n)$$

$$+ \sum_{j=1}^{m} \begin{pmatrix} w_{1\,j}(n) \\ w_{2\,j}(n) \\ \vdots \\ w_{m\,j}(n) \end{pmatrix} * f_j(n) + \begin{pmatrix} \underline{\tilde{w}}_1(n) \\ \underline{\tilde{w}}_2(n) \\ \vdots \\ \underline{\tilde{w}}_m(n) \end{pmatrix} \tag{6.100}$$

$$= \sum_{j=1}^{m} \underline{q}_j(n) * f_j(n) + \underbrace{\sum_{j=1}^{m} \underline{w}_{c_j}(n) * f_j(n) + \underline{\tilde{w}}(n)}_{\underline{w}(n)}$$

where

$$\underline{w}_{c_j}(n) = \begin{pmatrix} w_{1\,j}(n) \\ w_{2\,j}(n) \\ \vdots \\ w_{m\,j}(n) \end{pmatrix} \tag{6.101}$$

represents the clutter vector induced by the jth transmitter location received at the various receiver locations. In (6.100), $\underline{\tilde{w}}(n)$ represents the additive noise at the receivers. Thus in general, each transmitter location induces a clutter vector at the receivers, depending on their relative spatial locations. These clutter vectors can be correlated in general. As a result, the clutter vector associated with each transmitter generates its own autocorrelation matrix in addition to the various cross-correlation matrices generated between the different clutter vectors.

If we assume these clutter vector processes to be jointly wide-sense-stationary, then we have

$$E\{\underline{w}_{c_i}(n)\underline{w}^*_{c_j}(n+m)\} = \mathbf{R}_{i\,j}(m), \quad i = 1 \to m, \quad j = 1 \to m \tag{6.102}$$

represent the $m \times m$ cross-correlation matrices, and let $\mathbf{G}_{i\,j}(\omega)$ represent the $m \times m$ cross-power spectral matrices associated with the $m \times m$ cross-correlation matrices $\mathbf{R}_{i\,j}(m)$. Thus

$$\mathbf{R}_{i\,j}(m) \leftrightarrow \mathbf{G}_{i\,j}(\omega), \quad i = 1 \to m, \quad j = 1 \to m \tag{6.103}$$

and the total clutter/interference power spectrum matrix ($m^2 \times m^2$) is given by the block matrix

$$\mathbf{S}(\omega) = \begin{pmatrix} \mathbf{G}_{11}(\omega) & \mathbf{G}_{12}(\omega) & \cdots & \mathbf{G}_{1\,m}(\omega) \\ \mathbf{G}_{21}(\omega) & \mathbf{G}_{22}(\omega) & \cdots & \mathbf{G}_{2\,m}(\omega) \\ \vdots & \vdots & \ddots & \vdots \\ \mathbf{G}_{m1}(\omega) & \mathbf{G}_{m2}(\omega) & \cdots & \mathbf{G}_{m\,m}(\omega) \end{pmatrix} > 0. \tag{6.104}$$

In terms of $\mathbf{G}_{i\,j}(\omega)$ in (6.103) and (6.104), it follows that the total clutter/interference power in the received vector $\underline{w}(n)$ in (6.100) is given by

$$\mathbf{S}_F(\omega) = \sum_{i=1}^{m} \sum_{j=1}^{m} \mathbf{G}_{i\,j}(\omega) F_i(e^{j\omega}) F_j^*(e^{j\omega}) > 0. \tag{6.105}$$

Notice that the clutter spectrum $\mathbf{S}_F(\omega)$ is transmit signal-dependent. Let $\mathbf{R}_n(m) = E\,\{\underline{\tilde{w}}(n)\underline{\tilde{w}}^*(n+m)\}$ represent the autocorrelation process of the noise process $\underline{\tilde{w}}(n)$ in (6.100). With $\mathbf{R}_n(m) \leftrightarrow \mathbf{G}_n(\omega)$ with $\mathbf{G}_n(\omega)$ representing the power spectral density associated with the independent noise process $\underline{\tilde{w}}(n)$ in (6.100)

$$\mathbf{S}_o(\omega) = \mathbf{S}_F(\omega) + \mathbf{G}_n(\omega) \tag{6.106}$$

represents the total interference plus noise power at the receiver complex. Let $\mathbf{L}_f(z)$ represent the minimum phase left-spectral factor matrix function associated with the total interference/noise spectrum $\mathbf{S}_o(\omega)$, then

$$\mathbf{S}_o(\omega) = \mathbf{L}_f(e^{j\omega}) \mathbf{L}_f^*(e^{j\omega}) > 0. \tag{6.107}$$

Let

$$\underline{f}(n) = [f_1(n), \ \ f_2(n), \ \ \cdots \ \ f_m(n)]^T \tag{6.108}$$

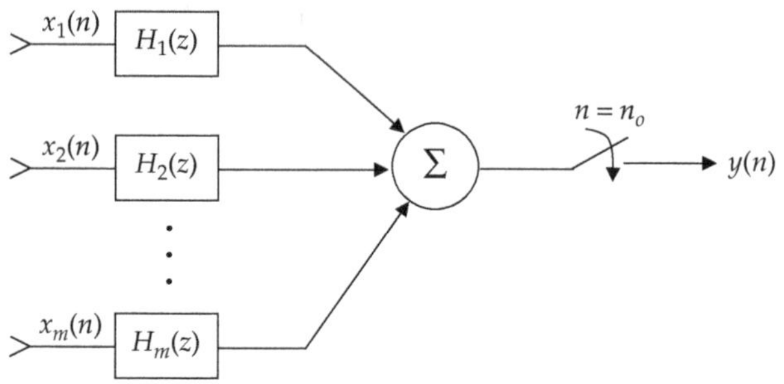

FIGURE 6.20 Multichannel receivers.

represent the transmit signal set. Notice that $\mathbf{L}_f(z)$ depends on the transmit signal set $\underline{f}(n)$, and for a given set of transmit signals, the left spectral factor matrix $\mathbf{L}_f(z)$ can be evaluated, using (6.105)–(6.107) using any matrix spectral factorization procedure.

At the receiver, data is passed through a set of receivers $H_j(z)$ to generate a combined output signal $y(n)$ as shown in Fig. 6.20.

If we write

$$y(n) = s(n) + w(n) \tag{6.109}$$

to denote the signal and clutter/interference plus noise components respectively, then

$$s(n) \leftrightarrow S(\omega) = \underline{H}(e^{j\omega})\mathbf{Q}(e^{j\omega})\underline{F}(e^{j\omega}). \tag{6.110}$$

Here the row vector

$$\underline{H}(z) = [H_1(z), \quad H_2(z), \quad \cdots \quad H_m(z)] \tag{6.111}$$

represents the unknown receiver bank of filters and $\mathbf{Q}(e^{j\omega})$ is the transfer function matrix associated with the target matrix function, and it is given by

$$[\underline{q}_1(n) \quad \underline{q}_2(n) \quad \cdots \quad \underline{q}_m(n)] \leftrightarrow \mathbf{Q}(e^{j\omega}) \tag{6.112}$$

where $\underline{q}_j(n)$ are as defined in (6.97), and $\underline{F}(e^{j\omega})$ represents the transform vector of the transmit set of signals in (6.108). Thus

$$\underline{f}(n) \leftrightarrow \underline{F}(e^{j\omega}). \tag{6.113}$$

The average interference plus noise power is given by

$$E\{|w(n)|^2\} = \mathbf{R}_w(0) = \frac{1}{2\pi}\int_{-\pi}^{\pi} \underline{H}(e^{j\omega})\left(\mathbf{S}_F(\omega) + \mathbf{G}_n(\omega)\right)\underline{H}^*(e^{j\omega})\,d\omega. \tag{6.114}$$

Following Sec. 6.2, the output *SINR* at the decision instant $n = n_o$ can be used as an optimization criterion in this case as well. This gives

$$\rho(n_o) = \frac{|s(n_o)|^2}{E\left\{|w(n)|^2\right\}} = \frac{\left|\frac{1}{2\pi}\int_{-\pi}^{\pi} \underline{H}(e^{j\omega})\mathbf{Q}(e^{j\omega})\underline{F}(e^{j\omega})e^{j\omega n_o}\,d\omega\right|^2}{\frac{1}{2\pi}\int_{-\pi}^{\pi} \underline{H}(e^{j\omega})\,\mathbf{S}_o(\omega)\underline{H}^*(e^{j\omega})d\omega}. \tag{6.115}$$

Proceeding as in (6.13)–(6.31), we can simplify the above *SINR* to

$$\rho(n_o) = \frac{\left|\frac{1}{2\pi}\int_{-\pi}^{\pi} \underline{H}(e^{j\omega})\mathbf{L}_f(e^{j\omega})\underline{K}^*(e^{j\omega})\,d\omega\right|^2}{\frac{1}{2\pi}\int_{-\pi}^{\pi} \underline{H}(e^{j\omega})\mathbf{L}_f(e^{j\omega})\mathbf{L}_f^*(e^{j\omega})\underline{H}^*(e^{j\omega})\,d\omega} \tag{6.116}$$

$$\leq \frac{1}{2\pi}\int_{-\pi}^{\pi} \underline{K}(e^{j\omega})\underline{K}^*(e^{j\omega})d\omega = \sum_{n=0}^{n_o} \underline{g}^*(n)\underline{g}(n) = SINR_{\max}.$$

With equality if and only if the row vector $\underline{H}_{opt}(z)$ satisfies

$$\underline{H}_{opt}(z) = \underline{K}(z)\mathbf{L}_f^{-1}(z). \tag{6.117}$$

Here

$$\underline{g}(n) \leftrightarrow \mathbf{L}_f^{-1}(e^{j\omega})\mathbf{Q}(e^{j\omega})\underline{F}(e^{j\omega}) \tag{6.118}$$

represents a column vector and

$$\underline{g}^*(n_o - n)u(n) \leftrightarrow \underline{K}(e^{j\omega}) \tag{6.119}$$

represents a row vector. Notice that both $\underline{g}(n)$ and $\underline{g}^*(n_o - n)u(n)$ represent causal signals. The optimum receivers in (6.117) can be interpreted as a whitening followed by a generalized matched filter represented by $\underline{K}(e^{j\omega})$. To simplify the $SINR_{\max}$ in (6.116), note that $\underline{g}(n)$ in (6.118) can be expressed as

$$\underline{g}(n) = \sum_{k=0}^{n_o} \mathbf{M}(n-k)\underline{f}(k) \tag{6.120}$$

where

$$\mathbf{M}(n) \leftrightarrow \mathbf{L}_f^{-1}(e^{j\omega})\mathbf{Q}(e^{j\omega}). \tag{6.121}$$

This gives

$$\begin{aligned} \rho(n_o) &= \sum_{n=0}^{n_o} \underline{g}^*(n)\underline{g}(n) = \sum_{k=0}^{n_o}\sum_{m=0}^{n_o} \underline{f}^*(k) \underbrace{\sum_{n=0}^{n_o} \mathbf{M}^*(n-k)\mathbf{M}(n-m)}_{\boldsymbol{\Omega}(k,\,m)} \underline{f}(m) \\ &= \underline{f}^*\boldsymbol{\Omega}\,\underline{f} \end{aligned} \tag{6.122}$$

where

$$\underline{f} = \begin{bmatrix} \underline{f}(0) \\ \underline{f}(1) \\ \vdots \\ \underline{f}(n_o) \end{bmatrix} \tag{6.123}$$

represents the column vector consisting of the transmit signal set. Notice that $\boldsymbol{\Omega}$ in (6.122) can be expressed as

$$\boldsymbol{\Omega} = \mathbf{M}^*\mathbf{M} \tag{6.124}$$

where

$$\mathbf{M} = \begin{pmatrix} \mathbf{M}(0) & 0 & \cdots & 0 \\ \mathbf{M}(1) & \mathbf{M}(0) & \cdots & 0 \\ \vdots & \vdots & \ddots & \vdots \\ \mathbf{M}(n_o) & \mathbf{M}(n_o-1) & \cdots & \mathbf{M}(0) \end{pmatrix}. \tag{6.125}$$

Equation (6.122) can be used to design the normalized optimum transmit signal set. From (6.122), the eigenvector $\underline{f}_{opt}$ associated with the largest eigenvalue of $\boldsymbol{\Omega}$ represents the normalized optimum transmit vector that maximizes the target echo while minimizing the clutter and noise components.

In general, (6.122) represents a highly nonlinear problem that can be solved iteratively as in (6.84)–(6.95). However, in the absence of clutter, the above formulation can be solved exactly since in that case

$$\mathbf{S}_o(\omega) = \mathbf{G}_n(\omega) = \mathbf{L}_n(e^{j\omega})\mathbf{L}_n^*(e^{j\omega}) \tag{6.126}$$

where $\mathbf{L}_n(z)$ represents the minimum phase factor associated with the noise spectral density matrix $\mathbf{G}_n(\omega)$. From (6.121) and (6.122), the optimum transmitter vector is given by the eigenvector associated with the largest eigenvalue of the known positive definite matrix

$$\boldsymbol{\Omega}(i, j) = \sum_{n=0}^{n_o} \mathbf{M}^*(n-i)\mathbf{M}(n-j) \tag{6.127}$$

where

$$\mathbf{M}(n) \leftrightarrow \mathbf{L}_n^{-1}(e^{j\omega})\mathbf{Q}(e^{j\omega}). \tag{6.128}$$

Figures 6.21 and 6.22 show a two-transmitter, two-receiver case with

$$\mathbf{Q}(n) = \begin{pmatrix} q_{11}(n) & q_{12}(n) \\ q_{21}(n) & q_{22}(n) \end{pmatrix} \tag{6.129}$$

and

$$\mathbf{G}_n(\omega) = \begin{pmatrix} G_n^{(11)}(\omega) & G_n^{(12)}(\omega) \\ G_n^{(21)}(\omega) & G_n^{(22)}(\omega) \end{pmatrix} > 0. \tag{6.130}$$

Figure 6.21 shows the target signal and its spectra and Fig. 6.22 shows the noise spectra. Notice that the noise has flat spectra in this case.

The optimum transmit signals $f_1(n)$ and $f_2(n)$ obtained as the eigenvector associated with the largest eigenvalue of (6.126) are shown in Fig. 6.23. In this case, the overall $SINR_{\max}$ in (6.116) corresponds to 21 dB while each single channel case gives 14.5 dB and 13.4 dB separately. Notice that in the single channel case, each transmit signal energy is

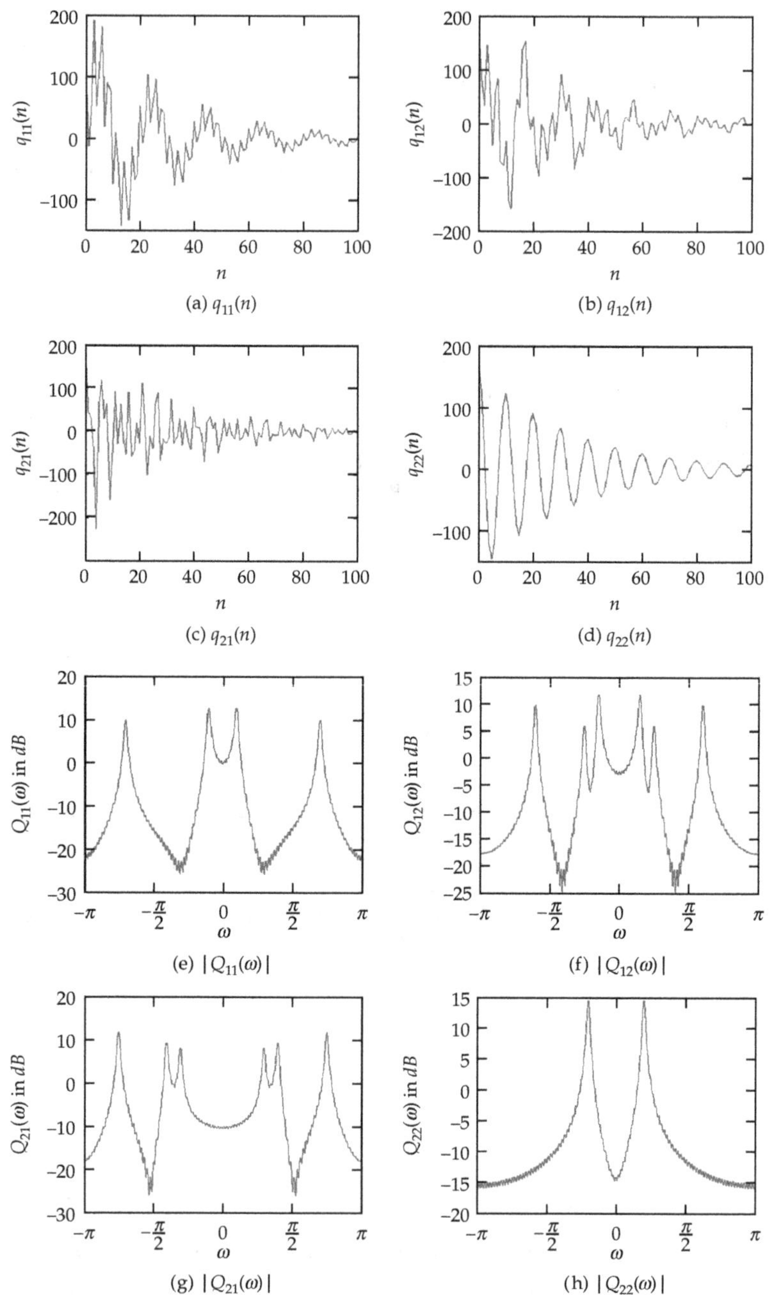

FIGURE 6.21 Target signals and their spectra for the two-transmitter, two-receiver case.

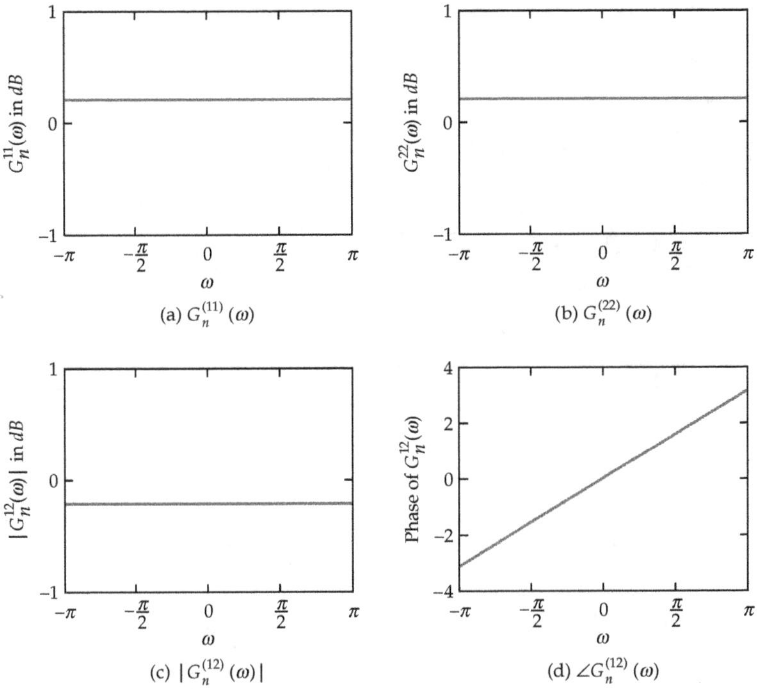

FIGURE 6.22 Noise spectra for the two-transmitter, two-receiver case.

the same as the combined energy in the two channel case. The noise is assumed to be white noise in the single channel cases. Thus the optimum transmit signals focus most of the energy at the frequency [the resonant frequency at $Q_{22}(\omega)$] where the target signal is strongest. As a result, the optimum solution distributes most of the transmit energy to transmit waveform $f_2(n)$. In this case, the energy ratio $\frac{E_{f_2}}{E_{f_1}} = 60.3$.

Figures 6.24 and 6.25 show a new set of noise spectra and the corresponding optimum transmit waveform spectra for the same target set as in Fig. 6.21 are used. The overall $SINR_{\max}$ in (6.116) corresponds to 17.5 dB. In this case, the optimum solution focuses most of the transmit energy at the second resonant frequency of $Q_{11}(\omega)$. The optimum solution distributes most of the transmit energy to transmit waveform $f_1(n)$. In this case, the energy ratio $\frac{E_{f_1}}{E_{f_2}} = 55.8$.

Returning to the general case, when clutter is present as in (6.84), let $\underline{F}_j^{(k)}(\omega)$, $j = 1 \to m$ represent the solutions at stage k. Let $\mathbf{L}_k(z)$

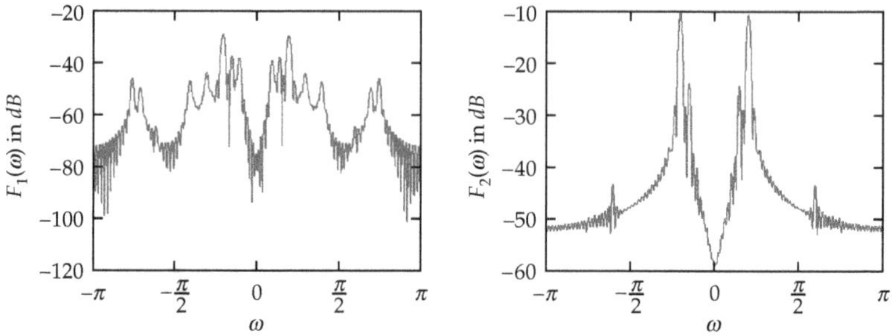

(a) $|F_1(\omega)|$ (b) $|F_2(\omega)|$

FIGURE 6.23 Optimum transmit signal transform magnitude for the 2 × 2 case.

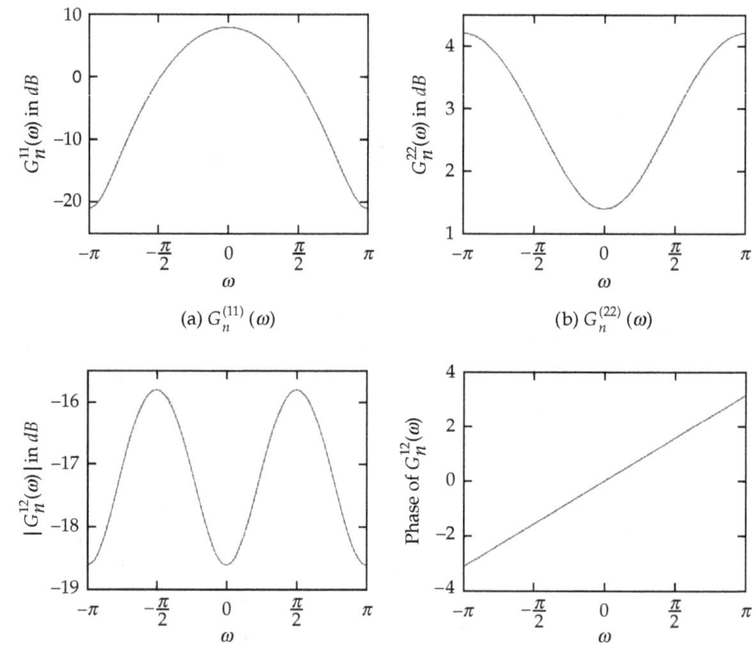

(a) $G_n^{(11)}(\omega)$ (b) $G_n^{(22)}(\omega)$ (c) $|G_n^{(12)}(\omega)|$ (d) $\angle G_n^{(12)}(\omega)$

FIGURE 6.24 Noise spectra for the two-transmitter, two-receiver case.

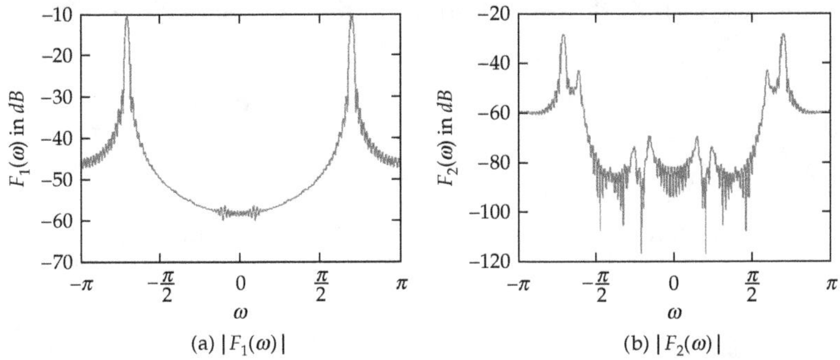

FIGURE 6.25 Optimum transmit signals.

represent the minimum phase factor associated with the equation

$$\mathbf{L}_k(e^{j\omega})\mathbf{L}_k^*(e^{j\omega}) = \sum_{i=1}^{m}\sum_{j=1}^{m} \mathbf{G}_{i\,j}(\omega)F_i^{(k)}(e^{j\omega})F_j^{*(k)}(e^{j\omega}) + \mathbf{G}_n(\omega) \quad (6.131)$$

and define

$$\mathbf{M}_k(n) \leftrightarrow \mathbf{L}_k^{-1}(e^{j\omega})\mathbf{Q}(e^{j\omega}). \quad (6.132)$$

Following (6.125), compute

$$\mathbf{M}_k = \begin{pmatrix} \mathbf{M}_k(0) & 0 & \cdots & 0 \\ \mathbf{M}_k(1) & \mathbf{M}_k(0) & \cdots & 0 \\ \vdots & \vdots & \ddots & \vdots \\ \mathbf{M}_k(n_o) & \mathbf{M}_k(n_o-1) & \cdots & \mathbf{M}_k(0) \end{pmatrix} \quad (6.133)$$

and define

$$\boldsymbol{\Omega}_k = \mathbf{M}_k^*\mathbf{M}_k > 0. \quad (6.134)$$

From (6.122) and (6.123), the normalized eigenvector $\underline{u}_1^{(k)}$ associated with the largest eigenvalue $\lambda^{(k)}$ of the equation

$$\boldsymbol{\Omega}_k\underline{u}_1^{(k)} = \lambda^{(k)}\underline{u}_1^{(k)} \quad (6.135)$$

should coincide with the optimum solution at some stage k. Consequently, the error vector

$$\underline{\varepsilon}_k = \underline{f}_k - \sqrt{E}\,\underline{u}_1^{(k)} \tag{6.136}$$

can be used to update $\underline{f}_k$ as in (6.91)–(6.94). Proceeding as there

$$\underline{f}_{k+1} = \frac{\underline{f}_k + \gamma_k \underline{u}_1^{(k)}}{\left\|\underline{f}_k + \gamma_k \underline{u}_1^{(k)}\right\|} \tag{6.137}$$

where

$$\gamma_k = \|\underline{\varepsilon}_k\| \tag{6.138}$$

can be used to update $\mathbf{L}_{k+1}(z)$ from $\mathbf{L}_k(z)$. This algorithm is repeated until the error $\|\underline{\varepsilon}_k\|$ is acceptably small. Hence, the optimum transmit signal vector is given by

$$\underline{f} = \lim_{k\to\infty} \underline{f}_k \tag{6.139}$$

and the maximum realizable *SINR* is given by

$$SINR_{\max} = \lim_{k\to\infty} \lambda_1(k). \tag{6.140}$$

Figures 6.26–6.28 show the simulation results for a 2×2 case when noise and clutter are present. In this case, the clutter to noise ratio is kept at 6 dB and target and noise spectra are as shown in Figs. 6.21 and 6.24, respectively. The clutter spectra are shown in Figs. 6.26 and 6.27 [see (6.104)]. Figure 6.26 shows the two-channel clutter spectra and Fig. 6.27 shows the cross clutter spectra. The optimum transmit signals obtained using (6.131)–(6.139) are shown in Fig. 6.28 and the final output *SINR* is at 14.7 dB. Notice the optimum solution distributes most of the transmit energy to transmit waveform $f_1(n)$. In this case, the energy ratio is $\frac{E_{f_1}}{E_{f_2}} = 51.7$.

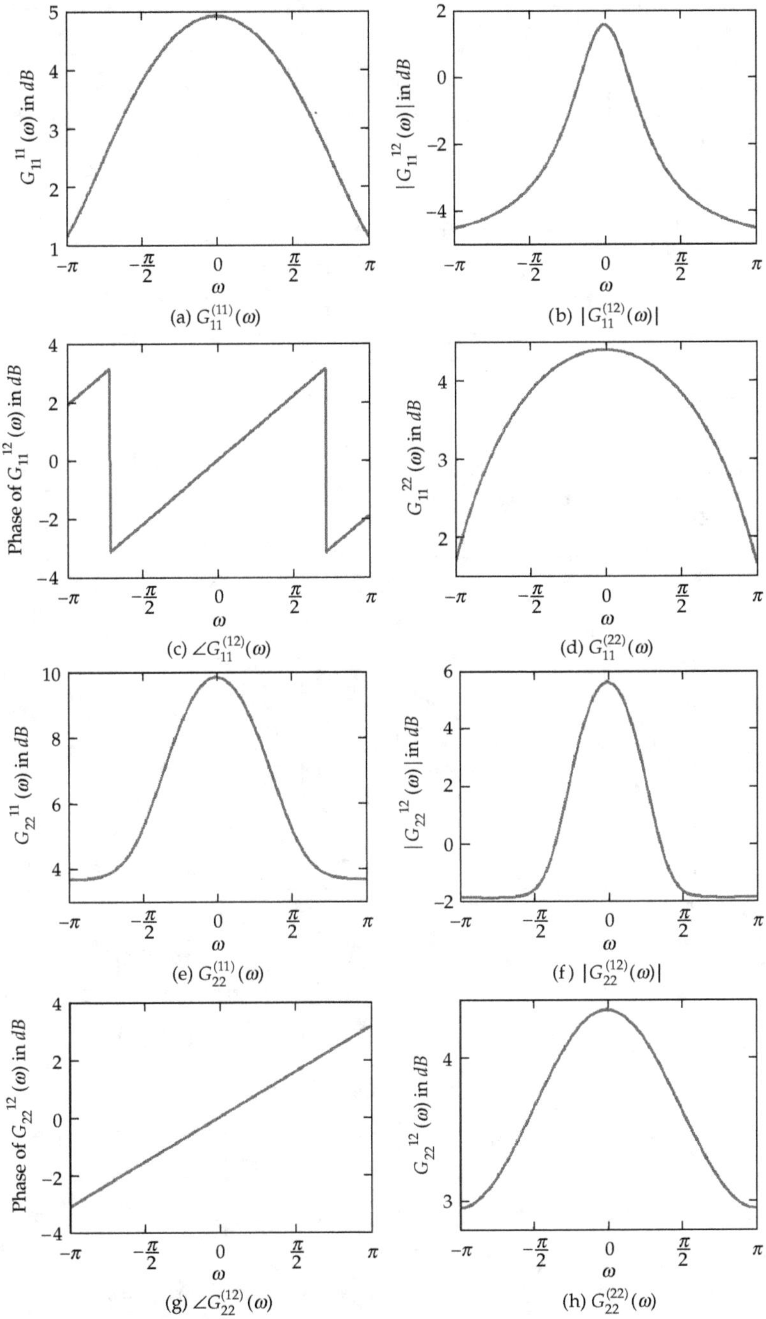

FIGURE 6.26 Clutter spectra for the 2×2 case.

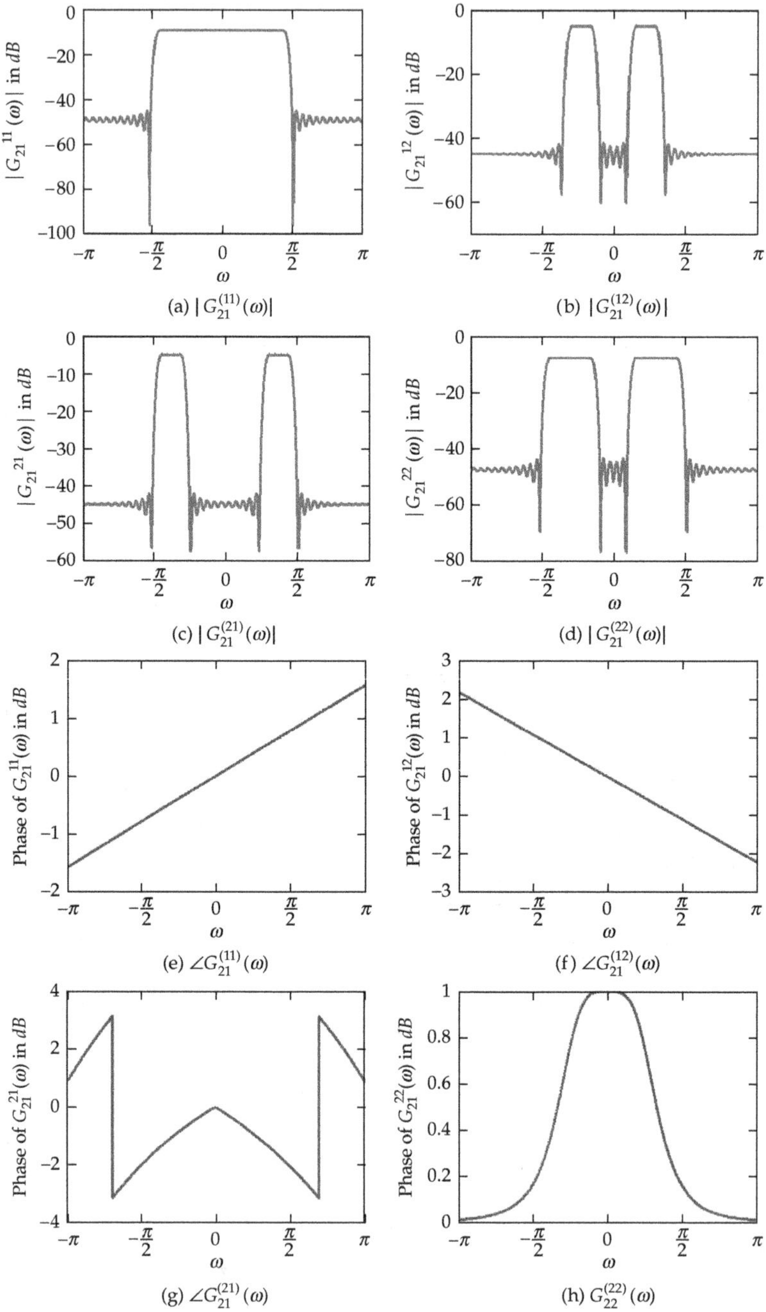

FIGURE 6.27 Cross clutter spectra for the 2 × 2 case.

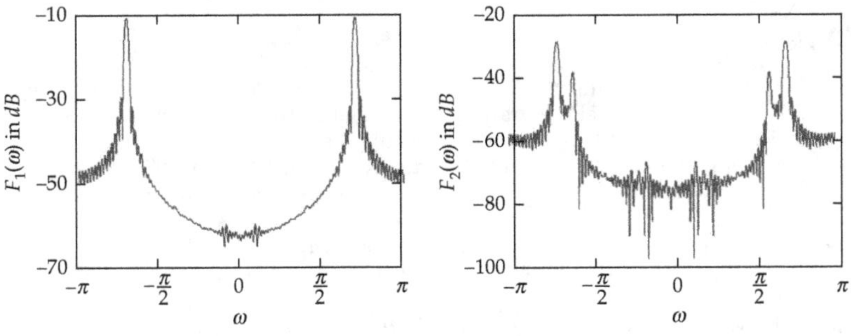

FIGURE 6.28 Optimum transit signal transform magnitude for the 2 × 2 case.

Several variations to the transmit waveform design problem are possible by imposing constraints, such as constant modulus, orthogonal input waveforms, equal energy waveforms, etc. These constraints make sense in various applications and in those cases, the design problem becomes more restrictive.

References

[1] D.C. Youla, N. Kazanjian, "Bauer-Type Factorization of Positive Matrices and the Theory of Matrix Polynomials Orthogonal on the Unit Circle," *IEEE Transactions on Circuits and Systems*, vol. 25, no. 2, Feb. 1978.

[2] F.R. Gantmacher, *The Theory of Matrices*. New York: Chelsea, 1977.

[3] R. Horn, C. Johnson, *Topics in Matrix Analysis*, New York: Cambridge University Press, 1991.

[4] P.F. Sammartino, C. Baker, "The Frequency Diverse Bistatic System," *IEEE International Waveform Diversity & Design Conference*, Orlando, FL, Feb. 8–13, 2009.

[5] P. Stoica, J. Li, X. Zhu, "Waveform Synthesis for Diversity-Based Transmit Beampattern Design," *IEEE Transactions on Signal Processing*, vol. 56, no. 6, June 2008.

[6] K.W. Forsythe, D.W. Bliss, "Waveform Correlation and Optimization Issues for MIMO Radar," *Proc. 39th Asilomar Conf. Signals, Systems, Computer*, Pacific Grove, CA, pp. 1306–1310, Nov. 2005.

[7] J. Klare, D. Cerutti-Maori, A. Brenner, J.H.G. Ender, "Image Quality Analysis of the Vibrating Sparse MIMO Antenna Array of the Airborne 3D Imaging Radar ARTINO," *Proc. IEEE Intl. Geoscience and Remote Sensing Symposium (IGARSS 2007)*, pp.5310–5314, 2007.

[8] Y. Chi, A. Pezeshki, R. Calderbank, S. Howard, "Range Sidelobe Suppression in a Desired Doppler Interval," *IEEE International Waveform Diversity & Design Conference*, Orlando, FL, Feb. 8–13, 2009.

[9] S. Sen, M. Hurtado, A. Nehorai, "Adaptive OFDM Radar for Detecting a Moving Target in Urban Scenarios," *IEEE International Waveform Diversity & Design Conference*, Orlando, FL, Feb. 8–13, 2009.

[10] Y. Li, W. Moran, S. Sira, A. Papandreou-Suppappola, "Adaptive Waveform Design in Rapidly Varying Radar Scenes," *IEEE International Waveform Diversity & Design Conference*, Orlando, FL, Feb. 8–13, 2009.

[11] H.I. Demirkiran, D. Weiner, A. Drozd, I. Kasperovich, "Equivalent Based Analysis for Narrow Weakly Nonlinear Bandpass Systems for Waveform Diversity Applications," *IEEE International Waveform Diversity & Design Conference*, Orlando, FL, Feb. 8–13, 2009.

[12] M. Stuff, B. Thelen, N. Subotic, J. Parker, J. Browning, "Optimization and Waveforms for Compressive Sensing Applications in the Presence of Interference," *IEEE International Waveform Diversity & Design Conference*, Orlando, FL, Feb. 8–13, 2009.

[13] H. Mendelson, P. Zulch, "Mitigation of Radar Range Ambiguities via Waveform Diversity," *IEEE International Waveform Diversity & Design Conference*, Orlando, FL, Feb. 8–13, 2009.

[14] J. Goodman, B. Miller, M. Herman, M. Wai, P. Monticciolo, "Extending the Dynamic Range of RF Receiver with Nonlinear Equalization," *IEEE International Waveform Diversity & Design Conference*, Orlando, FL, Feb. 8–13, 2009.

[15] M.R. Bell, "Information Theory and Radar Waveform Design," *IEEE Trans. Inform. Theory*, vol. 39, no. 5, pp. 1578–1597, Sep. 1993.

[16] S.U. Pillai, H.S. Oh, D.C. Youla, J.R. Guerci, "Optimum Transmit Receiver Design in the Presence of Signal-Dependent Interference and Channel Noise," *IEEE Trans. Inform. Theory*, vol. 46, no. 2, pp. 577–584, March 2000.

[17] S. Kay, "Optimal Signal Design for Detection of Gaussian Point Targets in Stationary Gaussian Clutter/Reverberation," *IEEE J. Sel. Topics in Sig. Proc. Mag.*, vol. 1, no. 1, pp. 31–41, Jun. 2007.

[18] J. Goodman, B. Miller, M. Herman, M. Wai, P. Monticciolo, "Extending the Dynamic Range of RF Receiver with Nonlinear Equalization," *IEEE International Waveform Diversity & Design Conference*, Orlando, FL, Feb. 8–13, 2009.

[19] J. Zhang, A. Papandreou-Suppappola, "MIMO Radar with Frequency Diversity," *IEEE International Waveform Diversity & Design Conference*, Orlando, FL, Feb. 8–13, 2009.

[20] Y. Paichard, M. Brooker, M. Inggs, "A Signal Level Simulator for Netted Radar Waveforms Evaluation," *IEEE International Waveform Diversity & Design Conference*, Orlando, FL, Feb. 8–13, 2009.

[21] I. Bradaric, G.T. Capraro, P. Zulch, "Signal Processing and Waveform Selection Strategies in Multistatic Radar Systems," *International Waveform Diversity and Design Conference*, Pisa, Italy, June 2007, invited paper.

[22] R. Bonneau, "A Coherence Gradient Method in Waveform Design for Apertures," *IEEE International Waveform Diversity & Design Conference*, Orlando, FL, Feb. 8–13, 2009.

[23] G. Dingley, C. Alabaster, "Radar Based Automatic Target System," *IEEE International Waveform Diversity & Design Conference*, Orlando, FL, Feb. 8–13, 2009.

[24] R. Romero, N. Goodman, "Improved Waveform Design for Target Recognition with Multiple Transmissions," *IEEE International Waveform Diversity & Design Conference*, Orlando, FL, Feb. 8–13, 2009.

[25] J. Guerci, "Ultra-Low MDV GMTI MIMO Radar," *IEEE International Waveform Diversity & Design Conference*, Orlando, FL, Feb. 8–13, 2009.

[26] J. Goodman, B. Miller, M. Herman, M. Wai, P. Monticciolo, "Extending the Dynamic Range of RF Receiver with Nonlinear Equalization," *IEEE International Waveform Diversity & Design Conference*, Orlando, FL, Feb. 8–13, 2009.

[27] I. Shirokov, I. Serdyuk, "The Investigation of Phase Synchronization of Reference Oscillators Through Atmospheric Channel," *IEEE International Waveform Diversity & Design Conference*, Orlando, FL, Feb. 8–13, 2009.

[28] I. Akduman, A. Yapar, F. Akleman, H. Sahinturk, "Electromagnetic Imaging of Objects Buried under a Dielectric Rough Surface," *IEEE International Waveform Diversity & Design Conference*, Orlando, FL, Feb. 8–13, 2009.

[29] M. Inggs, "Passive Coherent Location as Cognitive Radar," *IEEE International Waveform Diversity & Design Conference*, Orlando, FL, Feb. 8–13, 2009.

[30] E. Fishler, A. Haimovich, R. Blum, D. Chizhik, L. Cimini, R. Valenzuela, "MIMO Radar: An Idea Whose Time Has Come," in *Proceedings of IEEE radar conference, 2004*. p. 71–78, April 2004.

[31] Q. He, N.H. Lehmann, R.S. Blum, A.M. Haimovich, "MIMO Radar Moving Target Detection in Homogeneous Clutter," *IEEE Trans. on Aerospace and Electronic Systems (TAES)*, 2009

[32] H. Godrich, A.M. Haimovich, R.S. Blum, "Target Localization Techniques and Tools for MIMO Radar," submitted to *IET Radar, Sonar and Navigation*, 2008

[33] A.M. Haimovich, R.S. Blum, L.J. Cimini, "MIMO Radar with Widely Separated Antennas," *IEEE Signal Processing Magazine*, vol. 25, no. 1, pp. 116–129, 2008.

[34] N. Lehmann, E. Fishler, A.M. Haimovich, R.S. Blum, D. Chizhik, L.J. Cimini, R. Valenzuela, "Evaluation of Transmit Diversity in MIMO-Radar Direction Finding," *IEEE Trans. on Signal Processing*, vol. 55, no. 5, Part 2, pp 2215–2225, May 2007.

[35] E. Fishler, A.M. Haimovich, R.S. Blum, L.J. Cimini, D. Chizhik, R. Valenzuela, "Spatial Diversity in Radars–Models and Detection Performance," *IEEE Trans. on Signal Processing*, vol. 54, pp.823–838, March 2006.

[36] V.M. Chiriac, A.M. Haimovich, S.C. Schwartz, J.A. Dabin, "Performance Bound for Localization of a Near Field Source," *Asilomar Conf. Signals, Systems, and Computers*, Nov. 2009.

[37] Q. He, R.S. Blum, A.M. Haimovich, "Noncoherent MIMO Radar for Target Estimation: More Antennas Means Better Performance," *Conf. Information Sciences and Systems*, 2009.

[38] J. Li, P. Stoica, "MIMO Radar with Collocated Antennas," *IEEE Signal Processing Magazine*, pp.1053–5888, Sep. 2007.

[39] J. Li, P. Stoica (eds), *MIMO Radar Signal Processing*, John Wiley & Sons, 2009.
[40] Q. He, R.S. Blum, A.M. Haimovich, "NonCoherent MIMO Radar for Target Estimation: More Antennas Means Better Performance," *Conf. Information Sciences and Systems*, 2009.
[41] M.A. Haleem, A.M Haimovich, "Sidelobe Mitigation in MIMO Radar with Multiple Subcarriers," *IEEE Radar Conference*, May 2009.
[42] M.A. Haleem, A.M Haimovich, "On the Distribution of Ambiguity Levels in MIMO Radar," *Asilomar Conf. Signals, Systems, and Computers*, Oct. 2008.
[43] H. Godrich, A.M. Haimovich, R.S. Blum, "Target Localization Accuracy and Multiple Target Localization: Tradeoff in MIMO Radars," *Asilomar Conf. Signals, Systems, and Computers*, Oct. 2008.
[44] D.J. Rabideau, "Nonadaptive MIMO Radar Techniques for Reducing Clutter," *2008 IEEE Radar Conference*, Rome, Italy, pp.1–6, May 26–30, 2008.
[45] H. Godrich, A.M. Haimovich, R.S. Blum, "Target Localization Techniques and Tools for MIMO Radar," *IEEE Radar Conference*, Rome, Italy, May 2008.
[46] H. Godrich, A.M. Haimovich, R.S. Blum, "Cramer-Rao Bound on Target Localization Estimation in MIMO Radar Systems," *Conf. Information Sciences and Systems, (CISS '08)*, pp. 134–139, March 2008.
[47] Q. He, R.S. Blum, H. Godrich, A.M. Haimovich, "Cramer-Rao Bound for Target Velocity Estimation in MIMO Radar with Widely Separated Antennas," *Conf. Information Sciences and Systems, (CISS '08)*, pp. 123–127, March 2008.
[48] N.H. Lehmann, A.M. Haimovich, R.S. Blum, L.J. Cimini, "High resolution capabilities of MIMO radar," *Proc. 40th Asilomar Conf. Signals, Systems, and Computers*, pp. 25–30, Oct. 2006.
[49] N. Lehmann, A.M. Haimovich, R.S. Blum, L. Cimini, "MIMO-Radar Application to Moving Target Detection in Homogenous Clutter," presented at the *14th Annual Workshop on Adaptive Sensor Array Processing*, June 2006.
[50] D.R. fuhrmann, G. San Antonio, "Transmit Beamforming for MIMO Radar Systems Using Signal Cross-Correlation," *IEEE Transactions on Aerospace and Electronic Systems*, vol. 44. no. 1, pp. 171–186, Jan. 2008.
[51] P.F. Sammartino, C.J. Baker, H.D. Griffiths, "Target Model Effects on MIMO Radar Performance," *IEEE International Conference on Acustics, Speech and Signal Processing*, vol. 5, pp. 1129–1132, May 2006.e.
[52] E. Fishler, A.M. Haimovich, R.S. Blum, L.J. Cimini, D. Chizhik, R. Valenzuela, "Performance of MIMO Radar Systems: Advantages of Angular Diversity," *Proc. 38th Asilomar Conf. Signals, Systems, and Computers*, pp. 305–309, Nov. 2004.
[53] E. Fishler, A.M. Haimovich, R. Blum, L.J. Cimini, D. Chizhik, R. Valenzuela, "Statistical MIMO Radar," *Proc. Conf. on Adaptive Sensors Array Processing*, March 2004. http://www.stormingmedia.us/22/2263/A226334.html
[54] H. Godrich, A.M Haimovich, R.S. Blum, "Concepts and Applications of MIMO Radar System with Widely Separated Antenna," *MIMO Radar Signal Processing*, (ed) J. Li.
[55] Y. Yang, R.S. Blum, S. Sfar, "Antenna selection for MIMO systems with closely spaced antennas," *EURASIP Journal on Wireless Communications and Networking*, 2009.
[56] Z. Xu, S. Sfar, R.S. Blum, "Analysis of MIMO Systems with Receive Antenna Selection in Spatially Correlated Rayleigh Fading Channels," *IEEE Trans. on Vehicular Technology*, vol. 58, no. 1, pp. 251–262, Jan. 2009.
[57] D.C. Youla, "On the Factorization of Rational Matrices," *IRE Trans. Inform. Theory*, vol. IT-7, pp. 172–189, 1961.
[58] Y. Yang, R.S. Blum, "Minimax Robust MIMO Radar Waveform Design," *IEEE Journal of Selected Topics in Signal Processing*, special issue on *Adaptive Waveform Design for Agile Sensing and Communication*, vol. 1, pp. 147–155, June 2007.
[59] Y. Yang, R.S. Blum, "MIMO Radar Waveform Design Based on Mutual Information and Minimum Mean Square Error Estimation," *IEEE Transactions on Aeospace and Electronic System*, vol. 43, pp. 330–343, Jan. 2007.

[60] J. Li, L. Xu, P. Stoica, D. Bliss, K. Forsythe, "Range Compression and Waveform Optimization for MIMO Radar–A Cramer-Rao Bound Based Study," *IEEE Transactions on Signal Processing*, vol. 56, no. 1, pp. 218–232, Jan. 2008.
[61] F.C. Robey, S. Coutts, D. Werikle, C. Mcharg, K. Cuomo, "MIMO Radar Theory and Experimental Results," *Proc. 38th Asilomar Conf. Signal, System and Computers*, pp. 300–304, Nov. 2004.
[62] P.F. Sammartino, C.J. Baker, H.D. Griffiths, "Adaptive MIMO Radar Systems in Clutter," *2007 IEEE Radar Conference*, pp. 276–281, April 2007.
[63] A. Sadeque, M. Saquib, M. Rice, "Delay Diversity for Correlated MIMO Channel," *IEEE International Waveform Diversity & Design Conference*, Orlando, FL, Feb. 8–13, 2009.
[64] T. Qureshi, M. Zoltowski, R. Calderbank, "MIMO-OFDM Channel Estimation Using Golay Complementary Sequence," *IEEE International Waveform Diversity & Design Conference*, Orlando, FL, Feb. 8–13, 2009.
[65] R.S. Blum, "MIMO Capacity with Interference," *IEEE Journal on Selected Area in Communications, Special Issue on MIMO*, vol. 21, pp. 793–801, June 2003.
[66] R.S. Blum, J.H. Winters, "On Optimum MIMO with Antenna Selection," *IEEE Communications Letters*, pp. 322–324, Aug. 2002.
[67] A.M. Haimovich, R.S. Blum, "Spatial Coherence of Targets in MIMO Radar Applications," *Asilomar Conference on Signals, Systems, and Computers*, 2009.
[68] J. Zhang, G. Maddlouli, A. Papandreou-Suppappola, D. Morrell, "Cramer-Rao Lower Bounds for the Joint Estimation of Target Attributes Using MIMO Radars," *IEEE International Waveform Diversity & Design Conference*, Orlando, FL, Feb. 8–13, 2009.
[69] G. Wang, Y. Lu, "Clutter Rank of MIMO Radar with a Special Class of Waveforms," *IEEE International Waveform Diversity & Design Conference*, Orlando, FL, Feb. 8–13, 2009.
[70] F. Daum, J. Huang, "MIMO Radar" Snake Oil or Good Idea," *IEEE International Waveform Diversity & Design Conference*, Orlando, FL, Feb. 8–13, 2009.
[71] H. Godrich, A.M. Haimovich, R.S. Blum, "A MIMO Radar System Approach to Target Tracking," *2009 Asilomar Conference on Signals, Systems and Computers*, 2009.
[72] J.H.G. Ender, J. Klare, "System Architectures and Algorithms for Radar Imaging by MIMO-SAR," *2009 IEEE Radar Conference*, Pasadena, CA, pp. 1–6, May 4–8 2009.
[73] K.W. Forsythe, D.W. Bliss, "Waveform Correlation and Optimization Issues for MIMO Radar," *Proc. 39th Asilomar Conf. Signals, Systems, and Computers*, Pacific Grove, CA, pp.1306–1310, Nov. 2005.
[74] K.W. Forsythe, D.W. Bliss, G.S. Fawcett, "Multiple-Input Multiple-Output (MIMO) Radar: Performance Issues," *Proc. 38th Asilomar Conf. Signals, System, Computers*, Pacific Grove, CA, vol. 1, pp. 310–315, Nov. 2004.
[75] M. Xue, X. Zhu, J. Li, D. Vu, P. Stoica, "MIMO Radar Angle-Doppler Imaging via Iterative Space-Time Adaptive Processing," *IEEE International Waveform Diversity & Design Conference*, Orlando, FL, Feb. 8–13, 2009.
[76] D. Bliss. K.W. Forsythe, S.K. Davis, G.S. Fawcett. D.J. Rabideau, L.L. Horowitz, S. Kraut, "GMTI MIMO Radar," *IEEE International Waveform Diversity & Design Conference*, Orlando, FL, Feb. 8–13, 2009.
[77] J. Bergin, S. Mcneil, L. Fomundam, P. Zhulch, "MIMO Phase-Array for SMTI Radar," *IEEE International Waveform Diversity & Design Conference*, Orlando, FL, Feb. 8–13, 2009.
[78] H. Godrich, A. Haimovich, R. Blum, "A Comparative Study of Target Localization in MIMO Radar Systems," *IEEE International Waveform Diversity & Design Conference*, Orlando, FL, Feb. 8–13, 2009.

CHAPTER 7

Sparsity-Based Receivers

7.1 Introduction

The conventional receiver for a radar system consists of either a matched filter or other suitably designed linear filter. The goal of the receiver is to detect the presence and location of objects based on the received signal. Objects that are near other objects sometimes go undetected due to receiver sidelobes and noise.

The received signal $y(n)$ can be modeled as the convolution of the radar scene $x(n)$ and the transmit pulse $h(n)$ plus noise. We write the observed signal $\mathbf{y}$ as

$$\mathbf{y} = \mathbf{H}\mathbf{x} + \mathbf{w} \tag{7.1}$$

where $\mathbf{H}$ is a linear transform representing the convolution of $\mathbf{x}$ with $\mathbf{h}$ and $\mathbf{w}$ is additive noise. Generally, the system $\mathbf{H}$ may be either singular or near singular and therefore $\mathbf{x}$ cannot be recovered by inverting $\mathbf{H}$. Even if invertible, using the inverse directly may amplify noise and lead to poor results. In radar systems, $\mathbf{H}$ is often an LFM chirp waveform in which case the matched filter is an approximate inverse and is given by $\mathbf{H}^*$.

Nonlinear radar receivers that leverage recently developed sparse signal processing techniques provide an alternative to the matched filter. Sparse signal processing is useful for numerous problems in signal processing: denoising, deconvolution, interpolation, super-resolution, declipping, etc. [9].

The estimation of $\mathbf{x}$ from $\mathbf{y}$ can be viewed as a linear inverse problem. A standard approach to solve linear inverse problems is to define a suitable objective function $J(\mathbf{x})$ and to find the signal $\mathbf{x}$ minimizing $J(\mathbf{x})$. Generally, the chosen objective function is the sum of two terms:

$$J(\mathbf{x}) = D(\mathbf{y}, \mathbf{H}\mathbf{x}) + \lambda R(\mathbf{x}) \tag{7.2}$$

where $D(\mathbf{y}, \mathbf{H}\mathbf{x})$ measures the discrepancy between $\mathbf{y}$ and $\mathbf{H}\mathbf{x}$, and $R(\mathbf{x})$ is a regularization term (or penalty function). The parameter λ is called the regularization parameter and is used to adjust the tradeoff between the two terms. On one hand, we want to find a signal $\mathbf{x}$ so that $\mathbf{H}\mathbf{x}$ is similar to $\mathbf{y}$; that is, we want to find a signal $\mathbf{x}$ which is consistent with the observed data $\mathbf{y}$. For $D(\mathbf{y}, \mathbf{H}\mathbf{x})$, we can use the mean square error, namely

$$D(\mathbf{y}, \mathbf{H}\mathbf{x}) = \|\mathbf{y} - \mathbf{H}\mathbf{x}\|_2^2. \tag{7.3}$$

The notation $\|\mathbf{v}\|_2^2$ represents the sum of squares of the vector $\mathbf{v}$,

$$\|\mathbf{v}\|_2^2 \stackrel{\triangle}{=} v_1^2 + v_2^2 + \cdots + v_N^2. \tag{7.4}$$

Minimizing $D(\mathbf{y}, \mathbf{H}\mathbf{x})$ will generate a signal $\mathbf{x}$ that is as consistent with $\mathbf{y}$ as possible according to the square error criterion. We could try to minimize $D(\mathbf{y}, \mathbf{H}\mathbf{x})$ by setting $\mathbf{x} = \mathbf{H}^{-1}\mathbf{y}$; however, $\mathbf{H}$ may not be invertible in which case the signal $\mathbf{x}$ minimizing $J(\mathbf{x})$ is not unique. Even if $\mathbf{H}$ were invertible, it may be very ill-conditioned in which case this solution amplifies the noise, sometimes so much that the solution is useless. The role of the regularization term $R(\mathbf{x})$ is exactly to address this problem. The regularizer $R(\mathbf{x})$ should be chosen so as to penalize undesirable/unwanted behavior in $\mathbf{x}$. For example, if it is expected that $\mathbf{x}$ is a smooth signal, then $R(\mathbf{x})$ should be chosen to penalize nonsmooth signals. For example, one could set $R(\mathbf{x}) = \sum_n |x(n) - x(n-1)|^2$ or $R(\mathbf{x}) = \sum_n |x(n) - x(n-1)|$. (The second $R(\mathbf{x})$ is called the "total variation" of $\mathbf{x}$ and is zero only when $\mathbf{x}$ is a constant-valued signal. The more the signal $\mathbf{x}$ varies, the greater is its total variation.)

In this chapter, it is assumed that the signal of interest, $\mathbf{x}$, is known to be *sparse*. That is, $\mathbf{x}$ has relatively few nonzero values. For example, $\mathbf{x}$ may consist of a few impulses and is otherwise zero. In this case, ideally $R(\mathbf{x})$ should count the number of nonzero values of $\mathbf{x}$. Unfortunately, with $R(\mathbf{x})$ defined in that manner, the objective function $J(\mathbf{x})$ is very difficult to minimize. First, this $R(\mathbf{x})$ is not differentiable. Second, and more importantly, this $R(\mathbf{x})$ is not a *convex* function of $\mathbf{x}$, and therefore $J(\mathbf{x})$ will have many local minima. In order to develop robust numerical algorithms for the minimization of $J(\mathbf{x})$, it is best that $J(\mathbf{x})$ be a convex function of $\mathbf{x}$. We would like a function that measures sparsity, but which is also convex. For this reason, when $\mathbf{x}$ is known to be sparse, the regularization function $R(\mathbf{x})$ is often chosen to be the ℓ_1-norm,

$$\|\mathbf{x}\|_1 := |x_1| + |x_2| + \cdots + |x_N|. \tag{7.5}$$

Hence, the approach defines the objective function

$$J(\mathbf{x}) = \|\mathbf{y} - \mathbf{H}\mathbf{x}\|_2^2 + \lambda\|\mathbf{x}\|_1 \tag{7.6}$$

and an estimate $\mathbf{x}$ based on $\mathbf{y}$ is obtained by minimizing $J(\mathbf{x})$,

$$\hat{\mathbf{x}} = \arg\min_{\mathbf{x}} J(\mathbf{x}). \tag{7.7}$$

This is called an ℓ_1-norm regularized linear inverse problem.

The development of fast algorithms to minimize (7.6), and related functions, is an active research topic. An early and important algorithm is the iterated soft-thresholding algorithm (ISTA), also called the thresholded-Landweber (TL) algorithm. This algorithm was developed [7, 12] for the purpose of wavelet-based signal restoration, but the algorithm in a more general form appeared earlier in the optimization literature [2]. Since the development of ISTA, faster algorithms have been introduced for the minimization of (7.6), for example [2,3,6,8,10,11,13,14,18,26,28] and others. Some of these algorithms can be used for the minimization of more general or related objective functions as well.

This chapter describes the derivation of the ISTA. ISTA is a combination of the Landweber algorithm and soft-thresholding. The derivation below is based on majorization-minimization (a concept in optimization theory) and ℓ_1-norm regularized denoising (which leads to soft-thresholding).

7.2 Majorization-Minimization

Majoriziation-minimization (MM) replaces a difficult minimization problem by a sequence of easier minimization problems. The MM approach generates a sequence of vectors $\mathbf{x}_k$, $k = 0, 1, 2, \ldots$ which converge to the desired solution. The MM approach is useful for the minimization of $J(\mathbf{x})$ in (7.6) because this $J(\mathbf{x})$ cannot be easily minimized. There is no formula for the vector $\mathbf{x}$ that minimizes $J(\mathbf{x})$.

The MM idea can be described as follows. Suppose we have a vector $\mathbf{x}_k$, a "guess" for the minimum of $J(\mathbf{x})$. Based on $\mathbf{x}_k$, we would like to find a new vector, $\mathbf{x}_{k+1}$, which further decreases $J(\mathbf{x})$; that is, we want to find $\mathbf{x}_{k+1}$ such that

$$J(\mathbf{x}_{k+1}) < J(\mathbf{x}_k). \tag{7.8}$$

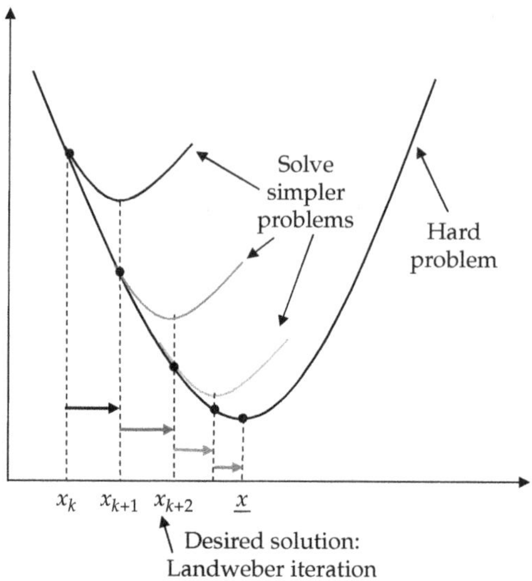

FIGURE 7.1 Illustration of the majorization-minimization (MM) method for convex optimization.

The MM approach first chooses a new function which majorizes $J(\mathbf{x})$, and then it minimizes the new function to obtain $\mathbf{x}_{k+1}$. MM puts some requirements on this new function, call it $G(\mathbf{x})$. We should choose $G(\mathbf{x})$ such that $G(\mathbf{x}) \geq J(\mathbf{x})$ for all $\mathbf{x}$ [it is said that $G(\mathbf{x})$ majorizes $J(\mathbf{x})$]. In addition, $G(\mathbf{x})$ should equal $J(\mathbf{x})$ at $\mathbf{x}_k$. We find $\mathbf{x}_{k+1}$ by minimizing $G(\mathbf{x})$. For this method to be useful, we should choose $G(\mathbf{x})$ to be a function we can minimize easily. The function $G(\mathbf{x})$ will be different at each iteration, so we denote it $G_k(\mathbf{x})$. Figure 7.1 illustrates the MM method.

Figure 7.2 illustrates the MM procedure with an actual example. For clarity, the figure illustrates the minimization of a function of one variable. However, the MM procedure works in the same way for the minimization of multivariate functions, and it is in the multivariate case where the MM procedure is especially useful. (In the following sections, MM will be used for the minimization of multivariate functions.)

To summarize, the MM algorithm for the minimization of a function $J(\mathbf{x})$ is given by the following iteration:

1. Set $k = 0$. Initialize $\mathbf{x}_0$.
2. Choose $G_k(\mathbf{x})$ such that

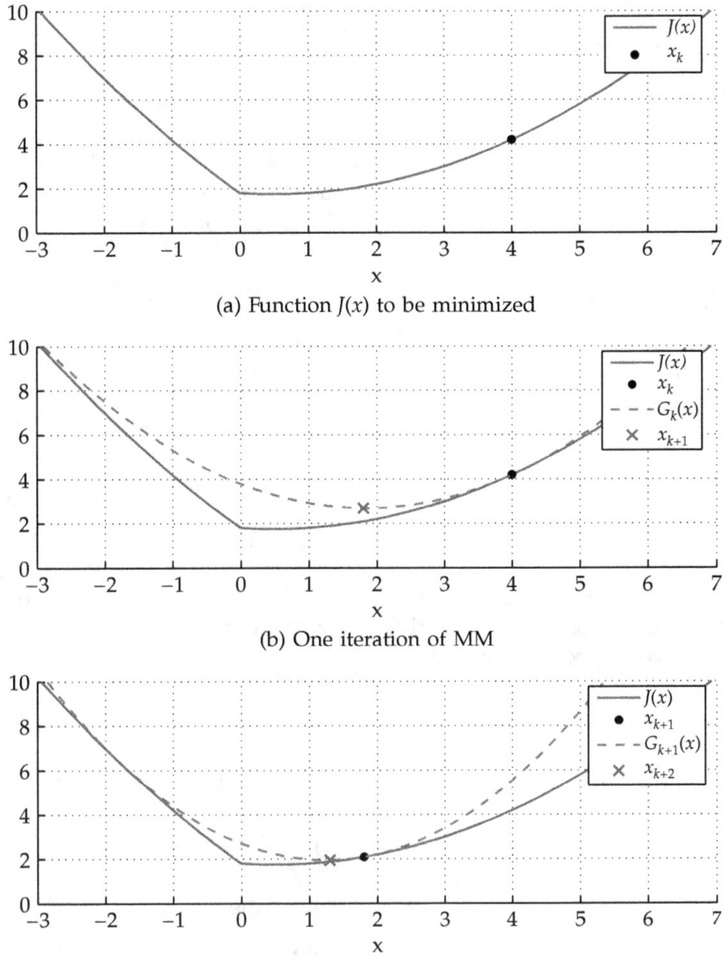

(a) Function $J(x)$ to be minimized

(b) One iteration of MM

(c) Subsequent iteration

FIGURE 7.2 Illustration of MM procedure. Plot (a) illustrates the function to be minimized, $J(x)$, and a "guess" of the minimizer, x_k. Plot (b) illustrates the majorizing function $G_k(x)$ which coincides with $J(x)$ at $x = x_k$ but is otherwise greater than $J(x)$. The minimizer of $G_k(x)$, shown as a cross in (b), is taken as x_{k+1}. Plot (c) illustrates the subsequent iteration. In (c) the majorizer coincides with $J(x)$ at $x = x_{k+1}$. The iterates x_k converge to the minimum of $J(x)$.

(a) $G_k(\mathbf{x}) \geq J(\mathbf{x})$ for all $\mathbf{x}$
(b) $G_k(\mathbf{x}_k) = J(\mathbf{x}_k)$

3. Set $\mathbf{x}_{k+1}$ as the minimizer of $G_k(\mathbf{x})$.
4. Set $k = k + 1$ and go to step (2).

More details about the MM procedure are provided in [11] and the references therein.

Note that the MM procedure is known as an "optimization transfer" method or as "surrogate" functionals. For example, optimization transfer methods are applied to SAR image restoration in [20].

7.2.1 The Landweber Iteration

Although we are interested in minimizing $J(\mathbf{x})$ in (7.6), let us consider first the minimization of the simpler quadratic objective function

$$J(\mathbf{x}) = \|\mathbf{y} - \mathbf{H}\mathbf{x}\|_2^2 \tag{7.9}$$

$$= (\mathbf{y} - \mathbf{H}\mathbf{x})^*(\mathbf{y} - \mathbf{H}\mathbf{x}) \tag{7.10}$$

$$= \mathbf{y}^*\mathbf{y} - 2\mathbf{y}^*\mathbf{H}\mathbf{x} + \mathbf{x}^*\mathbf{H}^*\mathbf{H}\mathbf{x}. \tag{7.11}$$

Because $J(\mathbf{x})$ in (7.9) is differentiable and convex, we can obtain its minimizer by setting the derivative with respect to $\mathbf{x}$ to zero. The derivative of $J(\mathbf{x})$ is given by

$$\frac{\partial}{\partial \mathbf{x}} J(\mathbf{x}) = -2\mathbf{H}^*\mathbf{y} + 2\mathbf{H}^*\mathbf{H}\mathbf{x}. \tag{7.12}$$

Setting the derivative to zero gives a system of linear equations:

$$\frac{\partial}{\partial \mathbf{x}} J(\mathbf{x}) = \mathbf{0} \quad \Longrightarrow \quad (\mathbf{H}^*\,\mathbf{H})\,\mathbf{x} = \mathbf{H}^*\mathbf{y}. \tag{7.13}$$

So the minimizer of $J(\mathbf{x})$ in (7.9) is given by the classic solution

$$\mathbf{x} = (\mathbf{H}^*\,\mathbf{H})^{-1}\mathbf{H}^*\mathbf{y}. \tag{7.14}$$

Thus, $J(\mathbf{x})$ in (7.9) can be minimized by solving a linear system of equations. However, we may not be able to solve these equations easily. For example, if $\mathbf{x}$ is a very long signal, then $\mathbf{H}$ will be a very large matrix and solving the system of equations may require too much memory and computation time. Additionally, the matrix $\mathbf{H}^*\mathbf{H}$ might not be invertible, or it may be very ill-conditioned.

By using the MM approach to minimize $J(\mathbf{x})$ in (7.9), we can avoid solving a system of linear equations. As described in Sec. 7.2, at each iteration k of the MM approach, we should find a function $G_k(\mathbf{x})$ that coincides with $J(\mathbf{x})$ at $\mathbf{x}_k$ but otherwise upper-bounds $J(\mathbf{x})$. We should choose a majorizer $G_k(\mathbf{x})$ which can be minimized more easily (without having to solve a system of equations).

We will find a function $G_k(\mathbf{x})$ that majorizes $J(\mathbf{x})$ by adding a nonnegative function to $J(\mathbf{x})$,

$$G_k(\mathbf{x}) = J(\mathbf{x}) + \text{ nonnegative function of } \mathbf{x}. \tag{7.15}$$

In order that $G_k(\mathbf{x})$ coincides with $J(\mathbf{x})$ at $\mathbf{x} = \mathbf{x}_k$, the nonnegative function we add to $J(\mathbf{x})$ should be equal to zero at $\mathbf{x}_k$. With this in mind, we choose $G_k(\mathbf{x})$ to be:

$$G_k(\mathbf{x}) = \|\mathbf{y} - \mathbf{H}\mathbf{x}\|_2^2 + \underbrace{(\mathbf{x} - \mathbf{x}_k)^*(\alpha\,\mathbf{I} - \mathbf{H}^*\mathbf{H})(\mathbf{x} - \mathbf{x}_k)}_{\text{nonnegative}}. \tag{7.16}$$

The function we have added to $J(\mathbf{x})$ is clearly zero at $\mathbf{x}_k$ so we have $G_k(\mathbf{x}) = J\,(\mathbf{x}_k)$ as desired.

To ensure the function we have added to $J(\mathbf{x})$ is nonnegative for all $\mathbf{x}$, the matrix $\alpha\,\mathbf{I} - \mathbf{H}^*\mathbf{H}$ must be positive semidefinite. Therefore, the scalar parameter α must be chosen to be equal to or greater than the maximum eigenvalue of $\mathbf{H}^*\mathbf{H}$. Thus

$$\alpha \geq \text{maxeig}(\mathbf{H}^*\mathbf{H}). \tag{7.17}$$

Following the MM procedure, we need to minimize $G_k(\mathbf{x})$ to obtain $\mathbf{x}_{k+1}$. Expanding $G_k(\mathbf{x})$ in (7.16) gives

$$G_k(\mathbf{x}) = \mathbf{y}^*\mathbf{y} - 2\mathbf{y}^*\mathbf{H}\mathbf{x} + \mathbf{x}^*\mathbf{H}^*\mathbf{H}\mathbf{x} + (\mathbf{x} - \mathbf{x}_k)^*(\alpha\,\mathbf{I} - \mathbf{H}^*\mathbf{H})(\mathbf{x} - \mathbf{x}_k) \tag{7.18}$$

$$= \mathbf{y}^*\mathbf{y} + \mathbf{x}_k{}^*\,(\alpha\,\mathbf{I} - \mathbf{H}^*\mathbf{H})\,\mathbf{x}_k - 2\,(\mathbf{y}^*\mathbf{H} + \mathbf{x}_k{}^*(\alpha\,\mathbf{I} - \mathbf{H}^*\mathbf{H}))\,\mathbf{x} + \alpha\,\mathbf{x}^*\mathbf{x}. \tag{7.19}$$

Note that the quadratic term in (7.19) is simply $\alpha\,\mathbf{x}^*\mathbf{x}$ instead of $\mathbf{x}^*\mathbf{H}^*\mathbf{H}\mathbf{x}$ as it was in (7.11). Therefore we can minimize $G_k(\mathbf{x})$ more easily as follows:

$$\frac{\partial}{\partial \mathbf{x}} G_k(\mathbf{x}) = -2\mathbf{H}^*\mathbf{y} - 2\,(\alpha\,\mathbf{I} - \mathbf{H}^*\mathbf{H})\,\mathbf{x}_k + 2\alpha\mathbf{x} \tag{7.20}$$

$$\frac{\partial}{\partial \mathbf{x}} G_k(\mathbf{x}) = \mathbf{0} \quad \Longrightarrow \quad \mathbf{x} = \mathbf{x}_k + \frac{1}{\alpha}\mathbf{H}^*(\mathbf{y} - \mathbf{H}\mathbf{x}_k) \tag{7.21}$$

Hence, we obtain through the MM procedure, the update equation

$$\mathbf{x}_{k+1} = \mathbf{x}_k + \frac{1}{\alpha}\mathbf{H}^*(\mathbf{y} - \mathbf{H}\mathbf{x}_k) \tag{7.22}$$

which is known as the "Landweber" iteration. It is guaranteed that $J(\mathbf{x})$ in (7.9) decreases with each iteration, due to the properties of the MM procedure. Note that (7.22) does not require the solution to a linear system of equations. It only requires multiplying by $\mathbf{H}$ and $\mathbf{H}^*$.

To emphasize the functional dependence of $G_k(\mathbf{x})$ on $\mathbf{x}$ we can write $G_k(\mathbf{x})$ in (7.19) as

$$G_k(\mathbf{x}) = \alpha\,(-2\mathbf{b}^*\mathbf{x} + \mathbf{x}^*\mathbf{x}) + c \tag{7.23}$$

where

$$\mathbf{b} = \frac{1}{\alpha}\left[\mathbf{H}^*\mathbf{y} + (\alpha\,\mathbf{I} - \mathbf{H}^*\mathbf{H})\,\mathbf{x}_k\right] = \mathbf{x}_k + \frac{1}{\alpha}\mathbf{H}^*(\mathbf{y} - \mathbf{H}\mathbf{x}_k) \tag{7.24}$$

and where c consists of the first two terms of (7.19) which do not depend on $\mathbf{x}$. Note that for any vectors $\mathbf{b}$ and $\mathbf{x}$,

$$\mathbf{b}^*\mathbf{b} - 2\mathbf{b}^*\mathbf{x} + \mathbf{x}^*\mathbf{x} = \|\mathbf{b} - \mathbf{x}\|_2^2, \tag{7.25}$$

hence from (7.23) we can write $G_k(\mathbf{x})$ as

$$G_k(\mathbf{x}) = \alpha\,\|\mathbf{b} - \mathbf{x}\|_2^2 - \alpha\mathbf{b}^*\mathbf{b} + c. \tag{7.26}$$

That is, using (7.24), the majorizer $G_k(\mathbf{x})$ in (7.16) can be written as

$$G_k(\mathbf{x}) = \alpha\left\|\mathbf{x}_k + \frac{1}{\alpha}\mathbf{H}^*(\mathbf{y} - \mathbf{H}\mathbf{x}_k) - \mathbf{x}\right\|_2^2 + K \tag{7.27}$$

where K does not depend on $\mathbf{x}$. This shows that $G_k(\mathbf{x})$ is shaped like a bowl with circular level sets. A key point in the Landweber iteration is that in $J(\mathbf{x})$ in (7.9) the variables (the elements of $\mathbf{x}$) are coupled: x_1 multiplies x_2, etc. On the other hand, in the majorizer $G_k(\mathbf{x})$ the variables are uncoupled: x_1 is not multiplied with any other x_i, etc.

7.3 Soft-Thresholding

As in Sec. 7.2.1, consider the minimization of an objective function simpler than (7.6). Thus consider the minimization of the function

$$J(\mathbf{x}) = \|\mathbf{y} - \mathbf{x}\|_2^2 + \lambda\|\mathbf{x}\|_1. \tag{7.28}$$

This is a special simple case of (7.6) with $\mathbf{H} = \mathbf{I}$. The function $J(\mathbf{x})$ in (7.28) is convex, but not differentiable (because $\|\mathbf{x}\|_1$ is not differentiable). Nevertheless, the minimizer of this $J(\mathbf{x})$ is given by a simple formula. Expanding $J(\mathbf{x})$ in (7.28) gives

$$J(\mathbf{x}) = |y_1 - x_1|^2 + \lambda|x_1| + |y_2 - x_2|^2 + \lambda|x_2| + \cdots + |y_N - x_N|^2 + \lambda|x_N|. \tag{7.29}$$

Note that the variables, x_i, are *uncoupled*. That is, the function $J(\mathbf{x})$ can be minimized by minimizing each term $|y_i - x_i|^2 + \lambda|x_i|$ individually

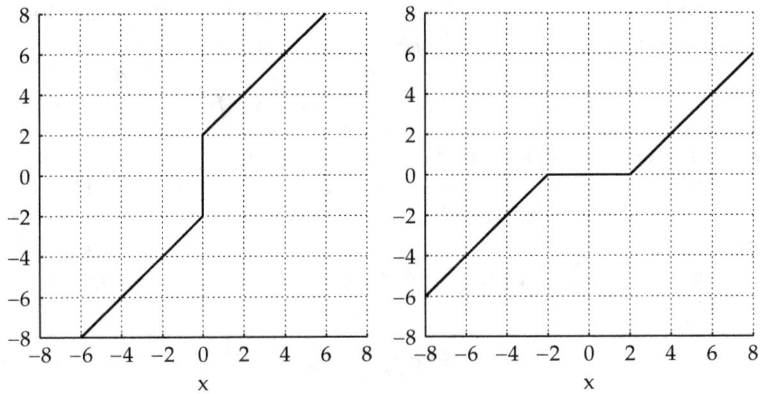

(a) (b) Soft (x, T)

FIGURE 7.3 The soft-threshold rule (7.34) with threshold $T = 2$.

to get x_i, for $1 \leq i \leq N$. Therefore we need only consider the scalar minimization of the function

$$f(x) = |y - x|^2 + \lambda|x|. \tag{7.30}$$

Taking the derivative, assuming x and y are real-valued,

$$f'(x) = -2(y - x) + \lambda \text{sign}(x), \tag{7.31}$$

setting $f'(x) = 0$ gives

$$y = x + \frac{\lambda}{2}\text{sign}(x), \tag{7.32}$$

which is shown in Fig. 7.3(a). Solving for x gives the graph shown in Fig. 7.3(b) with threshold $\lambda/2$. Thus, the minimizer of $f(x)$ is obtained by applying the soft-threshold rule to y with threshold $\lambda/2$.

The soft-threshold rule is the nonlinear function defined as

$$\text{soft}(x, T) := \begin{cases} x + T & x \leq -T \\ 0 & |x| \leq T \\ x - T & x \geq T \end{cases} \tag{7.33}$$

or more compactly, as

$$\text{soft}(x, T) \triangleq \text{sign}(x)\max(0, |x| - T). \tag{7.34}$$

The soft-threshold rule is illustrated in Fig. 7.3(b).

When y is complex-valued and x is allowed to be complex-valued, then the minimizer of $f(x) = |y - x|^2 + \lambda|x|$ is given by

$$\text{soft}(x, T) \stackrel{\Delta}{=} \max(0, |x| - T)\exp(j\angle x) \tag{7.35}$$

which is the same as (7.34) when x is real. This extension of the soft-threshold function to complex-valued x essentially applies the soft-thresold function to the modulus of x and leaves the phase of x as unchanged.

In terms of the soft-threshold rule, the minimization of (7.30) is given by

$$x = \text{soft}(y,\ \lambda/2). \tag{7.36}$$

Because the variables in the function $J(\mathbf{x})$ in (7.28) are uncoupled and the solution is obtained by minimizing with respect to each x_i individually, the minimizer of $J(\mathbf{x})$ is obtained by applying the soft-thresholding rule to each element of $\mathbf{x}$:

$$\mathbf{x} = \text{soft}\left(\mathbf{y},\ \lambda/2\right). \tag{7.37}$$

The minimization of (7.28) does not require an iterative algorithm. It is minimized simply by soft-thresholding each element of $\mathbf{y}$ with threshold $\lambda/2$.

7.3.1 Iterated Soft-Thresholding Algorithm

In this section, the minimization of the objective function $J(\mathbf{x})$ in (7.6) is considered, with

$$J(\mathbf{x}) = \|\mathbf{y} - \mathbf{H}\mathbf{x}\|_2^2 + \lambda\|\mathbf{x}\|_1. \tag{7.38}$$

It can be minimized by combining the results of Secs. 7.2.1 and 7.3.

Applying the MM approach, the goal is to find a majorizer of $J(\mathbf{x})$ which coincides with $J(\mathbf{x})$ at $\mathbf{x}_k$, and which is easily minimized. We can add to $J(\mathbf{x})$ the same nonnegative function as in (7.16) in Sec. 7.2.1,

$$G_k(\mathbf{x}) = J(\mathbf{x}) + (\mathbf{x} - \mathbf{x}_k)^*(\alpha\,\mathbf{I} - \mathbf{H}^*\mathbf{H})(\mathbf{x} - \mathbf{x}_k). \tag{7.39}$$

By design, $G_k(\mathbf{x})$ coincides with $J(\mathbf{x})$ at $\mathbf{x}_k$. We need to minimize $G_k(\mathbf{x})$ to obtain $\mathbf{x}_{k+1}$. Using (7.38), we write

$$G_k(\mathbf{x}) = \|\mathbf{y} - \mathbf{H}\mathbf{x}\|_2^2 + (\mathbf{x} - \mathbf{x}_k)^*(\alpha\,\mathbf{I} - \mathbf{H}^*\mathbf{H})(\mathbf{x} - \mathbf{x}_k) + \lambda\|\mathbf{x}\|_1 \tag{7.40}$$

which is the same as (7.16) except here we have the additional term $\lambda\|\mathbf{x}\|_1$. Using (7.27), we can write $G_k(\mathbf{x})$ as

$$G_k(\mathbf{x}) = \alpha\left\|\mathbf{x}_k + \frac{1}{\alpha}\mathbf{H}^*(\mathbf{y} - \mathbf{H}\mathbf{x}_k) - \mathbf{x}\right\|_2^2 + \lambda\|\mathbf{x}\|_1 + K \tag{7.41}$$

```
function [x,J] = ista(y,H,lambda,alpha,Nit)
% [x, J] = ista(y, H, lambda, alpha, Nit)
% L1-regularized signal restoration using the iterated
% soft-thresholding algorithm (ISTA)
% Minimizes J(x) = sum(abs(y-H*x).^2) + lambda*sum(abs(x))
%   INPUT
%    y - observed signal
%    H - matrix or operator
%    lambda - regularization parameter
%    alpha - alpha >= max(eig(H'*H))
%    Nit - number of iterations
% OUTPUT
%    x - estimated signal
%    J - objective function

J = zeros(1, Nit);          % Objective function
x = 0*H'*y;                 % Initialize x
T = lambda/(2*alpha);
for k = 1:Nit
    Hx = H*x;
    J(k) = sum(abs(Hx(:)-y(:)).^2) + lambda*sum(abs(x(:)));
    x = soft(x + (H'*(y - Hx))/alpha, T);
end
```

LISTING 7.1 A MATLAB program, `ista`, implementing the iterated soft-thresholding algorithm (ISTA). The main update equation is (7.43).

where K is a constant with respect to $\mathbf{x}$. Minimizing $G_k(\mathbf{x})$ is equivalent to minimizing $(1/\alpha)\, G_k(\mathbf{x})$, so $\mathbf{x}_{k+1}$ is obtained by minimizing

$$\left\| \mathbf{x}_k + \frac{1}{\alpha}\mathbf{H}^*(\mathbf{y} - \mathbf{H}\mathbf{x}_k) - \mathbf{x} \right\|_2^2 + \frac{\lambda}{\alpha}\|\mathbf{x}\|_1. \tag{7.42}$$

We omit the additive constant term K because the vector $\mathbf{x}$ minimizing $G_k(\mathbf{x})$ is not influenced by it. Note that (7.42) has exactly the same form as (7.28) which is minimized by the formula (7.37). Therefore, minimizing (7.42) is achieved by the following soft-thresholding equation:

$$\mathbf{x}_{k+1} = \text{soft}\left(\mathbf{x}_k + \frac{1}{\alpha}\mathbf{H}^*(\mathbf{y} - \mathbf{H}\mathbf{x}_k),\ \frac{\lambda}{2\alpha} \right) \tag{7.43}$$

where $\alpha \geq \text{maxeig}(\mathbf{H}^*\mathbf{H})$. This gives the iterated soft-thresholding algorithm (ISTA). Also, note that ISTA requires no matrix inversion nor the solution to any linear system of equations—the algorithm requires only the matrix-vector products. The MATLAB program `ista` in Listing 7.1 implements the iterated soft-thresholding algorithm.

Being an iterative algorithm, the convergence in ISTA is an important issue. Each iteration of ISTA is guaranteed to decrease the objective function, $J(\mathbf{x})$, from one iteration to the next. Moroever, it can be shown [2] that the sequence $\mathbf{x}_k$ converges linearly to the minimizer $\mathbf{x}_*$,

$$J(\mathbf{x}_k) - J(\mathbf{x}_*) \leq \frac{C}{k} \tag{7.44}$$

where the constant C depends the initialization $\mathbf{x}_0$ and $\mathbf{H}$. However, ISTA converges (very) slowly for some problems and more recently developed algorithms exhibit faster convergence, as discussed in Sec. 7.3.2.

Example 7.1 Figure 7.4 illustrates an example of ISTA. A clean 100-point sparse signal, $\mathbf{x}$, is illustrated in Fig. 7.4(a). The observed signal $\mathbf{y}$ is obtained by convolving $\mathbf{x}$ by the impulse response

$$\mathbf{h} = [1,\ 2,\ 3,\ 4,\ 3,\ 2,\ 1]/16 \tag{7.45}$$

and adding white Gaussian noise with standard deviation 0.05.

The result of 500 iterations of ISTA is illustrated in Fig. 7.4(c). The estimated signal is very similar to the original signal (although not exactly the same). Note that the objective function $J(\mathbf{x})$, also illustrated, decays monotonically.

The following commands implement the above example in MATLAB:

```
h = [1 2 3 4 3 2 1]/16;
N = 100;
H = convmtx(h',N);
lambda = 0.1;
alpha = 1;
Nit = 500;
[x, J] = ista(y, H, lambda, alpha, Nit);
```

Note that the selection of α in ISTA depends on the largest eigenvalue of $\mathbf{H}^*\mathbf{H}$. For a very large matrix, the computation of this eigenvalue may be overly time-consuming. However, when $\mathbf{H}$ represents discrete-time convolution, then

$$\text{maxeig}(\mathbf{H}^*\mathbf{H}) \leq \max_{\omega} |H(e^{j\omega})|^2 \tag{7.46}$$

and we can therefore set α equal to the maximum value of $|H(e^{j\omega})|^2$. For other problems, the *power method* or a related method can be used to compute the largest eigevalue; or one can use a version of ISTA that employs back-tracking for which the maximum eigenvalue of $\mathbf{H}^*\mathbf{H}$ is not needed [2].

Avoiding Large Matrices

The formulation of ISTA calls for matrix-vector multiplications; however, for many problems of interest the required matrices can be very large. For example, if $\mathbf{x}$ represents the pixel values of a 500 by 500 pixel image, then the matrix $\mathbf{H}$ will have 25,000 columns. Often, the explicit computation and storage of the matrix $\mathbf{H}$ can be avoided. For example, if $\mathbf{H}$ represents discrete-time convolution, then it is necessary to

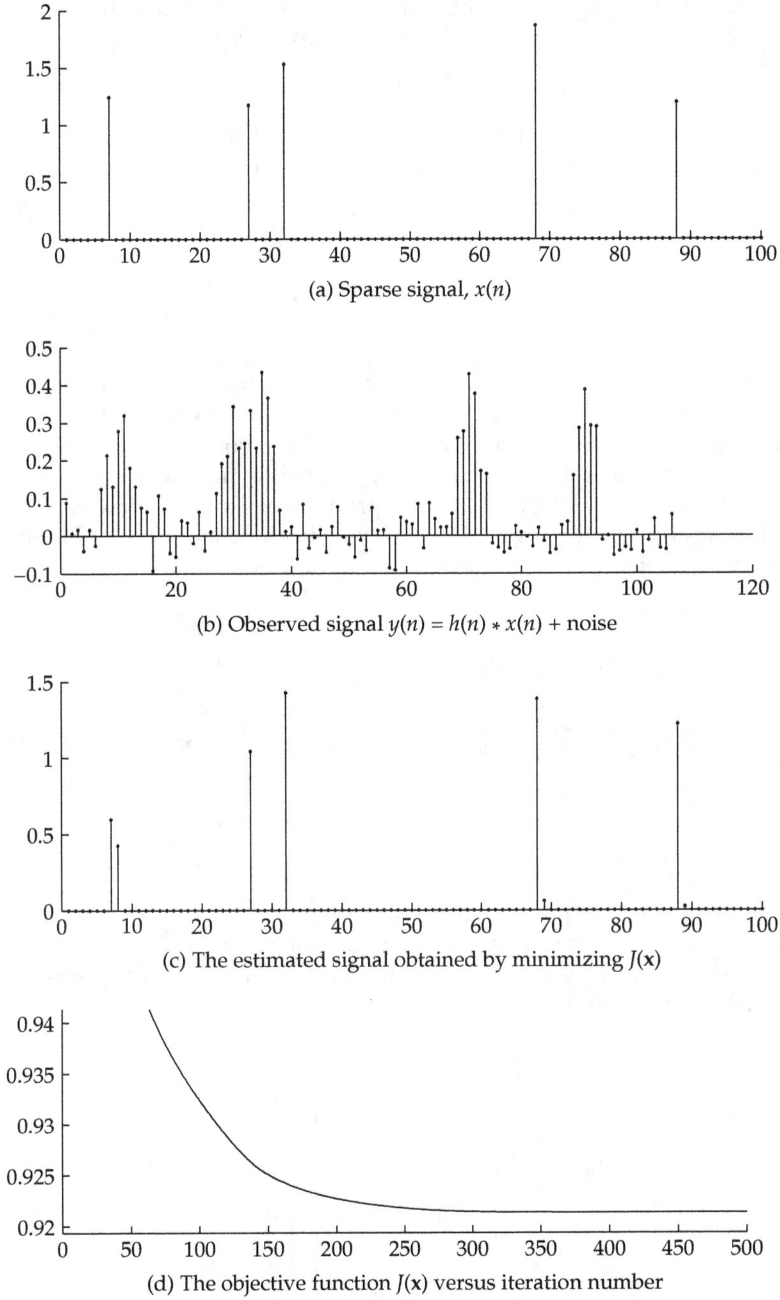

(a) Sparse signal, $x(n)$

(b) Observed signal $y(n) = h(n) * x(n) + \text{noise}$

(c) The estimated signal obtained by minimizing $J(\mathbf{x})$

(d) The objective function $J(\mathbf{x})$ versus iteration number

FIGURE 7.4 Sparse signal estimation using ISTA.

store only the impulse response $\mathbf{h}$. Specifically, although convolution can be represented by matrix-vector multiplication,

$$\mathbf{H}\mathbf{x} = \begin{bmatrix} h_0 & & & & \\ h_1 & h_0 & & & \\ h_2 & h_1 & h_0 & & \\ & h_2 & h_1 & h_0 & \\ & & h_2 & h_1 & h_0 \\ & & & h_2 & h_1 \\ & & & & h_2 \end{bmatrix} \begin{bmatrix} x_0 \\ x_1 \\ x_2 \\ x_3 \\ x_4 \end{bmatrix}, \tag{7.47}$$

we do not need to perform explicit matrix-vector multiplication. We can use a nonrecursive difference equation or zero-padded FFTs to implement the matrix-vector product in (7.47). For example, in MATLAB, the command `conv(h,x)` can be used to implement (7.47).

ISTA also calls for matrix-vector products with the matrix $\mathbf{H}^*$. For discrete-time convolution, $\mathbf{H}^*$ is similar but not identical to $\mathbf{H}$. For example, using (7.47), we have

$$\mathbf{H}^*\mathbf{g} = \begin{bmatrix} h_0^* & h_1^* & h_2^* & & & & \\ & h_0^* & h_1^* & h_2^* & & & \\ & & h_0^* & h_1^* & h_2^* & & \\ & & & h_0^* & h_1^* & h_2^* & \\ & & & & h_0^* & h_1^* & h_2^* \end{bmatrix} \begin{bmatrix} g_0 \\ g_1 \\ g_2 \\ g_3 \\ g_4 \\ g_5 \\ g_6 \end{bmatrix}. \tag{7.48}$$

Note that this is not exactly the same as the discrete-time convolution of two finite-length sequences. [The command `conv` in MATLAB will not implement (7.48) exactly.] On the other hand, if the matrix $\mathbf{H}^*$ is vertically extended as in (7.49):

$$\left[\begin{array}{ccccccc} h_2^* & & & & & & \\ h_1^* & h_2^* & & & & & \\ \hline h_0^* & h_1^* & h_2^* & & & & \\ & h_0^* & h_1^* & h_2^* & & & \\ & & h_0^* & h_1^* & h_2^* & & \\ & & & h_0^* & h_1^* & h_2^* & \\ & & & & h_0^* & h_1^* & h_2^* \\ \hline & & & & & h_0^* & h_1^* \\ & & & & & & h_0^* \end{array}\right] \begin{bmatrix} g_0 \\ g_1 \\ g_2 \\ g_3 \\ g_4 \\ g_5 \\ g_6 \end{bmatrix}, \tag{7.49}$$

```
function b = convt(h, g)
% b = convt(h, g)
% Conjugate transpose convolution: b = H' g.
Nh = length(h);
Ng = length(g);
b = conv(conj(h(Nh:-1:1)), g);
b = b(Nh:Ng);
```

LISTING 7.2 A MATLAB program, `convt`, for conjugate transpose convolution as in (7.48).

then a convolution matrix is obtained, representing the impulse response $h^*(-n)$. Therefore, the matrix-vector product $\mathbf{H}^*\mathbf{g}$ in (7.48) can be seen as discrete-time convolution of $\mathbf{g}$ with the reversed complex-conjugated version of $\mathbf{h}$, followed by a truncation [discarding the first and last $(N_h - 1)$ samples where N_h denotes the length of $\mathbf{h}$]. The MATLAB program `convt` in Listing 7.2 implements $\mathbf{H}^*\mathbf{g}$ when $\mathbf{H}$ represents convolution.

The MATLAB program `ista2` in Listing 7.3 is a second program for ISTA that uses function handles for $\mathbf{H}$ and $\mathbf{H}^*$ instead of matrices, so it is not necessary to create the matrix $\mathbf{H}$ in MATLAB.

```
function [x,J] = ista2(y,H,Ht,lambda,alpha,Nit)
% [x, J] = ista2(y, H, lambda, alpha, Nit)
% L1-regularized signal restoration using the iterated
% soft-thresholding algorithm (ISTA)
% Minimizes J(x) = sum(abs(y-H*x).^2) + lambda*sum(abs(x))
%  INPUT
%   y - observed signal
%   H - function handle
%   Ht - function handle for H'
%   lambda - regularization parameter
%   alpha - alpha >= max(eig(H'*H))
%   Nit - number of iterations
% OUTPUT
%   x - estimated signal
%   J - objective function

J = zeros(1, Nit);          % Objective function
x = 0*Ht(y);                % Initialize x
T = lambda/(2*alpha);
for k = 1:Nit
    Hx = H(x);
    J(k) = sum(abs(Hx(:)-y(:)).^2) + lambda*sum(abs(x(:)));
    x = soft(x + (Ht(y - Hx))/alpha, T);
end
```

LISTING 7.3 A MATLAB program, `ista2`, implementing the iterated soft-thresholding algorithm (ISTA). This program uses function handles for $\mathbf{H}$ and $\mathbf{H}^*$ so as to avoid the use of large matrices.

Instead, it is necessary to supply a function for each of **H** and $\mathbf{H}^*$. In the preceding example where **H** represents convolution, we can specify function handles as follows:

```
h = [1 2 3 4 3 2 1]/16;
H = @(x) conv(h,x);
Ht = @(y) convt(h,y);
[x, J] = ista2(y, H, Ht, lambda, alpha, Nit);
```

The output **x** is exactly the same as using the `ista` program in Listing 7.1, but with `ista2` no large matrices need be created or stored.

Target Detection Example

Let us compare matched filtering and the ℓ_1 receiver. [The output of the "ℓ_1 receiver" being the result of minimizing the objective function $J(\mathbf{x})$ in (7.38).] Figure 7.6 illustrates a target scene with several targets at different ranges and powers. The transmit waveform is the LFM chirp illustrated in Fig. 7.5. Figure 7.7 illustrates the outputs of

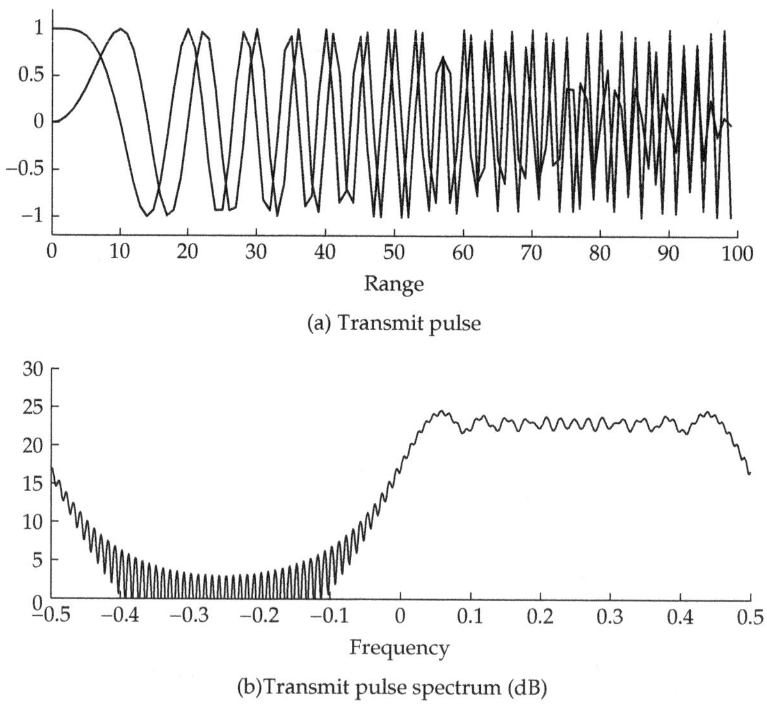

(a) Transmit pulse

(b)Transmit pulse spectrum (dB)

FIGURE 7.5 The transmit pulse is an LFM chirp signal.

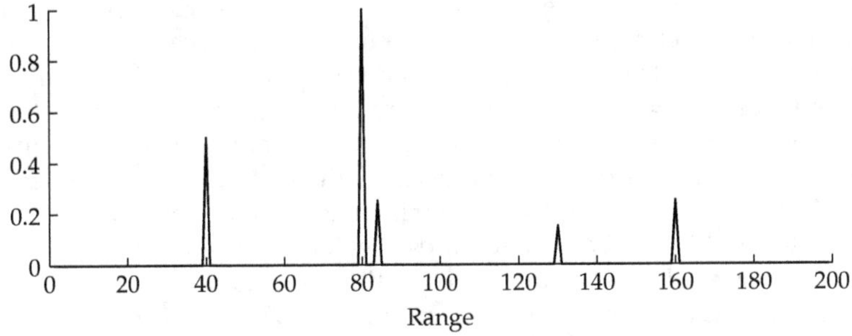

FIGURE 7.6 Target scene range profile.

the matched filter and of the ℓ_1 receiver. The ℓ_1 receiver has superior range resolution, accurately resolves the targets, and does not exhibit the sidelobes of the matched filter. The ℓ_1 receiver works well in this example because the target scene was truly sparse and thus a sparsity-based receiver can effectively exploit that prior knowledge.

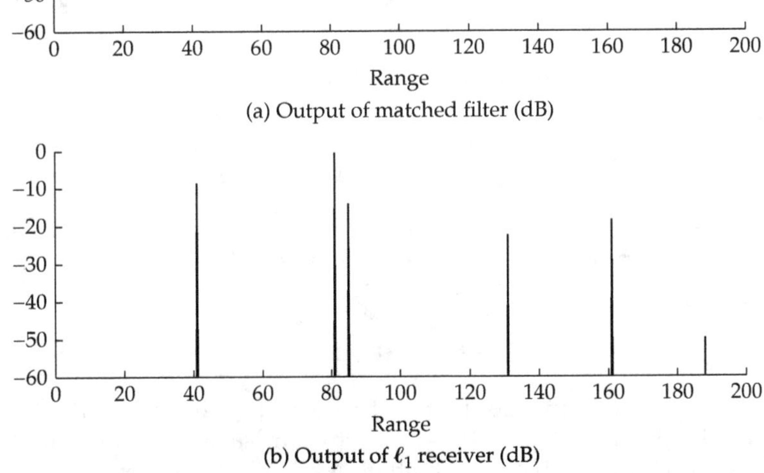

FIGURE 7.7 Comparison of matched filter and ℓ_1 receiver. (a) Output of matched filter. (b) Output of ℓ_1 receiver.

There are several issues that need to be addressed in using an ℓ_1 receiver in practice: The performance of the ℓ_1 receiver depends on the target scene truly being sparse and on the accuracy of the model of the received radar data. The performance can be sensitive to calibration errors (e.g., if the transmit waveform is not precisely known). Another important point is that because the ℓ_1 receiver is nonlinear, the influence of noise on its output signal is less easily predicted.

Note that the ℓ_1 receiver requires a value for λ, whereas no such parameter is required of the matched filter (or of other receivers implemented as linear receivers). An appropriate value for λ will depend on the noise power. Therefore, the ℓ_1 receiver may be more sensitive to noise than the matched filter. However, under the appropriate conditions, the output of the ℓ_1 receiver can have better resolution and greater noise immunity.

The ℓ_1 receiver can be generalized by modifying the regularizer in (7.6). The same sparsity-based approach applies to more general radar imaging applications, such as SAR imaging.

7.3.2 Fast ISTA (FISTA)

Due to its importance in a number of sparse signal processing problems of current interest, several faster algorithms for minimizing $J(\mathbf{x})$ in (7.6) have been proposed in the literature. One such algorithm is "Fast ISTA" (FISTA), introduced in [2]. FISTA is given by:

$$t_1 = 0, \quad \mathbf{z}_1 = \mathbf{x}_0 \tag{7.50}$$

$$t_{k+1} = \frac{1}{2} + \frac{1}{2}\sqrt{1 + 4t_k^2} \tag{7.51}$$

$$\mathbf{z}_{k+1} = \mathbf{x}_k + \left(\frac{t_k - 1}{t_{k+1}}\right)(\mathbf{x}_k - \mathbf{x}_{k-1}) \tag{7.52}$$

$$\mathbf{x}_{k+1} = \text{soft}\left(\mathbf{z}_{k+1} + \frac{1}{\alpha}\mathbf{H}^*(\mathbf{y} - \mathbf{H}\mathbf{z}_{k+1}),\ \frac{\lambda}{2\alpha}\right) \tag{7.53}$$

FISTA is particularly attractive because it maintains the simplicity of ISTA, yet is proven to converge quadratically to the minimizer $\mathbf{x}_*$, since

$$J(\mathbf{x}_k) - J(\mathbf{x}_*) \leq \frac{C}{k^2}. \tag{7.54}$$

We have found in practice that FISTA is substantially more effective for radar problems than is ISTA. For the example in Sec. 7.3.1, Fig. 7.8 illustrates the convergence of both ISTA and FISTA. Although FISTA does not guarantee that the objective function is strictly decreasing like ISTA, the asymptotic convergence of FISTA is superior to that of ISTA.

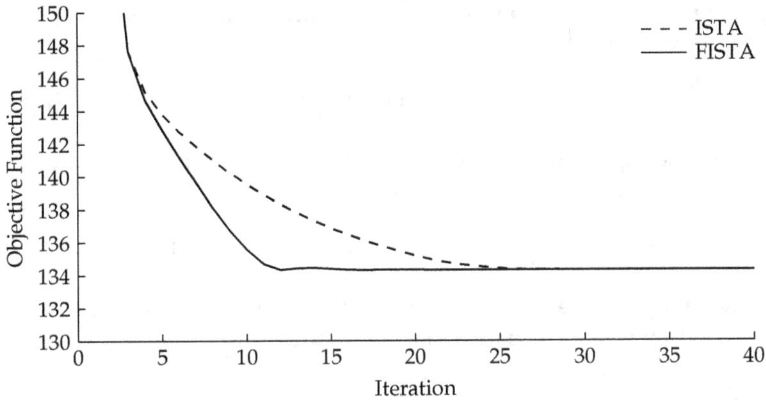

FIGURE 7.8 Comparison: convergence of ISTA and FISTA. FISTA converges faster than ISTA.

7.4 Nonlinear Receiver as a Bayesian Estimator

This chapter began with the objective function (7.6); however, starting with a specific estimation problem based upon Bayesian estimation theory leads to the same objective function. It is informative to consider the Bayesian formulation because it provides insight as to how the objective function should be modified to take into account prior information regarding the behavior of the signal of interest. Additionally, the Bayesian formulation can give some insight into the relationship between different methods.

In the following, we illustrate how the objective function (7.6) can be obtained using Bayesian estimation theory. Consider the observation of a linearly distorted signal $\mathbf{x}$ in additive independent noise $\mathbf{w}$,

$$\mathbf{y} = \mathbf{H}\mathbf{x} + \mathbf{w}, \tag{7.55}$$

the observed signal being $\mathbf{y}$. The maximum a posteriori (MAP) estimate of $\mathbf{x}$ is defined as

$$\hat{\mathbf{x}}_{\text{MAP}} := \arg\max_{\mathbf{x}} p_{\mathbf{x}|\mathbf{y}}(\mathbf{x}|\mathbf{y}) \tag{7.56}$$

where $p_{\mathbf{x}|\mathbf{y}}(\mathbf{x}|\mathbf{y})$ represents the a posteriori probability density function of $\mathbf{x}$ given $\mathbf{y}$. Using rule, as

$$\hat{\mathbf{x}}_{\text{MAP}} = \arg\max_{\mathbf{x}} p_{\mathbf{y}|\mathbf{x}}(\mathbf{y}|\mathbf{x})\, p_{\mathbf{x}}(\mathbf{x}). \tag{7.57}$$

Using the observation model (7.55), we have that $p_{\mathbf{y}|\mathbf{x}}(\mathbf{y}|\mathbf{x}) = p_w(\mathbf{y} - \mathbf{Hx})$, so we can write

$$\hat{\mathbf{x}}_{\text{MAP}} = \arg\max_{\mathbf{x}} p_w(\mathbf{y} - \mathbf{Hx})\, p_{\mathbf{x}}(\mathbf{x}). \tag{7.58}$$

Taking the logarithm of a function does not affect the location of its maximum; therefore, we can write

$$\hat{\mathbf{x}}_{\text{MAP}} = \arg\max_{\mathbf{x}} \left[\log(p_n(\mathbf{y} - \mathbf{Hx})) + \log(p_{\mathbf{x}}(\mathbf{x}))\right]. \tag{7.59}$$

Now, if the additive noise, $\mathbf{w}$, is known to follow a zero-mean Gaussian distribution with covariance matrix Σ_n, then

$$\hat{\mathbf{x}}_{\text{MAP}} = \arg\min_{\mathbf{x}} \left[\frac{1}{2}(\mathbf{y} - \mathbf{Hx})^* \Sigma_n^{-1}(\mathbf{y} - \mathbf{Hx}) + \log(p_{\mathbf{x}}(\mathbf{x}))\right]. \tag{7.60}$$

Therefore, the MAP estimator of $\mathbf{x}$ given $\mathbf{y}$ is found by minimizing the objective function

$$J(\mathbf{x}) = \frac{1}{2}\|\mathbf{y} - \mathbf{Hx}\|_{\Sigma_n^{-1}} + \log(p_{\mathbf{x}}(\mathbf{x})) \tag{7.61}$$

where $\|\mathbf{v}\|_{\mathbf{R}}$ denotes $\mathbf{v}^*\mathbf{R}\mathbf{v}$.

Gaussian Signal Model

If, in addition, the signal of interest, $\mathbf{x}$, is known to follow a zero-mean Gaussian distribution with covariance matrix Σ_x, then (7.60) can be written as

$$\hat{\mathbf{x}}_{\text{MAP}} = \arg\min_{\mathbf{x}} \left[\frac{1}{2}(\mathbf{y} - \mathbf{Hx})^* \Sigma_n^{-1}(\mathbf{y} - \mathbf{Hx}) + \frac{1}{2}\mathbf{x}^*\Sigma_x^{-1}\mathbf{x}\right] \tag{7.62}$$

when $\mathbf{w} \sim \mathcal{N}(\mathbf{0}, \Sigma_n)$, $\mathbf{x} \sim \mathcal{N}(\mathbf{0}, \Sigma_x)$. In this case, the MAP estimator is found by minimizing the objective function,

$$J(\mathbf{x}) = \frac{1}{2}\|\mathbf{y} - \mathbf{Hx}\|_{\Sigma_n^{-1}} + \frac{1}{2}\|\mathbf{x}\|_{\Sigma_x^{-1}}. \tag{7.63}$$

Setting to zero the derivative with respect to $\mathbf{x}$ gives

$$-\mathbf{H}^*\Sigma_n^{-1}(\mathbf{y} - \mathbf{Hx}) + \Sigma_x^{-1}\mathbf{x} = \mathbf{0} \tag{7.64}$$

or

$$\left(\mathbf{H}^*\Sigma_n^{-1}\mathbf{H} + \Sigma_x^{-1}\right)\mathbf{x} = \mathbf{H}^*\Sigma_n^{-1}\mathbf{y}. \tag{7.65}$$

Solving for $\mathbf{x}$ gives the MAP estimator in explicit form:

$$\hat{\mathbf{x}}_{\text{MAP}} = \left(\mathbf{H}^*\,\Sigma_n^{-1}\,\mathbf{H} + \Sigma_x^{-1}\right)^{-1}\mathbf{H}^*\,\Sigma_n^{-1}\,\mathbf{y} \tag{7.66}$$

when $\mathbf{w} \sim \mathcal{N}(\mathbf{0}, \Sigma_n)$, $\mathbf{x} \sim \mathcal{N}(\mathbf{0}, \Sigma_x)$. This is the Wiener filter for deconvolution.

As a special simple case, suppose that the noise $\mathbf{w}$ and signal $\mathbf{x}$ are known to be white Gaussian; that is, $\Sigma_n = \sigma_n^2 \mathbf{I}$ and $\Sigma_x = \sigma_x^2 \mathbf{I}$. Then, from (7.62),

$$\hat{\mathbf{x}}_{\text{MAP}} = \arg\min_{\mathbf{x}} \left[\frac{1}{2\sigma_n^2} \|\mathbf{y} - \mathbf{H}\mathbf{x}\|_2^2 + \frac{1}{2\sigma_x^2} \|\mathbf{x}\|_2^2 \right]. \tag{7.67}$$

In this case, the MAP estimate is found by minimizing

$$J(\mathbf{x}) = \|\mathbf{y} - \mathbf{H}\mathbf{x}\|_2^2 + \lambda \|\mathbf{x}\|_2^2 \quad \text{where} \quad \lambda = \frac{\sigma_n^2}{\sigma_x^2}, \tag{7.68}$$

which has the explicit solution

$$\hat{\mathbf{x}}_{\text{MAP}} = \left(\mathbf{H}^*\mathbf{H} + \lambda\, \mathbf{I}\right)^{-1} \mathbf{H}^*\mathbf{y}. \tag{7.69}$$

In (7.69), the same constant λ is added to each element of the diagonal of $\mathbf{H}^*\mathbf{H}$, a step known as *diagonal loading*. This guarantees that the matrix is invertible.

Note that the objective functions (7.63) and (7.68) are *quadratic* functions of $\mathbf{x}$ and that their minimizers in (7.66) and (7.69) are *linear* functions of $\mathbf{y}$.

Laplace Signal Model

The estimators (7.66) and (7.69), derived based on Gaussian signal models, are generally far from being sparse. If it is known or expected that the signal of interest $\mathbf{x}$ is sparse, then the Gaussian signal model is not an appropriate model for $\mathbf{x}$. Indeed, the samples of sparse signals follow distributions that are more peaked at zero and more heavy-tailed than is the Gaussian distribution. (These are sometimes called "super-Gaussian" distributions.) There are several such well-known distributions. A simple one is the Laplace (or "double-exponential") distribution,

$$p(x) = \frac{1}{\sqrt{2}\sigma} \exp\left(-\frac{\sqrt{2}}{\sigma} |x|\right). \tag{7.70}$$

If the elements of $\mathbf{x}$ are assumed or known to be independently and identically distributed (IID) and follow the Laplace distribution with variance σ_x^2 [denoted $\mathbf{x} \sim \mathcal{L}(\mathbf{0}, \sigma_x^2 \mathbf{I})$], then

$$p_{\mathbf{x}}(\mathbf{x}) = \frac{1}{(\sqrt{2}\sigma)^N} \exp\left(-\frac{\sqrt{2}}{\sigma_x} \|\mathbf{x}\|_1\right). \tag{7.71}$$

In this case, the MAP estimator (7.60) is given by

$$\hat{\mathbf{x}}_{\mathrm{MAP}} = \arg\min_{\mathbf{x}} \left[\frac{1}{2}(\mathbf{y} - \mathbf{H}\mathbf{x})^{*} \Sigma_n^{-1} (\mathbf{y} - \mathbf{H}\mathbf{x}) + \frac{\sqrt{2}}{\sigma_x} \|\mathbf{x}\|_1 \right]. \tag{7.72}$$

If the noise is white, $\Sigma_n = \sigma_n^2 \mathbf{I}$, then (7.72) becomes

$$\hat{\mathbf{x}}_{\mathrm{MAP}} = \arg\min_{\mathbf{x}} \left[\frac{1}{2\sigma_n^2} \|\mathbf{y} - \mathbf{H}\mathbf{x}\|_2^2 + \frac{\sqrt{2}}{\sigma_x} \|\mathbf{x}\|_1 \right]. \tag{7.73}$$

The MAP estimator is therefore found by minimizing

$$J(\mathbf{x}) = \|\mathbf{y} - \mathbf{H}\mathbf{x}\|_2^2 + \lambda \|\mathbf{x}\|_1 \tag{7.74}$$

where $\lambda = 2\sqrt{2}\,\sigma_n^2/\sigma_x$. This is the same objective function as $J(\mathbf{x})$ in (7.6). Therefore, the ℓ_1-norm regularized objective function introduced in (7.6) at the beginning of this chapter, corresponds to the MAP estimator of a linearly distorted IID Laplace signal in additive white Gaussian noise. This derivation also provides a value for the regularization parameter λ.

Note that the objective function (7.74) is a nonquadratic function of $\mathbf{x}$ and that the estimate $\hat{\mathbf{x}}$ (minimizer of $J(\mathbf{x})$) is a nonlinear function of $\mathbf{y}$. This is in contrast with the all-Gaussian case where the optimal estimate is a linear function of $\mathbf{y}$. In order to utilize knowledge that the signal of interest is sparse, this approach forgoes linearity—and ISTA, FISTA, or another iterative numerical algorithm must be used instead.

The Bayesian formulation suggests how the ℓ_1-norm in the objective function (7.6) can be modified so as to more accurately reflect prior knowledge of the signal to be estimated. For example, prior distribution functions other than the Laplace distribution (generalized Gaussian, BesselK, mixture models, etc.) may better reflect the statistics of the signal. New objective functions can be derived based on the chosen prior distribution; moreover, the prior distribution need not be IID.

7.5 Sparsity-Based Range-Doppler Processing

The estimation of the velocity of a moving target in a radar scene can be accomplished by transmitting a sequence of M high-frequency pulses and by suitably processing the subsequently acquired radar data. This method utilizes the Doppler effect [24]. This section describes the measurement of velocity by conventional Doppler processing and illustrates an alternative method based on sparse-signal models. The accuracy of the sparsity model in this case depends on there being few moving objects within range of the radar moving with distinct velocities. We can assume that the transmitted pulse is given by

$$s(t) = a(t)\, e^{j(2\pi f_o t+\phi)} \tag{7.75}$$

where

$$f_o = c/\lambda, \qquad c = 3 \cdot 10^8\,\text{m/s}. \tag{7.76}$$

A pulse is transmitted every T seconds, where T is large enough so that all returns due to each pulse arrive at the receiver before the next pulse is transmitted. For example, if $T = 0.0016$ seconds (PRF = 625 pulses/s), then each pulse would travel $c \cdot T = 480$ km during the interpulse period.

Received Data

If an object is D_o meters away, then the received signal due to the first transmitted pulse is

$$r_0(t) = s(t - 2\,\tau_0) \tag{7.77}$$

where we set $t = 0$ as the time the first pulse is transmitted. Then τ_0, the time required for the transmitted pulse to reach the object, is given by $\tau_0 = D_o/(c+v)$ where v is the velocity of the object toward the transmitter/receiver base. After T seconds, when another pulse is transmitted, the distance between the base and the object is $D_1 = D_0 - v\,T$ due to the motion of the object toward the base. The received signal due to the second transmitted pulse is therefore

$$r_1(t) = s(t - 2\,\tau_1 - T) \tag{7.78}$$

where

$$\tau_1 = \frac{D_1}{c+v} = \frac{D_0 - v\,T}{c+v} = \tau_0 - \frac{v\,T}{c+v}. \tag{7.79}$$

Similarly,

$$r_m(t) = s(t - 2\,\tau_m - m\,T), \quad m = 0, \ldots, M-1, \tag{7.80}$$

where

$$\tau_m = \tau_0 - m\frac{v\,T}{c+v}. \tag{7.81}$$

The set of signals $r_m(t)$, $m = 0, \ldots, M-1$, is the received signal data. Because $v \ll c$, from (7.81),

$$\tau_m \approx \tau_0 - m\frac{vT}{c}. \tag{7.82}$$

Therefore the received radar data is given by

$$r_m(t) \approx s\left(t - 2\tau_0 + 2m\frac{vT}{c} - mT\right), \quad m = 0, \ldots, M-1 \tag{7.83}$$

Demodulating and time-shifting gives

$$y_m(t) := r_m(t + mT)\, e^{-j2\pi f_o t} \tag{7.84}$$

$$\approx a(t - 2\tau_m)\, \exp\left(j\left[-4\pi f_o \tau_m + \phi\right]\right). \tag{7.85}$$

Using (7.82),

$$y_m(t) \approx a(t - 2\tau_m)\, \exp\left(j\left[-4\pi f_o(\tau_0 - mvT/c) + \phi\right]\right) \tag{7.86}$$

$$\approx a(t - 2\tau_m)\, \exp\left(j\left[-4\pi f_o \tau_0 + \phi\right]\right) \exp\left(j4m\pi f_o vT/c\right) \tag{7.87}$$

Assuming $a(t)$ does not vary too rapdily, we have

$$a(t - 2\tau_0) \approx a(t - 2\tau_m), \quad \text{for all } t \in \mathbf{R} \text{ and } m = 1, \ldots, M-1, \tag{7.88}$$

so

$$y_m(t) \approx A(t - 2\tau_0)\, \exp\left(j\frac{4 f_o T v}{c}\pi m\right), \quad m = 0, \ldots, M-1, \tag{7.89}$$

where $A(t) = a(t)\, e^{j(-4\pi f_o \tau_0 + \phi)}$. To avoid ambiguity (aliasing) in measuring v from $y_m(t)$, we need

$$\left|\frac{4 f_o v T}{c}\right| < 1 \tag{7.90}$$

or equivalently

$$|v| < \frac{c}{4 f_o T} =: v_{\max}. \tag{7.91}$$

For example, if the pulse repetition rate (PRF) is 625 pulses/s and $f_o = 435$ MHz, then $v_{\max} = 107.7$ m/s.

With the normalized Doppler frequency, ω_d, defined as

$$\omega_d := \frac{v}{v_{\max}}, \tag{7.92}$$

the received data is written as

$$y_m(t) \approx A(t - 2\,\tau_0)\,\exp(j\omega_d \pi m), \qquad m = 0, \ldots, M-1. \tag{7.93}$$

Assuming that only one object is present in range of the radar, the received radar data at time $t = t_p$ is given by the M complex values $y_m(t_p)$, for $m = 0, \ldots, M-1$. Equation (7.93) reveals that this M-point sequence is a complex exponential sequence with frequency $\omega_d\,\pi$.

Conventional Doppler Processing

When one object is in range of the radar, the estimation of v from $y_m(t_p)$, $m = 0, \ldots, M-1$, requires estimating the frequency of a complex exponential from M samples. This can be achieved by computing the Fourier transform of the M-point sequence and locating the peak in the spectrum. In practice, a zero-padded FFT is used to compute the Fourier transform. Once the peak ω_d is estimated, then the velocity v can be found from (7.92) as $v = \omega_d \cdot v_{\max}$.

When several moving objects are present, then the Fourier transform of the M-point sequence $y_m(t_p)$ is expected to have several peaks, each corresponding to the velocity of an object. However, the presence of noise, the unequal power reflected from each object, and the similarity between object velocities can make the accurate detection of object velocities challenging.

Discretization

Discretize the time variable t so that

$$t_i = t_0 + i\,T_s, \qquad i \in \mathbb{Z} \tag{7.94}$$

where T_s is the sampling period and i is the range bin index. Let us define the discretized received data,

$$\mathbf{y}(m, i) := A(t_i - 2\,\tau_0)\,\exp(j\omega_d \pi m), \qquad m = 0, \ldots, M-1, \quad i \in \mathbb{Z}, \tag{7.95}$$

$$= A(t_0 + i\,T_s - 2\,\tau_0)\,\exp(j\omega_d \pi m), \qquad m = 0, \ldots, M-1, \quad i \in \mathbb{Z}. \tag{7.96}$$

Assume that τ_0 is discretized as

$$\tau_0 = \frac{1}{2}t_0 + l\frac{T_s}{2}, \quad \text{for some } l \in \mathbb{Z}, \tag{7.97}$$

then

$$\mathbf{y}(m, i) = A(t_0 + i\,T_s - t_0 - l T_s)\exp(j\omega_d \pi m), \tag{7.98}$$

$$= A((i - l)\,T_s)\exp(j\omega_d \pi m), \tag{7.99}$$

$$= \mathbf{f}(i - l)\exp(j\omega_d \pi m), \tag{7.100}$$

where $\mathbf{f}(i) := A(i\,T_s)$ is the transmit pulse.[1]

Discretize the normalized Doppler ω_d into N_d Doppler bins (with $N_d \geq M$),

$$[\omega_d]_k = \frac{2k}{N_d}, \qquad 0 \leq k \leq N_d - 1. \tag{7.101}$$

Then, from (7.100), the received radar data due to a unit-amplitude object in range bin l and normalized Doppler $2k/N_d$ (with $0 \leq k \leq N_d - 1$) is given by

$$\mathbf{y}(m, i) = \mathbf{f}(i - l)\exp\left(j\frac{2\pi}{N_d}km\right), \qquad m = 0, \ldots, M-1, \quad i \in \mathbb{Z}. \tag{7.102}$$

The total radar scene will have numerous objects and responses at multiple range bins and multiple Dopplers. Denote the Doppler-range profile of the input scene as $\mathbf{x}(k, i)$,

$$\mathbf{x}(k, i), \qquad k = 0, \ldots, N_d - 1, \qquad i = 0, \ldots, N_r - 1, \tag{7.103}$$

where N_r is the number of range bins. Then the total received data will be the sum of received data due to each Doppler-range bin

$$\mathbf{y}(m, i) = \sum_{k=0}^{N_d-1}\sum_{l=0}^{N_r-1} \mathbf{x}(k, l)\,\mathbf{f}(i - l)\exp\left(j\frac{2\pi}{N_d}km\right),$$
$$m = 0, \ldots, M-1, \quad i \in \mathbb{Z}. \tag{7.104}$$

Let us assume that $\mathbf{f}$ is of finite duration and causal,

$$\mathbf{f}(i), \quad i = 0, \ldots, N_f - 1 \tag{7.105}$$

where N_f is the length of $\mathbf{f}$ in range bins. Then $\mathbf{y}(m, i)$ is defined for

$$m = 0, \ldots, M-1, \quad i = 0, \ldots, N_r + N_f - 1. \tag{7.106}$$

[1] There is an additional range-dependent phase term (τ_0) in the definition of $A(t)$ which we neglect here.

We can write (7.104) as

$$\mathbf{y}(m,i) = \sum_{k=0}^{N_d-1} \left(\sum_{l=0}^{N_r-1} \mathbf{x}(k,l)\,\mathbf{f}(i-l) \right) \exp\left(j\frac{2\pi}{N_d}km \right) \qquad (7.107)$$

where we can recognize the inner sum as the one-dimensional (1D) convolution of $\mathbf{x}(k,i)$ and $\mathbf{f}(i)$ (with respect to i). Defining

$$\mathbf{g}(k,i) := \sum_{l=0}^{N_r-1} \mathbf{x}(k,i)\,\mathbf{f}(i-l), \qquad (7.108)$$

we have

$$\mathbf{y}(m,i) = \sum_{k=0}^{N_d-1} \mathbf{g}(k,i)\exp\left(j\frac{2\pi}{N_d}km \right) \qquad (7.109)$$

which we can recognize as the N_d-point inverse DFT of $\mathbf{g}(k,i)$ (with respect to k) scaled by N_d.

$$\mathbf{y}(m,i) = N_d \cdot \mathrm{IDFT}_{N_d}\{\mathbf{g}(k,i),\ k = 0,\ldots,N_d-1\}. \qquad (7.110)$$

Because the received data consists of M pulses ($m = 0,\ldots,M-1$), only the first M coefficients of the inverse DFT are retained.

Sparse Doppler Processing[2]

With the time (range) and Doppler discretization given above, the Doppler-range profile of the radar scene can be taken to be a 2D array $\mathbf{x}$ of size $N_d \times N_r$. The received radar data is a 2D array $\mathbf{y}$ of size $M \times (N_r + N_f - 1)$. From (7.110) the received radar data is a linear function of the scene $\mathbf{x}$ of interest. We can therefore define the linear operator $\mathbf{H}$ such that $\mathbf{y} = \mathbf{Hx}$. The operator $\mathbf{H}$ comprises matched filtering along the rows of $\mathbf{x}$ followed by truncated inverse DFT along the columns.

Note that the conjugate transpose $\mathbf{H}^*$ performs conventional range compression (matched filtering) and Doppler processing (DFT). On the other hand, with the Doppler-range profile, $\mathbf{x}$ can be estimated by minimizing the ℓ_1-norm regularized objective function (7.6) so as to obtain a sparse estimate $\hat{\mathbf{x}}$.

An example comparing matched processing and sparse processing is illustrated in Fig. 7.9. Matched processing involves conventional range compression (matched filtering) along with FFT in the pulse domain. The result obtained by sparse processing does not have the

[2]©2010 IEEE. Reprinted with permission, from "Angle-Doppler Processing Using Sparse Regularization" by I.W. Selesnick, S. U. Pillai, K. Y. Li, and B. Himed, *Proc. IEEE Int. Conf. Acoust., Speech, Signal Processing (ICASSP)*, March 2010.

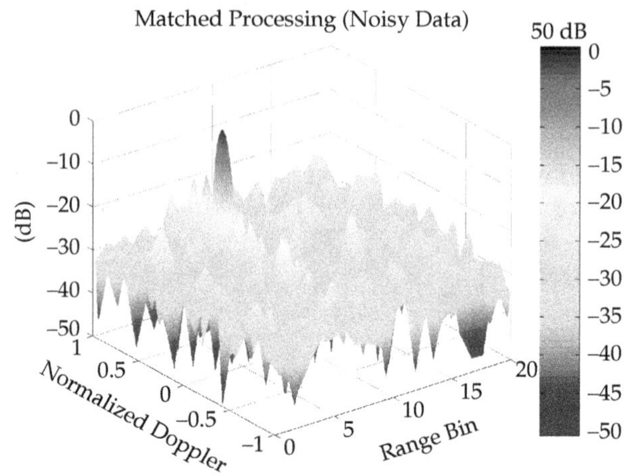

(a) Matched Processing

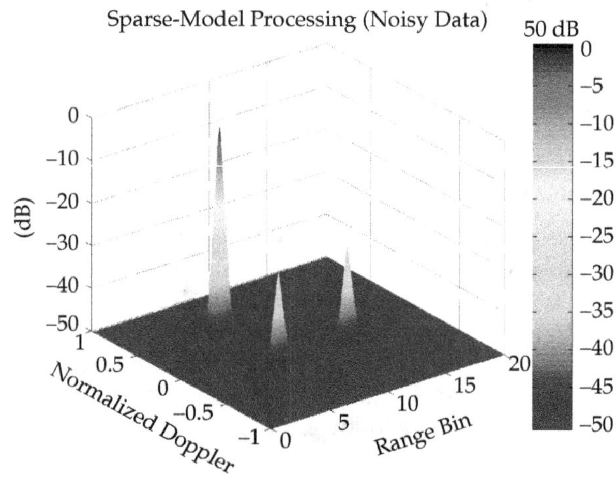

(b) Sparse Range-Doppler processing

FIGURE 7.9 Range-Doppler processing: estimation of Doppler-range profile.

sidelobes nor noise behavior present in the result obtained by conventional matched processing approach.

7.6 Space-Time Adaptive Processing

Space-time adaptive processing (STAP) simultaneously processes spatial (sensor array outputs) and temporal data (pulse returns) acquired by a set of sensors so as to detect and/or identify targets present in noisy observations [27]. Spatial data is collected using a set of sensors. Temporal data is collected using a sequence of pulses with a suitably adjusted pulse repetition frequency. With N sensors, M pulses from each of L range bins, the received radar data can be formed into the space-time data cube.

STAP is used in applications such as ground moving target indication (GMTI) and airborne moving target indication (AMTI), where it is desired to estimate both the position and velocity of the targets. Conventional STAP algorithms require the estimation of covariance matrices using space-time data from numerous range bins. Due to the nonstationary behavior of clutter present in the acquired data, algorithms for STAP, needing fewer range bins yet maintaining acceptable resolution and signal to noise ratio (SNR) levels, have been of particular interest [16,23].

Sparse-STAP

In contrast to conventional STAP algorithms, we can view the problem as an ℓ_1-norm regularized linear inverse problem as in [21, 22, 30]. Unlike conventional STAP, sparsity-based STAP (sparse-STAP) seeks to exploit the expectation that moving targets in the scene are isolated from each other and constitute a small percentage of the total angle-Doppler profile area. In sparse-STAP, the scene $\mathbf{x}$ is estimated by defining an operator $\mathbf{H}$ to model the received radar data (7.1), defining a cost function (7.6) or (7.72) and minimizing it (7.7) to obtain $\mathbf{x}$.

Angle-Doppler Data

Space-time adaptive processing (STAP) provides estimates of both the angle and velocity of targets [15]. Assume M consecutive pulses with interpulse period T and N sensors spaced $\tilde{d}$ apart along a linear array, as illustrated in Fig. 7.10. The transmitted signal can be modeled as

$$s(t) = a(t)\exp(j[2\pi f_o t + \phi]). \tag{7.111}$$

The signal received at the nth sensor in response to the mth pulse is denoted as $r_{m,n}(t)$, with $0 \le m \le M-1$ and $0 \le n \le N-1$.

The variable t (time) can be quantized into range bins $0 \le i \le N_r - 1$ (where N_r is the number of range bins). The normalized Doppler ω_d can be quantized into N_d Doppler bins (with $N_d \ge M$),

$$[\omega_d]_{k_1} = \frac{2k_1}{N_d} \quad \text{with } 0 \le k_1 \le N_d - 1, \tag{7.112}$$

and $d\cos(\theta)$ can be quantized into N_a angle bins (with $N_a \ge N$)

$$[d\,\cos(\theta)]_{k_2} = \frac{2k_2}{N_a} \quad \text{with } 0 \le k_2 \le N_a - 1. \tag{7.113}$$

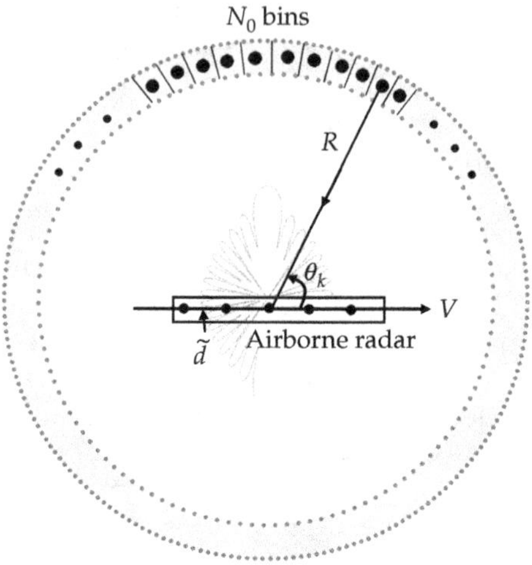

FIGURE 7.10 Airborne side-looking radar, as described in [15].

Then the received radar data for a unit-point target in range bin l, normalized Doppler $2k_1/N_d$, and angle index $2k_2/N_a$ is given by

$$\mathbf{y}(m, n, i) = \delta(i - l) \exp\left(j\frac{2\pi}{N_d}k_1 m - j\frac{2\pi}{N_a}k_2 n\right), \tag{7.114}$$

for $0 \leq m \leq M - 1$, $0 \leq n \leq N - 1$, and $0 \leq i \leq N_r - 1$.

Denoting the Doppler-angle-range profile of the input scene as $\mathbf{x}(k_1, k_2, i)$, the total received data will be the sum of received radar data due to each Doppler-angle-range bin

$$\mathbf{y}(m, n, i) = \sum_{k_1=0}^{N_d-1} \sum_{k_2=0}^{N_a-1} \mathbf{x}(k_1, k_2, i) \exp\left(j\frac{2\pi}{N_d}k_1 m - j\frac{2\pi}{N_a}k_2 n\right), \tag{7.115}$$

or equivalently,

$$\mathbf{y}(m, n, i) = N_d \cdot \mathrm{DFT}_{N_a}\{\mathrm{IDFT}_{N_d}\{\mathbf{x}(k_1, k_2, i),\ k_1\},\ k_2\}. \tag{7.116}$$

The received radar data consists of the first M coefficients of the inverse DFT along dimension k_1 and N coefficients of the DFT along dimension k_2.

Note that the acquired radar data (7.116) can be written as $\mathbf{y} = \mathbf{Hx}$ as (7.116) is a linear function of the input scene $\mathbf{x}$. That is, the operator $\mathbf{H}$ is defined through (7.116).

Sparsity and the Clutter Ridge

The effectiveness of sparsity-based signal estimation (in the case of STAP, the signal being a function of Doppler and angle) depends on the signal of interest being, in fact, sparse. If moving objects are relatively small and well isolated, then the sparsity assumption appears to be satisfied. However, for airborne radar, the relative motion of the terrain and the radar platform generates an additional secondary component in the angle-Doppler profile of the scene, and moreover this component is spread throughout a substantial fraction of the angle-Doppler plane, and is usually stronger than the returns from targets. That is, for airborne STAP, the ground return (clutter) cannot be modeled as being sparse.

Specifically, for a radar platform moving with velocity V performing ground moving target indication (GMTI), a stationary ground patch in the direction θ has a relative velocity $V\cos\theta$ with respect to the moving platform and hence, it generates a Doppler frequency

$$\omega_d = \frac{2VT}{\lambda/2}\cos\theta = \beta(d\cos\theta) \tag{7.117}$$

where

$$\beta = \frac{2VT}{\tilde{d}} = \frac{VT}{\tilde{d}/2} \tag{7.118}$$

refers to the Doppler foldover factor (Brennan factor). When $\beta = 1$, the clutter fills the angle-Doppler space exactly once; when $\beta > 1$, the clutter folds over into the Doppler space creating Doppler ambiguity. The ground return gives rise to the so-called *clutter ridge* in the angle-Doppler plane. While temporal or spatial processing cannot cancel the clutter ridge, space-time adaptive processing (STAP) algorithms can cancel the clutter.

In [22], this clutter is taken into account by modeling it as an additive noise/intereference term; hence the objective function (7.72) is utilized where the noise-covariance matrix is the covariance matrix estimated from multiple range bins. The following section describes another method, following [25], for taking the clutter ridge into account for sparse-STAP. Namely, the clutter ridge is omitted from the regularization term—the new cost function is

$$J(\mathbf{x}) = \|\mathbf{y} - \mathbf{H}\mathbf{x}\|_2^2 + \lambda\|\mathbf{M}\mathbf{x}\|_1 \tag{7.119}$$

where $\mathbf{M}$ is a diagonal operator (mask). The value of $\mathbf{M}$ is zero on the clutter ridge, and unity off the clutter ridge. The result of minimizing $J(\mathbf{x})$ will yield an angle-Doppler profile $\mathbf{x}$ consisting of (1) an estimate for the sparse angle-Doppler profile away from the clutter ridge, and (2) an estimate for the nonsparse clutter itself on the clutter ridge.

Examples

A comparison between sparse-STAP and two conventional STAP algorithms is illustrated in Figs. 7.11 and 7.12. The simulations are based on a system having 14 sensors ($N = 14$) with 16 pulses ($M = 16$) acquired at each sensor. In each example, the simulated clutter to noise ratio (CNR) is 40 dB, while the SNR is listed in the caption of each figure. In each simulation, $\beta = 1$ and $d = 1$, and it is assumed that the array characteristics are ideal and known (no calibration errors).

The angle-Doppler profiles illustrated in each figure are computed using three STAP algorithms: SMI-DL (sample matrix inverse with diagonal loading), HTPSASPFB (Hung-Turner projection with subarray subpulse forward-back smoothing) [15, 24], and the sparse-STAP method described above (ℓ_1-norm regularized linear inverse problem). The sparse-STAP has improved resolution and fewer sidelobes, yet it is based on a single range bin. In the simulation, the SMI-DL and HTPSASPFB STAP algorithms estimate covariance matrices using a total of 80 range bins and 10 range bins respectively, in order to estimate the covariance matrix needed for STAP.

Conventional STAP algorithms use radar data acquired from numerous range bins in order to estimate covariance matrices. In contrast, sparse-STAP does not require covariance matrices, and therefore, in comparison with conventional STAP, an advantage of sparse-STAP [21, 22, 25, 30] is that it can be performed with high resolution using only the range bin of interest. In principle, sparse-STAP does not need to make assumptions regarding the stationarity of clutter statistics and can therefore be used in highly nonstationary environments.

The sparse-STAP approach does have some limitations, however. First, the sparse-STAP, as described above, requires knowledge of the extent of clutter ridge support in the angle-Doppler plane. Second, sparse-STAP is more computationally costly in its implementation than conventional methods. Third, sparse-STAP may be sensitive to calibration and system model errors. (The simulations here assume no such miscalibration or errors.)

The use of sparse regularization also plays an important role in *compressed sensing*, which has also been applied as a new approach for radar signal processing [1, 5, 17, 19, 29]. Estimation and signal reconstruction using the compressed sensing framework relies on the ability to solve sparse-regularized inverse problems and on random or pseudo-random observations. The sparse-STAP, as described here employs sparse regularization; however, it does not utilize or require random sampling of the radar data, and hence, does not necessarily constitute a compressed-sensing method. The use of sparse (nonquadratic) regularization for radar imaging using the inverse problem formulation is well illustrated in earlier research [4].

(a)

(b)

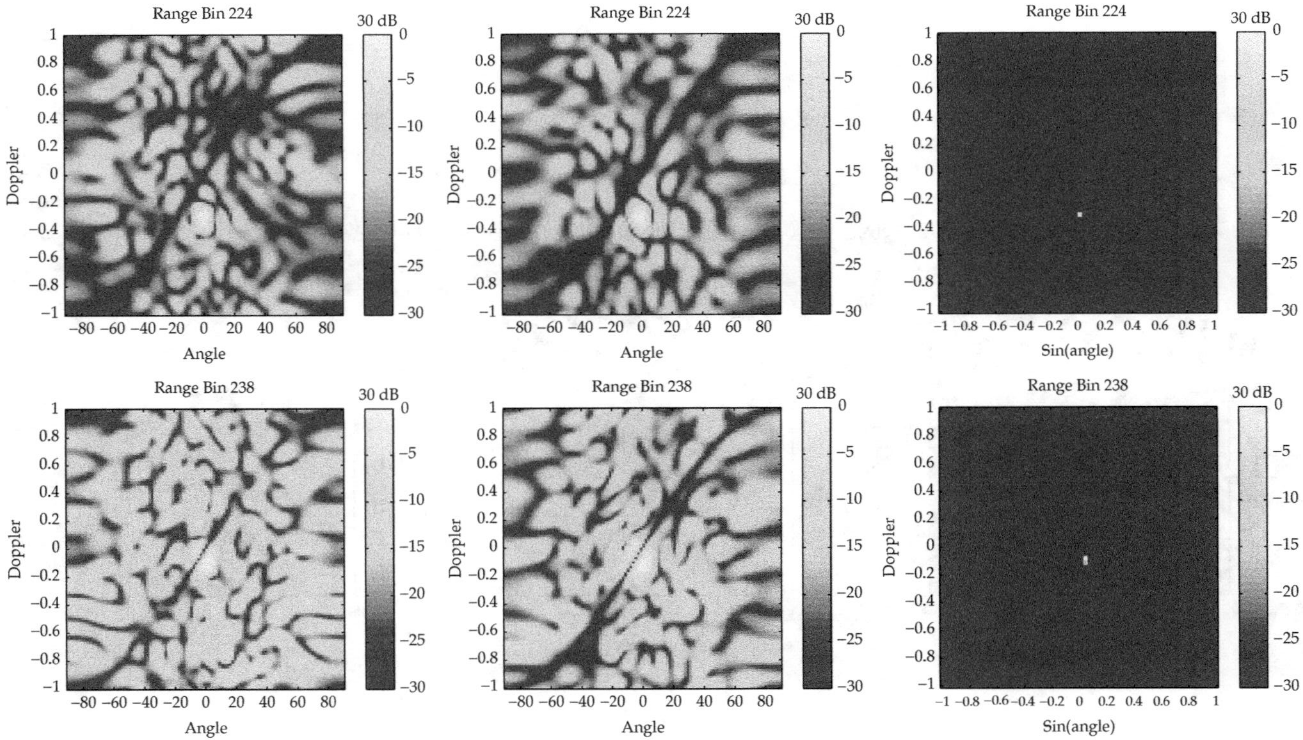

FIGURE 7.11 STAP comparison: (a) 1 target, SNR = 0 dB, CNR = 40 dB. SMI-DL and HTPSASPFB STAP algorithms work well. (b) 1 target close to clutter ridge, SNR = 0 dB, CNR = 40 dB. SMI-DL and HTPSASPFB STAP yield sidelobes around clutter notch.

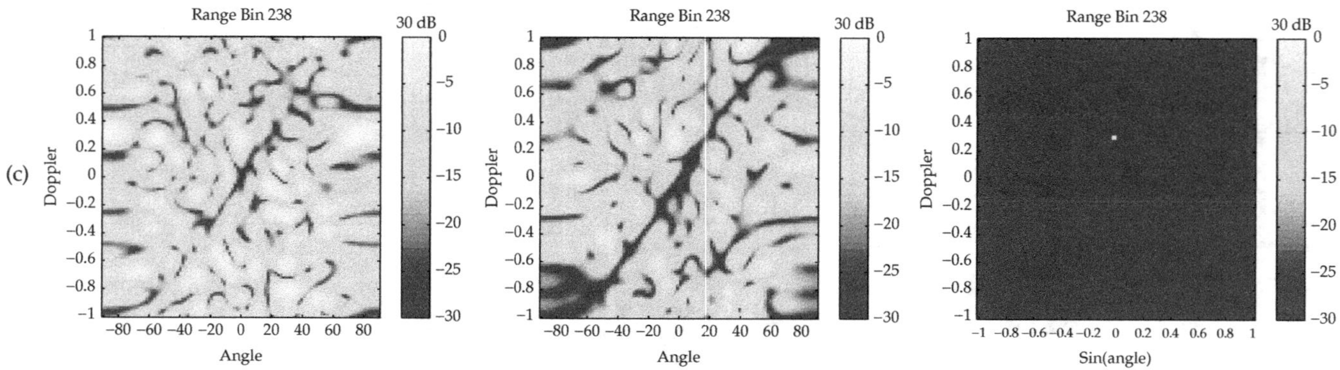

FIGURE 7.11 (continued). STAP comparison: (c) 1 target with low SNR, SNR = −10 dB, CNR = 40 dB. SMI-DL and HTPSASPFB STAP suffer from low SNR.

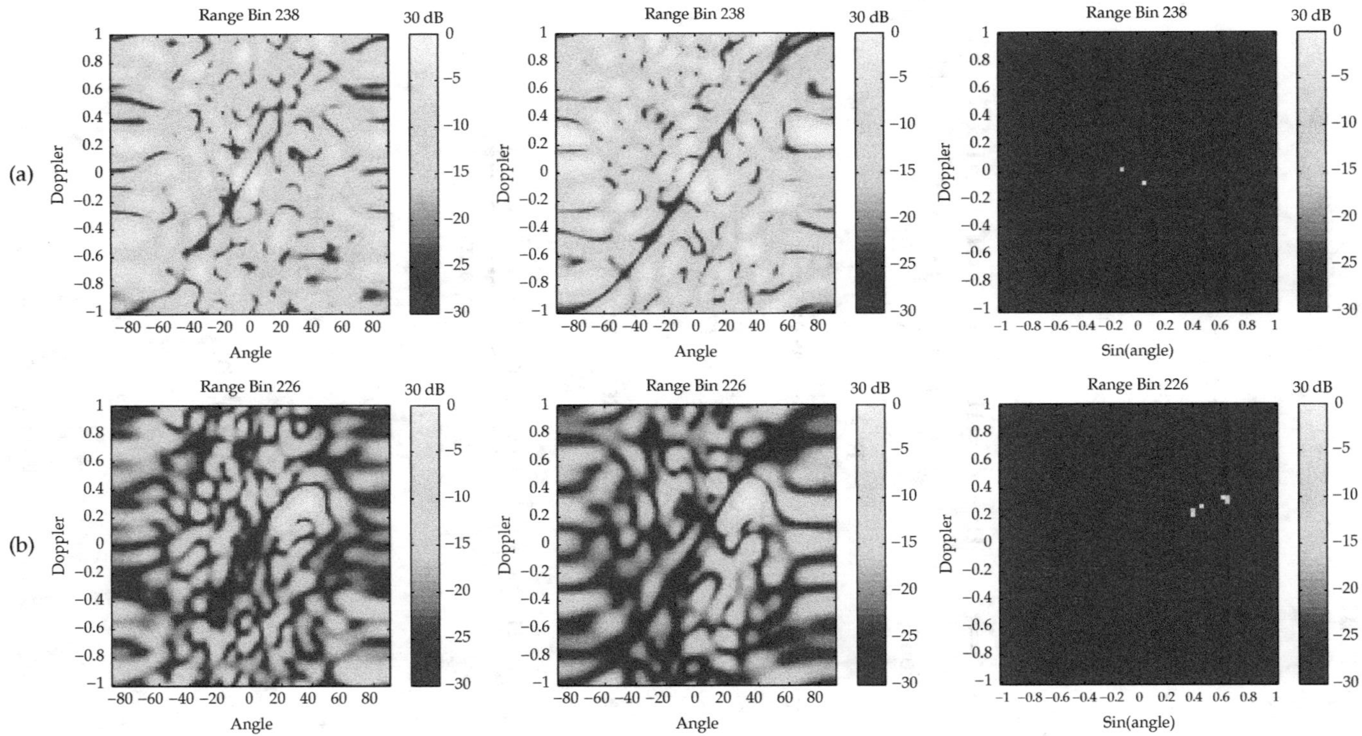

FIGURE 7.12 STAP comparison (a) 1 target very close to clutter ridge, SNR = 0 dB, CNR = 40 dB. Target is too close to clutter ridge for SMI-DL and HTPSASPFB STAP. Sparse-STAP yields one extraneous weak target. (b) STAP comparison (3 closely spaced targets, SNR = [0, −2.2, 1.5] dB, CNR = 40 dB). The targets are too close together for conventional STAP algorithms to distinguish.

7.7 ℓ_2 Regularization

Another standard linear inverse problem formulation uses ℓ_2-norm regularization:

$$J(\mathbf{x}) = \|\mathbf{y} - \mathbf{H}\mathbf{x}\|_2^2 + \lambda \|\mathbf{x}\|_2^2 \tag{7.120}$$

where $\lambda > 0$. Taking the derivative gives

$$\frac{\partial}{\partial \mathbf{x}} J(\mathbf{x}) = 2\mathbf{H}^*(\mathbf{H}\mathbf{x} - \mathbf{y}) + 2\lambda \mathbf{x}. \tag{7.121}$$

Setting the derivative to zero,

$$\frac{\partial}{\partial \mathbf{x}} J(\mathbf{x}) = \mathbf{0} \quad \Longrightarrow \quad \mathbf{H}^*\mathbf{H}\mathbf{x} + \lambda\mathbf{x} = \mathbf{H}^*\mathbf{y} \quad \Longrightarrow \quad (\mathbf{H}^*\mathbf{H} + \lambda\mathbf{I})\,\mathbf{x} = \mathbf{H}^*\mathbf{y}. \tag{7.122}$$

So the solution is given by

$$\mathbf{x} = (\mathbf{H}^*\mathbf{H} + \lambda\mathbf{I})^{-1}\mathbf{H}^*\mathbf{y}. \tag{7.123}$$

This is referred to as "diagonal loading" because a constant, λ, is added to the diagonal elements of $\mathbf{H}^*\mathbf{H}$. The approach also avoids the problem of rank deficiency because $\mathbf{H}^*\mathbf{H}+\lambda\mathbf{I}$ is invertible even if $\mathbf{H}^*\mathbf{H}$ is not.

Equation (7.123) is to be compared with (7.14): if $\lambda = 0$, then (7.123) gives (7.14).

Landweber Iteration for ℓ_2 Regularized Inverse Problem

Construct the majorizer

$$G_k(\mathbf{x}) = \|\mathbf{y} - \mathbf{H}\mathbf{x}\|_2^2 + \lambda \|\mathbf{x}\|_2^2 + \underbrace{(\mathbf{x} - \mathbf{x}_k)^*(\alpha\,\mathbf{I} - \mathbf{H}^*\mathbf{H})(\mathbf{x} - \mathbf{x}_k)}_{\text{nonnegative}}. \tag{7.124}$$

Expanding

$$G_k(\mathbf{x}) = \mathbf{x}^*\mathbf{H}^*\mathbf{H}\mathbf{x} - 2\mathbf{y}^*\mathbf{H}\mathbf{x} + \mathbf{y}^*\mathbf{y} + \lambda\mathbf{x}^*\mathbf{x} + (\mathbf{x} - \mathbf{x}_k)^*(\alpha\,\mathbf{I} - \mathbf{H}^*\mathbf{H})(\mathbf{x} - \mathbf{x}_k) \tag{7.125}$$

$$= (\alpha + \lambda)\,\mathbf{x}^*\mathbf{x} - 2(\mathbf{y}^*\mathbf{H} + \mathbf{x}_k{}^*(\alpha\,\mathbf{I} - \mathbf{H}^*\mathbf{H}))\mathbf{x} + \mathbf{y}^*\mathbf{y} + \mathbf{x}_k{}^*(\alpha\,\mathbf{I} - \mathbf{H}^*\mathbf{H})\mathbf{x}_k. \tag{7.126}$$

The leading term of $G_k(\mathbf{x})$ is $(\alpha + \lambda)\mathbf{x}^*\mathbf{x}$, so we can minimize $G_k(\mathbf{x})$ easily

$$\frac{\partial}{\partial \mathbf{x}} G_k(\mathbf{x}) = 2(\alpha + \lambda)\mathbf{x} - 2\mathbf{H}^*\mathbf{y} - 2(\alpha\,\mathbf{I} - \mathbf{H}^*\mathbf{H})\mathbf{x}_k \tag{7.127}$$

$$\frac{\partial}{\partial \mathbf{x}} G_k(\mathbf{x}) = \mathbf{0} \quad \Longrightarrow \quad \mathbf{x} = \frac{\alpha}{\alpha + \lambda}\mathbf{x}_k + \frac{1}{\alpha + \lambda}\mathbf{H}^*(\mathbf{y} - \mathbf{H}\mathbf{x}_k) \qquad (7.128)$$

so we obtain the iteration

$$\mathbf{x}_{k+1} = \frac{\alpha}{\alpha + \lambda}\mathbf{x}_k + \frac{1}{\alpha + \lambda}\mathbf{H}^*(\mathbf{y} - \mathbf{H}\mathbf{x}_k). \qquad (7.129)$$

Note that if $\lambda = 0$, then (7.129) becomes (7.22).

References

[1] R. Baraniuk, P. Steeghs, "Compressive Radar Imaging," In *IEEE Radar Conf.*, pp. 128–133, April 2007.

[2] A. Beck, M. Teboulle, "A Fast Iterative Shrinkage-Thresholding Algorithm for Linear Inverse Problems," *SIAM J. Imag. Sci.*, vol. 2, no. 1, pp. 183–202, 2009.

[3] J.M. Bioucas-Dias, M.A.T. Figueiredo, "A New TwIST: Two-Step Iterative Shrinkage—Thresholding Algorithms for Image Restoration," *IEEE Trans. Image Process.*, vol. 16, no. 12, pp. 2992–3004, Dec. 2007.

[4] M. Cetin, W.C. Karl, "Feature-Enhanced Synthetic Aperture Radar Image Formation Based on Nonquadratic Regularization," *IEEE Trans. Image Process.*, vol. 10, no. 4, pp. 623–631.

[5] Chun-Yang Chen, P.P. Vaidyanathan, "Compressed Sensing in MIMO Radar," In *Asilomar Conf. on Signals, Systems and Comp.*, Oct. 2008.

[6] P.L. Combettes, J.C. Pesquet, "Proximal Thresholding Algorithm for Minimization over Orthonormal Bases," *SIAM J. on Optimization*, vol. 18, no. 4, pp. 1351–1376, 2007.

[7] I. Daubechies, M. Defriese, C. De Mol, "An Iterative Thresholding Algorithm for Linear Inverse Problems with a Sparsity Constraint," *Commun. Pure Appl. Math*, vol. LVII, pp. 1413–1457, 2004.

[8] I. Daubechies, M. Fornasier, I. Loris, "Accelerated Projected Gradient Method for Linear Inverse Problems with Sparsity Constraints," *Journal of Fourier Analysis and Applications*, vol. 14, 764–792, Dec. 2008.

[9] I. Daubechies, G. Teschke, L. Vese, "On Some Iterative Concepts for Image Restoration," In P.W. Hawkes, editor, *Advances in Imaging and Electron Physics*, vol. 150, pp. 2–51. Elsevier, 2008.

[10] M. Elad, B. Matalon, M. Zibulevsky, "Coordinate and Subspace Optimization Methods for Linear Least Squares with Nonquadratic Regularization," *J. of Appl. and Comp. Harm. Analysis*, vol. 23, pp. 346–367, 2007.

[11] M. Figueiredo, J. Bioucas-Dias, R. Nowak, "Majorization-Minimization Algorithms for Wavelet-Based Image Restoration," *IEEE Trans. Image Process.*, vol. 16, no. 12, pp. 2980–2991, Dec. 2007.

[12] M. Figueiredo, R. Nowak, "An EM Algorithm for Wavelet-Based Image Restoration," *IEEE Trans. Image Process.*, vol. 12, no. 8, pp. 906–916, Aug. 2003.

[13] M.A.T. Figueiredo, R.D. Nowak, S. J. Wright, "Gradient Projection for Sparse Reconstruction: Application to Compressed Sensing and Other Inverse Problems," *IEEE J. Sel. Top. Signal Process.*, vol. 1, no. 4, pp. 586–598, Dec. 2007.

[14] T. Goldstein, S. Osher, "The Split Bregman Method for L1-Regularized Problems," *SIAM J. Imag. Sci.*, vol. 2, no. 2, pp.323–343, 2009.

[15] J.R. Guerci, *Space-Time Adaptive Processing for Radar*, Artech House, 2003.

[16] J.R. Guerci, J.S. Goldstein, I. S. Reed, "Optimal and Adaptive Reduced-Rank STAP," *IEEE Trans. on Aerospace and Electronic Systems*, vol. 36, no. 2, pp. 647–663, April 2000.

[17] A.C. Gurbuz, J.H. McClellan, W. R. Scott. "A Compressive Sensing Data Acquisition and Imaging Method for Stepped Frequency GPRs," *IEEE Trans. Signal Process.*, vol. 57, no. 7, pp. 2640–2650, July 2009.

[18] E.T. Hale, W. Yin, Y. Zhang, "Fixed-Point Continuation for ℓ_1-Minimization: Methodology and Convergence," *SIAM J. on Optimization*, vol. 19, no. 3, pp. 1107–1130, 2008.

[19] M.A. Herman, T. Strohmer, "High-Resolution Radar via Compressed Sensing," *IEEE Trans. Signal Process.*, vol. 57, no. 6, pp. 2275–2284, June 2009.

[20] T.J. Kragh, A.A. Kharbouch, "Monotonic Iterative Algorithms for SAR Image Restoration," In *Proc. IEEE Int. Conf. Acoust., Speech, Signal Processing (ICASSP)*, Oct. 2006.

[21] S. Maria, J.J. Fuchs, "Application of the Global Matched Filter to STAP Data an Efficient Algorithmic Approach," In *Proc. IEEE Int. Conf. Acoust., Speech, Signal Processing (ICASSP)*, May 2006.

[22] J. Parker, L. Potter, "A Bayesian Perspective on Sparse Regularization for STAP Post-Processing," In *Proc. IEEE Int. Radar Conf.*, May 2010.

[23] S.U. Pillai, J.R. Guerci, Y. L. Kim. "Generalized Forward/Backward Subaperture Smoothing Techniques for Sample Starved STAP," In *Proc. IEEE Int. Conf. Acoust., Speech, Signal Processing (ICASSP)*, vol. 4, May 1998.

[24] S.U. Pillai, K.Y. Li, B. Himed, *Space Based Radar: Theory & Applications*, McGraw-Hill, 2007.

[25] I.W. Selesnick, S.U. Pillai, K.Y. Li, B. Himed, "Angle-Doppler Processing Using Sparse Regularization," In *Proc. IEEE Int. Conf. Acoust., Speech, Signal Processing (ICASSP)*, March 2010.

[26] Y. Wang, J. Yang, W. Yin, Y. Zhang, "A New Alternating Minimization Algorithm for Total Variation Image Reconstruction," *SIAM J. on Imaging Sciences*, vol. 1, no. 3, pp. 248–272, 2008.

[27] J. Ward, Space-Time Adaptive Processing for Airborne Radar, Technical Report 1015, MIT Lincoln Laboratory, 1994.

[28] S.J. Wright, R.D. Nowak, M.A.T. Figueiredo, "Sparse Reconstruction by Separable Approximation," *IEEE Trans. Signal Process.*, vol. 57, no. 7, pp. 2479–2493, July 2009.

[29] Y. Yu, A.P. Petropulu, H.V. Poor, "Compressive Sensing for MIMO Radar," In *Proc. IEEE Int. Conf. Acoust., Speech, Signal Processing (ICASSP)*, pp. 3017–3020, April 2009.

[30] H. Zhang, G. Li, H. Meng, "A Class of Novel STAP Algorithms Using Sparse Recovery Technique," Preprint, http://arxiv.org/abs/0904.1313v1, 2009.

Index

f denotes figures

J

L

M

N

T

www.ingramcontent.com/pod-product-compliance
Lightning Source LLC
LaVergne TN
LVHW011526200926
848484LV00001B/165
9780071622899